Protecting natural capital and biodiversity in the agri-food sector

It is widely recognised that agriculture is a significant contributor to global warming and climate change. Agriculture needs to reduce its environmental impact and adapt to current climate change whilst still feeding a growing population, i.e. become more 'climate-smart'. Burleigh Dodds Science Publishing is playing its part in achieving this by bringing together key research on making the production of the world's most important crops and livestock products more sustainable.

Based on extensive research, our publications specifically target the challenge of climate-smart agriculture. In this way we are using 'smart publishing' to help achieve climate-smart agriculture.

Burleigh Dodds Science Publishing is an independent and innovative publisher delivering high quality customer-focused agricultural science content in both print and online formats for the academic and research communities. Our aim is to build a foundation of knowledge on which researchers can build to meet the challenge of climate-smart agriculture.

For more information about Burleigh Dodds Science Publishing simply call us on +44 (0) 1223 839365, email info@bdspublishing.com or alternatively please visit our website at www.bdspublishing.com.

Related titles:

Achieving sustainable cultivation of oil palm Volume 2: Diseases, pests, quality and sustainability
Print (ISBN 978-1-78676-108-8); Online (ISBN 978-1-78676-111-8, 978-1-78676-110-1)

Assessing the environmental impact of agriculture
Print (ISBN 978-1-78676-228-3); Online (ISBN 978-1-78676-231-3, 978-1-78676-230-6)

Reconciling agricultural production with biodiversity conservation
Print (ISBN 978-1-78676-348-8); Online (ISBN 978-1-78676-350-1, 978-1-78676-351-8)

Promoting pollination and pollinators in farming
Print (ISBN 978-1-80146-098-9); Online (ISBN 978-1-80146-101-6, 978-1-80146-100-9)

Chapters are available individually from our online bookshop:
https://shop.bdspublishing.com

climate•SMART•publishing
IN AGRICULTURAL SCIENCE

BURLEIGH DODDS SERIES IN AGRICULTURAL SCIENCE
NUMBER 143

Protecting natural capital and biodiversity in the agri-food sector

Edited by Professor Jill Atkins, Cardiff University, UK

burleigh dodds
SCIENCE PUBLISHING

Published by Burleigh Dodds Science Publishing Limited
82 High Street, Sawston, Cambridge CB22 3HJ, UK
www.bdspublishing.com

Burleigh Dodds Science Publishing, 1518 Walnut Street, Suite 900, Philadelphia, PA 19102-3406, USA

First published 2024 by Burleigh Dodds Science Publishing Limited

Library of Congress Control Number: 2023951809

British Library Cataloguing in Publication Data
A catalogue record for this book is available from the British Library

ISBN 978-1-80146-351-5 (print)
ISBN 978-1-80146-353-9 (PDF)
ISBN 978-1-80146-352-2 (ePub)
ISSN 2059-6936 (print)
ISSN 2059-6944 (online)

DOI: 10.19103/AS.2023.0128

Typeset by Deanta Global Publishing Services, Dublin, Ireland

Contents

Series list

Title	Series number
Achieving sustainable cultivation of maize - Vol 1 From improved varieties to local applications Edited by: Dr Dave Watson, CGIAR Maize Research Program Manager, CIMMYT, Mexico	001
Achieving sustainable cultivation of maize - Vol 2 Cultivation techniques, pest and disease control Edited by: Dr Dave Watson, CGIAR Maize Research Program Manager, CIMMYT, Mexico	002
Achieving sustainable cultivation of rice - Vol 1 Breeding for higher yield and quality Edited by: Prof. Takuji Sasaki, Tokyo University of Agriculture, Japan	003
Achieving sustainable cultivation of rice - Vol 2 Cultivation, pest and disease management Edited by: Prof. Takuji Sasaki, Tokyo University of Agriculture, Japan	004
Achieving sustainable cultivation of wheat - Vol 1 Breeding, quality traits, pests and diseases Edited by: Prof. Peter Langridge, The University of Adelaide, Australia	005
Achieving sustainable cultivation of wheat - Vol 2 Cultivation techniques Edited by: Prof. Peter Langridge, The University of Adelaide, Australia	006
Achieving sustainable cultivation of tomatoes Edited by: Dr Autar Mattoo, USDA-ARS, USA and Prof. Avtar Handa, Purdue University, USA	007
Achieving sustainable production of milk - Vol 1 Milk composition, genetics and breeding Edited by: Dr Nico van Belzen, International Dairy Federation (IDF), Belgium	008
Achieving sustainable production of milk - Vol 2 Safety, quality and sustainability Edited by: Dr Nico van Belzen, International Dairy Federation (IDF), Belgium	009
Achieving sustainable production of milk - Vol 3 Dairy herd management and welfare Edited by: Prof. John Webster, University of Bristol, UK	010
Ensuring safety and quality in the production of beef - Vol 1 Safety Edited by: Prof. Gary Acuff, Texas A&M University, USA and Prof. James Dickson, Iowa State University, USA	011
Ensuring safety and quality in the production of beef - Vol 2 Quality Edited by: Prof. Michael Dikeman, Kansas State University, USA	012
Achieving sustainable production of poultry meat - Vol 1 Safety, quality and sustainability Edited by: Prof. Steven C. Ricke, University of Arkansas, USA	013
Achieving sustainable production of poultry meat - Vol 2 Breeding and nutrition Edited by: Prof. Todd Applegate, University of Georgia, USA	014

Chapter 1

Protecting natural capital and biodiversity in agricultural supply chains: introduction

Jill Atkins and John Peirce, Cardiff Business School – Cardiff University, UK

1 Introduction

Since the dawn of human consciousness, humans have affected and altered the biodiversity of planet Earth. Prehistoric cave paintings provide us with the earliest records of wild species of animal being hunted for food. Palaeontologists, archaeologists and rock artist researchers have studied the evolution of the human race and the ways in which people evolved around the world from hunter gatherers to groups living in early settlements. In prehistoric times, as people began to shift from living nomadic lifestyles, the earliest forms of agriculture were born. Animals such as the prehistoric auroch, the original species of cattle from which all modern cattle evolved, were gradually domesticated over time (Atkins et al., 2023). Similarly, early civilisations began to cultivate plants for food. The earliest industrial activity involved food production and the development of agricultural practices in order for people to have the very basics for survival.

Researchers have noted that anthropologists have traced the path of human development through several stages, each portraying an increasing dominance over non-human species, from the earliest stages of evolution to 'elemental man', the advanced hunter and gatherer; the domesticator of plants and animals; the advanced food producer; industrial man; and atomic

http://dx.doi.org/10.19103/AS.2023.0128.25

man (Watson and Watson, 1969; Jones and Solomon, 2013). These successive stages of human evolution have positioned the human species as increasingly dominant over all other species of fauna. As human society has evolved and human populations have expanded around the world, people have had an increasing impact on the world around them, on the natural environment, on other species and on biodiversity.

By biodiversity, we refer to the variety and diversity of all life on earth. There are not only many similar but also competing definitions of 'biodiversity' but an appropriate definition for the subject of this book can be found on the World Wide Fund for Nature (WWF) website, as follows:

> Biodiversity is all the different kinds of life you'll find in one area–the variety of animals, plants, fungi, and even microorganisms like bacteria that make up our natural world. Each of these species and organisms work together in ecosystems, like an intricate web, to maintain balance and support life. Biodiversity supports everything in nature that we need to survive: food, clean water, medicine, and shelter.[1]

The previous definition is clearly anthropocentric in focus; in other words, it focuses on the importance and relevance of biodiversity to humans. As this book addresses the agricultural industry and its impacts on, and links to, nature and biodiversity, the WWF's definition is especially useful as it highlights how species of flora and fauna, whilst varied, are also interconnected and are also critically important to the survival of people on the planet.

If all was in balance in nature and if natural ecosystems and webs of species were healthy and functioning, as they have been for millennia, then this book would be unnecessary. The uncomfortable fact is that this is not the case. In the twenty-first century, humanity faces massive threats and challenges from water shortages to climate change, to political turmoil and wars and also the growing threat associated with biodiversity collapse and species extinctions. The risks of extinctions to humanity and especially to businesses, industry and the financial system have been explored and explained in detail in recent years (Atkins and Atkins, 2019). Linkages between human systems of finance, accounting, business and nature are becoming increasingly evident as we witness damage to natural ecosystem services such as pollination (Atkins and Atkins, 2016, 2019; Atkins and Macpherson, 2022).

Given the focus on agriculture and biodiversity and also adopting an anthropocentric approach to our discussion of biodiversity, we can also introduce the concept of 'natural capital'. In essence, we refer to natural capital as one of the forms of 'capital' that is necessary for businesses and economic production to operate. From a multi-capital perspective, natural capital is grouped together with financial capital, manufactured capital, social and

[1] See the following link: What is biodiversity? | Pages | WWF (worldwildlife.org).

relationship capital, human capital and intellectual capital as one of the building blocks of business. They are also the elements used in multi-capital accounting or integrated reporting. Indeed, natural capital has been defined as:

> Natural capital – that part of nature which directly or indirectly underpins value to people, including ecosystems, species, freshwater, soils, minerals, the air and oceans, as well as natural processes and functions. Natural capital underpins the four types of capital set out below. In combination with other types of capital, natural capital forms part of our wealth; that is, our ability to produce actual or potential goods and services into the future to support our wellbeing. (NCC, 2013)

We can immediately see from this definition that biodiversity constitutes only a part of what we understand by the term 'natural capital'. Biodiversity includes ecosystems and species, as it refers to living things, flora and fauna, but does not include freshwater or soils or any of the other elements encapsulated by natural capital. Of course, healthy air, water, oceans, freshwater and soils are all of critical importance to the survival of species, to the health of ecosystems and to rich biodiversity.[2] So for the purposes of this book, we are concerned with the interrelationships between natural capital and biodiversity in the agricultural industry and especially the ways in which agriculture's involvement in impacts on biodiversity.

It is now acknowledged that one of the most significant threats to the future of the human race is global warming and anthropocentric climate change, that is climate change that has been caused by humans. Shifts in weather patterns, increasing volatility in climate, rising temperatures, rising sea levels and increases in severe weather-related events such as hurricanes, wild fires and floods are all on the rise worldwide. The industrial revolution ushered in a new era for humanity and for the planet, as human activity began to have a growing impact on air, water and land. However, it is also widely acknowledged by scientists but also increasingly by the international financial and business communities that challenges associated with biodiversity loss and species extinctions are at least as substantial as climate change impacts. In the last few years, the focus has shifted for business from efforts to address climate change to a two-pronged approach which also recognises and seeks to address biodiversity loss.

This book hones in on the agri-food industry worldwide as it is recognised that agriculture, and especially agricultural intensification, is one of the major drivers of biodiversity decline on land. A recent report on the 'State of Nature' in a UK context identifies agricultural intensification as *the* major driver of biodiversity decline (Burns et al., 2023). Food production in every part of the planet has intensified with an ever-growing focus on quantity, with mono-agriculture and intensive farming practices such as the use of fertilisers and

[2] Maroun and Atkins (2020) demonstrated the importance of soil health and how it may be reported on.

agro-chemicals, from pesticides to parasiticides, antibiotic use, herbicides, insecticides, to name but a few. The direct impacts on biodiversity of these intensified farming practices are becoming clear. There is less and less natural habitat available to wild species of animals, birds and insects, as well as for wild flowers and plants, as land is gobbled up at an alarming rate for agricultural growth. Rainforests continue to be cut down and cleared for pasture or for agricultural production to satisfy demand for soya beans, palm oil, grapes and avocados, in fact an endless demand for varied crops and food products. Latest estimates suggest that there will need to be a 30- 60% increase in food production by the middle of the twenty-first century to meet the demands of the global population who are growing in size and in wealth (van Dijk et al., 2021; Royal Society, 2023). Indeed, the latest forecast suggest that the global population will reach 10 billion by 2060.[3]

Later in this book, contributors focus in on some of these industries, e.g. the expanding avocado market. Current trends in consumer demand include a shift towards vegetarian and vegan diets in part fuelled by a desire for our food to have less impact on the environment, climate and biodiversity. There is a weight of evidence showing that large-scale intensive animal farming, especially beef and poultry production, is contributing substantially to global warming. There are also repercussions of course on biodiversity, as more and more land is taken up for farm animals but also for the production of animal feeds. However, and there is always a however, this shifting focus to the production of soya, for example, and the currently exploding demand for products such as avocados, is having similar problems. The chapter devoted to avocados demonstrates these contradictions and the dilemmas they pose. There is certainly no easy answer and the overriding issue is that people have to be fed but at the same time there is an urgent need to protect and enhance biodiversity. In relation to agriculture:

> The conversion of land for human settlement, industrial purposes or agricultural use is resulting in unprecedented habitat degradation. Flora and fauna are being adversely affected by unsustainable levels of harvesting, poor farming practices, pollution and the effects of climate change (IPBES 2019a,b,c). (Maroun and Atkins, 2020)

The global context on biodiversity loss and deterioration of natural capital is now well-known, with scientists stating we are in a sixth period of mass extinctions of species on planet earth and around a million species being at risk of extinction. Global biodiversity is shrinking at an alarming rate and there is an acknowledged urgency that substantial efforts need to be made the world over to curb and reverse biodiversity loss. The World Economic Forum's

[3] Ferri and Arnés García (2023) make reference to the latest research by the United Nations (2022) on population.

Annual Risk Report 2020 ranks biodiversity loss as the third most impactful risk facing the global economy, and the fourth most likely to occur (CISL, 2020). The Intergovernmental Science-Policy Platform on Biodiversity and Ecosystem Services (IPBES)[4] stated that around 1 million species of flora and fauna on the planet are at risk of extinction. Land use, from the perspective of agricultural land use (as well as other competing demands on land), is acknowledged as the most significant factor affecting biodiversity and species loss, such that land management is

> ... central to preventing and reversing ecosystem loss - anthropogenic habitat degradation and loss is the single greatest cause of extinction risk. (Royal Society, 2023, p. 12; CISL, 2020)

As well as identifying the challenges and negative impacts of the agricultural industry on biodiversity and natural capital, it is essential that we find and implement solutions. The authors contributing to this book also focus on solutions and mechanisms that can assist in addressing these challenges. From efforts to increase pollinator protection in agriculture (Haefeker) to a broader approach to protect and enhance insect biodiversity on agricultural landscapes (Longing and Kelly), we can see the diversity of initiatives and strategies that are being invented and implemented to try and reverse and halt biodiversity loss.

2 Addressing biodiversity loss: the international context

There are high-level international efforts to address biodiversity loss in the agricultural industry. The human-focused approach to biodiversity protection is naturally embedded in these efforts. The Food and Agriculture Organization (FAO) of the United Nations published their first report on the 'State of the World's Biodiversity for Food and Agriculture' in 2019, which highlights the interdependencies of food/non-food production and biodiversity/species. This publication subsequently led to the endorsement of a voluntary Framework for Action on Biodiversity for Food and Agriculture in 2021. Furthermore, 2021 saw the concept of 'nature-positive production' in food and agriculture being introduced at the United Nations Food Systems Pre-Summit and Summit. (Ferri and Arnés García, 2023)

In 2022, the FAO published a further updated Framework for Action (FAO, 2023). The FAO's underlying research for the framework found that changes in land and water use and management were identified as the primary drivers of negative impacts on biodiversity in relation to the agriculture industry. However, they also found that efforts to manage biodiversity and protect/

[4] The IPBES is the intergovernmental body which assesses the state of biodiversity and of the ecosystem services it provides to society, in response to requests from decision-makers.

support ecosystem services through implementing biodiversity-friendly practices were increasing in most countries, which is encouraging. The stated objectives of the FAO's 2022 Framework include promoting transition to more sustainable agrifood systems and contributing to the achievement of the SDGs[5] and the Convention on Biological Diversity's (CBD's) Global Biodiversity Framework. Their stated strategic priority areas include improving availability of, and access to, information on biodiversity (focusing on biodiversity for food and agriculture);[6] promoting sustainable use of biodiversity and integrated approaches to its management; and improving conservation and restoration of biodiversity. They also prioritise raising awareness of biodiversity and agricultural industry issues worldwide, as well as improving and increasing research and education in this area, and establishing/improving legal and policy frameworks for biodiversity.

So as we can see there is now a concerted effort to address these challenges by many parties worldwide. In the UK, the recent 'State of Nature' report focuses on solutions (Burns et al., 2023) by identifying the impact of what they term 'sustainable agricultural management'. The report demonstrates that farmers in the UK are taking steps to conserve wildlife and biodiversity on the lands under their management and point to several agri-environment schemes (AES) in place which encourage better environmental stewardship in the farming sector in relation to biodiversity and wild species. Similarly, the report discusses the Nature Friendly Farming Network (NFFN) which was founded in 2017 and which brings over 1600 farmers together in efforts to manage their land for wildlife and biodiversity as well as obviously to provide food. Current initiatives across Europe are focusing on rewilding as a means of nature restoration as is evidenced by the extensive information available on the Rewilding Europe website, which identifies numerous rewilding projects over immense tracts of European land.[7]

The chapters in the rest of this book explore the relationship and linkages between agriculture, biodiversity and natural capital from many angles. The approach is international with country studies covering the US, South Africa, Brazil, Latin America generally, Sweden, Finland and Italy. We also address many different parts of the agricultural industry around the world. Given the very international flavour of the contributions, we have decided to focus our introductory discussion on the UK context. The following section provides some information on the current state of biodiversity in the UK in relation to agriculture, the challenges faced by the sector and by farmers and also some

[5] The SDGs are the United Nation's Sustainable Development Goals (SDGs). SDGs 14 and 15 refer specifically to the protection of biodiversity on land and in water, respectively.

[6] The FAO provides a specific definition of biodiversity in relation to agriculture as follows: 'Biodiversity for food and agriculture comprises all the components of biodiversity that contribute to crop and livestock production, forestry, fisheries and aquaculture - both wild and domesticated, and at genetic, species and ecosystem levels. It is vital to food security and nutrition, livelihoods and the supply of a range of ecosystem services' (FAO, 2022, p. 56).

[7] See Rewilding Europe | Making Europe a Wilder Place.

of the strategies and initiatives being proposed and used as solutions. We then provide a short case study which demonstrates some of the most poignant issues relating to agriculture and biodiversity in a Welsh context, with a discussion of the critical situation characterising rivers in Wales and the agricultural practices that are leading to biodiversity collapse. We then conclude this chapter with a summary of all the chapters in the remainder of the book.

3 Agriculture, nature and biodiversity: the UK context

Given that the chapters in this book cover a wide range of geographic regions as well as dealing with a large array of topics relating to biodiversity and species, we thought it would be appropriate to focus on the UK as a case study for our introductory chapter, and specifically in the context of Wales, given that we both work in a Welsh university. Furthermore, we decided to look closely at the case of the River Wye. The issues relating to agriculture and biodiversity in the Welsh context, and in the case study of the River Wye, represent the problems arising around the globe. The Welsh context represents a microcosm of the worldwide issues relating to natural capital, biodiversity, agriculture and agri-supply chains. We turn now to the historical context in the UK relating to agriculture and biodiversity before outlining the current issues relating to the context.

It is always useful to delve into history to gain an appreciation of the causes underlying current debates in society. One of the main issues underlying the current problems relating to natural capital, biodiversity protection and agriculture is the apparent trade-off between land for nature and land for agriculture. This is not a new debate by any means. If we go back in history a couple of hundred years, we can see the same arguments playing out in conflicts between agrarians and those belonging to what is known as the British Aesthetic tradition. This was especially the case during the period of the Picturesque movement, where wild landscape was seen as a necessary part of any picture or framing of landscape for artistic purposes. Indeed, looking back through history, we can see a pattern of contest between nature and agriculture which has been expressed through opposing 'factions'.

Costelloe (2013) highlights the historical tradition underlying this sentiment of the English Picturesque movements. He shows that a call for natural wilderness arose in the works of Francis Bacon as early as the sixteenth century. There was a feeling that wild areas, left to nature, should form a part of any garden or estate. Costelloe also points out that Alexander Pope in 1731 expressed a need for gardens and landscapes to incorporate woods and space for 'nature' to thrive. Similarly the works of William Gilpin, the founder of the Picturesque movement, especially his *Essays on the Picturesque,* published in 1790s, made wild-ness and roughness central to the aesthetic movement.

There was a strong backlash against these aesthetic views which manifested in a movement around the late eighteenth century, at the time of the Picturesque movement, which advocated for all land to be used for agricultural purposes to feed the nation. This was effectively a socialist movement which opposed private landholding and the creation of picturesque, aesthetic estates for the wealthy and upper class in society. These radicals sought to represent the needs of the working classes, calling for expanding agriculture to prevent starvation, especially during this time when starvation and famine were rife (Worrall, 1994). These revolutionaries, led by the English 'Radical' Thomas Spence, advocated for common ownership of land. The movement was what may be termed anti-establishment, linked to revolutionary ideas of taking land from the aristocracy and from private landowners. Worrall expresses the dichotomous views between those who saw 'nature' in an artistic, romanticised way and those who perceived 'nature' as effectively 'natural capital' (as defined earlier) to be used for agricultural production alone:

> For the Spenceans, 'Nature' was quite simply the provider of 'natural' abundance, an energy of the land which was free to be tapped as a 'natural' rights with no one having exclusive ownership of either the land or the produce. With the loss of the Spencean purchase on 'Nature's' meaning, 'Nature' could become an ideality, a function of the aesthetic (as in the case of the Picturesque) which could be tweaked, teased or neglected into shape. (Worrall, 1994, p. 246)

With current pressures on the natural environment and biodiversity combined with the continuing need to feed the population, politicians and farmers are faced with an urgent need to try and achieve a balance between preserving and protecting 'nature' as wildlife and biodiversity for ecological reasons and 'using' 'nature' as natural capital for food production.

These sentiments and the debates around them are ever-present in British society and are alluded to in the recent Royal Society Report on 'Multifunctional Landscapes' and land use. This UK-focused report portrays in some detail the conflicts over land use in an era when biodiversity and nature are becoming so incredibly important to the future of people and planet. Certain politicians have recently emphasised the need to use all available land across the UK for food production (or house building). Others, however, are prioritising the need for farmers to set aside tracts of land for wild plants, biodiversity and nature restoration. There are a whole multitude of examples of rewilding schemes being implemented by farmers but also by land owners. The polarity of these priorities and approaches reaffirm the perennial battle of views between those who want to 'use' all the land for producing food (and accommodation) and those who are seeking to retain wilderness and restore nature. Agricultural activities have a challenging balance to achieve to satisfy these very differing pressures and demands.

Latest estimates suggest that between 5% and 8% of land in England would have to be released from agricultural production by 2035 to meet current climate and biodiversity targets,[8] which gives an idea of the scale of the problem in balancing increasing demand for food production against increasing concerns about the need to conserve, protect and restore biodiversity.

The UK is experiencing severe, and it is not an exaggeration to say critical, biodiversity decline. The recent State of Nature Report highlights the uncomfortable fact that the UK is now one of the most 'nature-depleted countries on Earth' (Burns et al., 2023, p. 2). They also summarise the change in species populations in terms of abundance and distribution since 1970 in the UK. For example, they state that the abundance of 753 terrestrial and freshwater species has fallen on average by 19%. Similarly we have witnessed a decrease in the distribution of invertebrates (insects) by on average 13%, which is a significant finding for pollinators particularly, and its impact on pollination as an ecosystem service. Further the distribution of flowering plant species has fallen by a massive 54% whilst around 16% (almost 1500) species are currently at risk of extinction across the UK. The UK's recent Dasgupta Review leaves us in no doubt of the urgent need to address biodiversity loss and species extinction. The new report from the Royal Society also elucidates the negative impacts of increasingly intensive agriculture in the UK on biodiversity and natural capital in the following way:

> [I]ntensive farming has led to the homogenisation of large areas of the countryside, reducing its biodiversity and recreational value. Overall, changes in land use and management since WWII have increased agricultural production but diminished natural capital. (Royal Society, 2023, p. 15)

The overall situation is pretty critical. As we can see from the previous discussion, more and more food is required to feed the ever-growing human population but biodiversity and ecosystems need to be protected. Further, simply trying to halt biodiversity loss and the degradation of ecosystems is not enough, as it is necessary to aim for more outcomes, as

> ... the ideal system we should be aiming at is expected not only to avoid the depletion of natural resources but also to actively improve the capacity of the biosphere to sustain terrestrial and marine habitats while providing food for everyone. (Ferri and Arnés García, 2023, p. 15)

We now turn to consider a brief case study, at a local level in Wales and along the Welsh border with England, to illustrate the types of problems for biodiversity arising from even small-scale agriculture.

[8] See Royal Society (2023), quoting the National Food Strategy.

4 Agriculture and biodiversity decline: the case of the River Wye

The River Wye has the upsetting privilege of being an exemplar of decline in the waters of the UK. Once the birthplace of British tourism, it now exists as a case study into what happens when pollution and its consequences are not curtailed enough. The Wye is a major habitat for the Atlantic Salmon, with, according to the Wye Salmon Association, a peak population of 50 000 fished per year (Wye Salmon Association, 2019). Now, they report a decline to around 2000–3000 salmon per year - a decline by 94–96% - and note that this species, and its endangerment in the river, is *'a recognised barometer of overall water quality and the health of the environment'*. What can be done? According to a report by the House of Commons (2022), around 60–70% of the pollution in the river can be attributed to agriculture in general, with 42% of that due to poultry farms and their soils - covered by artificial fertilisers - running off into the river. They also suggest that 28% of the pollution is caused by waste, a fault of the monopolistic water companies in England and Wales instead. Perhaps these groups are the ones that should be held responsible?

The notion of agriculture and its effects on nature is defined quite poetically by Moss (2007), who likens the costs and benefits of agriculture on the land as the actions of Kali, the Hindu goddess of destruction, and Ceres, the Roman god of the harvest. With a less deific sentiment, any agriculture that takes place does a combination of two things: It displaces and destroys natural habitat in order to create the farmland for food production, and thereby provides food for the human population at a minimum. In the context of the Wye, an article by the BBC (Griffith, 2023) notes that one quarter of all poultry production in the UK lies in the Wye's catchment area. Yet, despite this alarming statistic from the House of Commons report, a spokesperson from the National Farmer's Union in Wales (NFU Cymru) claims, in the BBC article, that their farmers adhere to the latest science in order to do no harm to the river. Who are we to believe, and, more importantly, is the balance between destruction and harvest being maintained?

Looking to the water companies, a pair of Roman deities fit the introductory poeticism of the companies' roles quite well: One could look to Cloacina, the goddess of the sewers, and Februus, the god of purification - the roles of the water industry are to both provide sewerage services to their customers and ensure that the effluent taken is purified and cleaned before returning to the waters. In context, we can observe the water quality measure introduced by Saal and Parker (2000) which is composed of a Bathing Water quality factor and, more pertinently, a River Water quality factor. The whole industry boasts an average compliance of above 99.9% for almost 20 years, as illustrated by Peirce (Peirce, 2024), which suggests only negligible issues in the rivers of

England and Wales overall. However, we can compare this near-perfect claim to quality with another BBC article (BBC, 2023), which states that the Lower Wye has mean phosphates 175% above the upper acceptable limit set by Natural Resources Wales (NRW). This alone is an exceptional problem, so who again do we believe: the water companies, whose duties are allegedly achieved, or the citizen scientists who reveal much starker figures about the state of the river?

The answer to these almost-rhetorical questions are surprisingly simple: regardless of who *should* be responsible for solving the problems this river faces, *something* needs to be done. Be it the water companies whose regulator could target their pollution through regulatory targets which, despite claims of legacy measures espousing near-fully compliant water quality, have clear room for improvement according to newer quality measures and recent research (Peirce, 2024) or other governmental bodies who restrict or evaluate the state of farms and their use of artificial fertilisers, someone needs to be helping the river to return to its natural state, lest we, like the animist Ainu people of North Japan, grieve the loss of our waters and their wildlife. *Someone* needs to be held accountable.

This mini case study based around the River Wye exemplifies the problems addressed throughout this book at industry level, local level and country level. We now turn to introducing the rest of the chapters.

5 The layout of this book

Part 1 explores the impacts of agriculture on natural capital and biodiversity from a wide variety of perspectives, providing case studies from different agricultural industries around the world. These range from very specific agricultural industries on land to an exploration of agricultural impacts in water, in other words aquaculture.

Chapter 2 by Lange and Maroun addresses a wide variety of biodiversity issues in the agri-food sector, and associated challenges, in a South African context. The authors seek to identify best practice in relation to protecting biodiversity through agri-supply chains. The book's focus then turns to industry-specific case studies to demonstrate the ways in which different parts of agricultural production are impacting on biodiversity. In Chapter 3, the same authors explore the impact of the avocado industry on biodiversity focusing on the contradictions between a consumer-led industry that is expanding rapidly and the negative effects the industry is having on nature and biodiversity. In a similar way in Chapter 4, Van Zijl analyses the biodiversity challenges arising in the soya bean industry, in a Brazilian context. To broaden the discussion further, Chapter 5 moves away from agriculture on land to explore the extensive challenges for biodiversity arising from marine agriculture or 'aquaculture'. In this chapter, Ecim and Maroun examine seafood production and the fisheries industry and their impacts on marine biodiversity. Following this in Chapter 6, in a rather chilling study, Lieberman provides a detailed and scientific assessment of the deleterious impacts of the agro-chemical industry on biodiversity and nature, highlighting the need for agricultural companies to address the problems associated with pesticide use for example.

The book continues in Part 2 by exploring the ways in which companies in the agricultural sector are seeking to address their impacts on natural capital and biodiversity. The main focus of Part 2 is to examine the role of accounting for biodiversity and natural capital accounting as mechanisms by which companies can discharge accountability to stakeholders for their impacts on nature and biodiversity. Researching and analysing accounting, reporting and disclosures produced by companies in the area of biodiversity is one of the most effective ways of assessing what companies are doing to address biodiversity issues and challenges. Of course, as is widely acknowledged in the academic accounting literature, corporate accounting for biodiversity often falls prey to greenwashing and impression management but biodiversity reporting remains one of the easiest and most readily available ways of understanding corporate biodiversity strategies and action plans.

To this end, Chapter 7 by Houdet summarises the ways in which natural capital accounting has evolved and is evolving to enhance corporate accountability for negative impacts on biodiversity as well as to allow companies to demonstrate the ways in which they are seeking to protect and enhance biodiversity through their environmental and ecological strategies. Chapter 8 analyses accounting for biodiversity and natural capital accounting by companies in the Swedish agri-food sector in order to assess corporate biodiversity practice and the ways in which agricultural companies in Sweden are seeking to protect biodiversity. In Chapter 9, Ecim and Maroun explore the impact of the fast food industry on biodiversity. As human populations continue to increase and there is an ever-growing need to provide nutritious food worldwide at low cost, the fast food industry is expanding substantially, especially in the production of beef and chicken for fast food outlets. Specifically the authors analyse accounting for biodiversity by companies in the fast-food industry to discover the ways in which they are addressing negative impacts.

In Chapter 10, Schadewitz addresses biodiversity and natural capital issues in relation to the dairy industry in a Finnish context by exploring accounting for biodiversity and natural capital accounting by one company in the region. Turning to the palm oil industry in Malaysia, in Chapter 11, Rimmel and Abdullah explore accounting and accountability for biodiversity and natural capital in order to assess the ways in which palm oil companies are seeking to protect biodiversity and discharge accountability for their impacts. In an Italian context, Doni et al. in Chapter 12 investigate the Italian winemaking industry, analysing accounting for biodiversity by Italian winemakers and addressing biodiversity challenges arising in this sector.

As mentioned earlier, the role of accounting in communicating information about biodiversity impacts and protection is critically important for companies to explain their strategies and natural capital use. Accounting has traditionally been produced and intended for shareholders and investors as primary financial stakeholders. As is widely acknowledged today, other stakeholder groups use and need accounting. The same applies specifically to accounting for biodiversity and natural capital accounting, as a broad range of stakeholders in companies want and need to know how companies are seeking to address their impacts on biodiversity and discharging their accountability for natural capital usage. However, the institutional investment community remain an extremely significant and influential group of stakeholders who hold substantial power over companies through their investment in the stock market and in specific companies.

Recent years have seen an explosion in what is sustainable investment, or responsible investment, whereby large institutional investors worldwide are using their influence on companies to drive forward sustainable strategies and approaches in relation to climate policies, as well as most recently biodiversity. As well as being able to vote at company annual general meetings, one of the most significant ways institutional investors can influence corporate activities and behaviour is through engagement and dialogue. This literally involves investors having direct dialogue and engagement with company directors in the form of face to face, or one-on-one, meetings. These meetings and this process of dialogue represents an important means of encouraging investee companies to adopt approaches in line with the investors' expectations.

The recent increase in interest in what are widely called 'ESG', or environmental, social and governance issues across the financial markets has led to an increase in engagement and dialogue between investors and companies and biodiversity is one of the most salient areas of focus by sustainable or ESG investors at the moment. Chapter 13 seeks to provide a flavour of the ways in which the financial markets and specifically the institutional investment community is engaging with companies in the agricultural sector on biodiversity issues. In this chapter, Macpherson explores the role of global investors in protecting natural capital and delivering ecosystem services within agriculture, food and other land use value chains. The last chapter in Part 2 brings together issues of biodiversity and species with the increasing challenges associated with climate change and global warming, specifically in relation to the increasing volatility in weather patterns. In Chapter 14, Atkins et al. analyse corporate disclosures in the food retail sector of the agricultural industry to assess the impacts climate change is having on the industry and on nature and biodiversity.

Part 3, the final part of the book, focuses in on a very specific area of biodiversity and natural capital. The authors explore ways in which the agricultural industry can address, and is addressing, some of the serious problems raised by Lieberman in Chapter 6, by focusing on the negative impacts of agricultural activity on pollinators such as bees and on insect biodiversity more broadly. Recent academic research has highlighted the financial risks associated with ongoing pollinator decline and especially worldwide decline in bee populations for the food industry but also for other sectors within agriculture (Atkins and Atkins, 2016). The baseline is quite simple: no pollinators, no pollination. Replacing natural pollination, a critical ecosystem service, by artificial forms of pollination is extremely complicated and expensive. Therefore finding ways of protecting insect and pollinator diversity throughout the agri-supply chain is a critically important issue for food production and for the food industry as a whole. In Chapter 15, Haefeker, former President of the European Beekeepers' Association, examines ways in which advanced technologies can assist in protecting pollinators. The book is rounded off with a chapter from Longing and Kelly, US entomologists, who provide an in-depth discussion how insect biodiversity may be protected in the agricultural industry.

6 Conclusion

Since the dawn of human consciousness, humans have affected and altered the biodiversity of Planet Earth. As a result, the world as we know it is on the brink of an extinction event, anthropogenically inspired by the development of farmland,industry, and ironically attempts to solve previous issues caused by those developments via healthy food industries. If biodiversity is said to support everything in nature that we need to survive, as in the WWF's definition, then our constant want for more exploitation of land to solve the issue of agricultural pollution - seeking to substitute meat and poultry farms for soya,avocado and the like - is, in an example of dismal science, causing more harm to the environment, its biodiversity, and the people occupying it.

As has been explored throughout the book, this kind of self-sufficient downward spiral is not limited to one place, or one industry: non-exhaustively, this book has covered issues of biodiversity across four continents, on land, sea, rivers, and forests, and throughout the species governed by these spaces. To try and understand what efforts are being made by companies, the study of accounting is used in almost every chapter to uncover and report on the frameworks used by and useful for companies whose actions require environmental responsibility, and the bodies that govern them.

The aim of this chapter was to provide some introduction to the current issues of biodiversity and natural capital, and in so doing, has begun to elucidate the problems across the globe, and how practices such as accounting can help collect and explain the information necessary for these issues to be fixed. As the previous section covered in summary, the remainder of the book will now aim to explore various pertinent themes centred around biodiversity and its accounting, starting with the South African agri-food sector.

7 Where to look for further information

The definition of Biodiversity, as cited at the start of this chapter, can be found on the WWF's website: https://www.worldwildlife.org/pages/what-is-biodiversity. The page also discusses many of the themes present in this chapter, such as how humanity effects biodiversity, using Borneo as an example ecosystem.

For a further introduction to the notion of Natural Capital, the Ecosystems Knowledge Network discusses the role of natural capital and provides further links to the Natural Capital Protocol, which is a framework developed for businesses to assess their natural capital: https://ecosystemsknowledge.net/resources/by-theme/natural-capital-an-introduction/.

For a more economic introduction to the role of natural capital, and the neighbouring subject of Ecological Economics, Daly and Farly (2004) provide a comprehensive textbook.

The IPBES website provides a wealth of further information on biodiversity, and one such webpage is their breakdown of their duties in Assessing Knowledge, which is a part of their Work Programme section of the site: https://www.ipbes.net/assessing-knowledge.

Lastly, those interested in reading more about the state of the river Wye specifically, the Wye Salmon Association's website provides more information about the issues plaguing the river: https://www.wyesalmon.com/.

Another organisation invested in the repair of the Wye is the Friends of the Upper Wye, whose website highlights the issues that they are trying to resolve. The website also interestingly employs Citizen Science projects, which bridge the gap between professional science on the river and volunteer data collection: https://www.fouw.org.uk/.

8 References

Atkins, J. and Atkins, B. C. (Eds.). (2016). *The Business of Bees, An Integrated Approach to Bee Decline*. Saltaire: Greenleaf Publishers.

Atkins, J. and Atkins, B. (Eds.). (2019). *Around the World in 80 Species: Exploring the Business of Extinction*. New York, NY: Routledge.

Atkins, J. and Macpherson, M. (Eds.). (2022). *Extinction Governance, Finance & Accounting: Implementing a Species Protection Action Plan for the Financial Markets*. New York, NY: Routledge.

Atkins, J., Atkins, B. C., Zhao, L., Norton, S. D. and Maroun, W. (2023). Accounting, accountability and finance for de-extinction, rewilding, nature restoration and biodiversity protection. Working Paper Presented as the Plenary at the University of Bococca, Milan, Sustainability EU Module, 19th July.

BBC. (2023). River Wye tributary pollution above targets, campaigners claim. *BBC News*. https://www.bbc.co.uk/news/uk-wales-66830873.

Burns, F., Mordue, S., al Fulaij, N., Boersch-Supan, P. H., Boswell, J., Boyd, R. J., Bradfer-Lawrence, T., de Ornellas, P., de Palma, A., de Zylva, P., Dennis, E. B., Foster, S., Gilbert, G., Halliwell, L., Hawkins, K., Haysom, K. A., Holland, M. M., Hughes, J., Jackson, A. C., Mancini, F., Mathews, F., McQuatters-Gollop, A., Noble, D. G., O'Brien, D., Pescott, O. L., Purvis, A., Simkin, J., Smith, A., Stanbury, A. J., Villemot, J., Walker, K. J., Walton, P., Webb, T. J., Williams, J., Wilson, R. and Gregory, R. D. (2023). State of nature 2023, the state of nature partnership. www.stateofnature.org.uk.

Costelloe, T. (2013). *The British Aesthetic Tradition: From Shaftesbury to Wittgenstein*. Cambridge: Cambridge University Press.

FAO. (2022). *Framework for Action on Biodiversity for Food and Agriculture*. Rome: FAO Commission on Genetic Resources for Food and Agriculture. https://doi.org/10.4060/cb8338en.

Ferri, M. and Arnés García, M. (2023). *From Nature-Negative to Nature-Positive Production - A Conceptual and Practical Framework for Agriculture Based on Thermodynamics*. Budapest: FAO. https://doi.org/10.4060/cc7414en.

Griffith, H. (2023). River Wye: pollution not caused by farming, says NFU. *BBC News*. https://www.bbc.co.uk/news/uk-wales-66440390.

House of Commons. (2022). *Water Quality in Rivers* (4th Report of Session 2021-22). House of Commons.

IPBES. (2019a). Assessment report on biodiversity and ecosystem services for Africa. Accessed 1 November 2019. https://ipbes.net/assessment-reports/africa.

IPBES. (2019b). Global assessment report on biodiversity and ecosystem services for the Americas. Accessed 1 November 2019. https://ipbes.net/assessment-reports/americas.

IPBES. (2019c). Land degradation and restoration assessment. Accessed 1 November 2019. https:// ipbes.net/assessment-reports/ldr.

Jones, M. J. and Solomon, J. F. (2013). Problematising accounting for biodiversity *Accounting, Auditing and Accountability Journal* 26(5), 668–687.

Maroun, W. and Atkins, J. (2020). A practical application of accounting for biodiversity: The case of soil health. *Social and Environmental Accountability Journal* 41(1-2), 37–65.

Moss, B. (2007). Water pollution by agriculture. *Philosophical Transactions of the Royal Society B* 363, 659–666.

NCC. (2013). The state of natural capital: Towards a framework for measurement and valuation. https://www.gov.uk/government/publications/natural-capital-committees-first-state-of-natural-capital-report.

Peirce, J. (2024). *Efficiency and Quality in the English and Welsh Water and Sewerage Industry* (PhD Thesis). Cardiff University, Cardiff, UK.

Royal Society. (2023). Multifunctional landscapes: Informing a long-term vision for managing the UK's land Issued. Royal Society, DES7483 ISBN: 978-1-78252-628-5.

Saal, D. S. and Parker, D. (2000). The impact of privatization and regulation on the water and sewerage industry in England and Wales: A translog cost function model. *Managerial and Decision Economics* 21(6), 253–268.

United Nations. (2022). World population prospects 2022: summary of results. In Population Division. Cited 11 July 2023. https://www.un.org/development/desa/pd/content/World-Population-Prospects-2022.

University of Cambridge Institute for Sustainability Leadership (CISL). (2020). *Biodiversity Loss and Land Degradation: An Overview of the Financial Materiality.*

van Dijk, M., Morley, T., Rau, M. L. and Saghai, Y. (2021). A meta-analysis of projected global food demand and population at risk of hunger for the period 2010–2050. *Nature Food* 2(7), 494–501. https://doi.org/10.1038/s43016-021-00322-9.

Watson, R. A. and Watson, J. (1969). *Man and Nature: An Anthropological Essay in Human Ecology*. San Diego, CA: Harcourt, Brace and World.

Worrall, D. (1994). Agrarians against the Picturesque: Ultra-radicalism and the revolutionary politics of land. In Copley S. and Garside P. (Eds.), *The Politics of the Picturesque: Literature, Landscape and Aesthetics Since 1770*. (pp. 240-260). New York, NY: Cambridge University Press.

Wye Salmon Association. (2019). *River Wye Salmon Action Plan, Bold and Urgent Action: A way forward.* Wye Salmon Conservation Society.

Part 1

The impact of agriculture on natural capital and biodiversity: international case studies

Chapter 2

Factoring biodiversity into the agri-food sector: international best practices and experiences from South Africa

Yvette Lange and Warren Maroun, University of the Witwatersrand, South Africa

1 Introduction

Agriculture contributes significantly to the global economy, especially in developing economies in Africa and South America (Food and Agriculture Organisation of the United Nations, 2018). The agri-sector is an important source of employment, foreign currency and tax revenue and is essential for ensuring food security especially since the world's population is projected to reach 9 billion by 2050 (Food and Agriculture Organisation of the United Nations, 2018; WWF, 2022).

Unfortunately, agriculture has a large environmental footprint. The demand for agricultural commodities has led to half of the planet's available land being put to use, directly or indirectly, for agricultural purposes (WWF, 2022). Intensive agricultural use is resulting in soil loss, decreasing soil fertility, water scarcity, pollution and biodiversity loss (Rockström et al., 2017). A more sustainable approach to agriculture is urgently required.

'Sustainable agriculture' refers to a 'way of farming that should mimic natural ecosystems' (Gomiero et al., 2011). It is a broad and complex topic covering issues ranging from ecology to agronomy and soil management

http://dx.doi.org/10.19103/AS.2023.0128.16

(Paoletti et al., 2011). Sustainable agriculture is characterised by the following principles:

- The incorporation of biological and ecological processes such as nutrient cycling and soil regeneration into agricultural and food production practices;
- Using decreased amounts of non-renewable and unsustainable inputs, particularly environmentally harmful ones;
- Relying on farmers' expertise to work the land productively and promote their self-reliance and self-sufficiency; and
- Solving agricultural and natural resource problems through the cooperation and collaboration of people with different skills.

(Pretty, 2008)

The principles stated earlier are consistent with those outlined by the Food and Agriculture Organisation of the United Nations (FAO UN) as shown in Fig. 1. These are intended to guide sustainable agricultural practices, supported by the development of relevant technical, policy, governance and financing frameworks to drive innovation.

Different practices have emerged to help achieve more sustainable ways of feeding the world's population. These include, for example, agroecology, integrated agriculture, organic agriculture and sustainable intensification (Gomiero et al., 2011). Each of these practices is explained briefly in Table 1.

The importance of sustainable agriculture cannot be understated. Responsible farming practices which address the growing demand for food

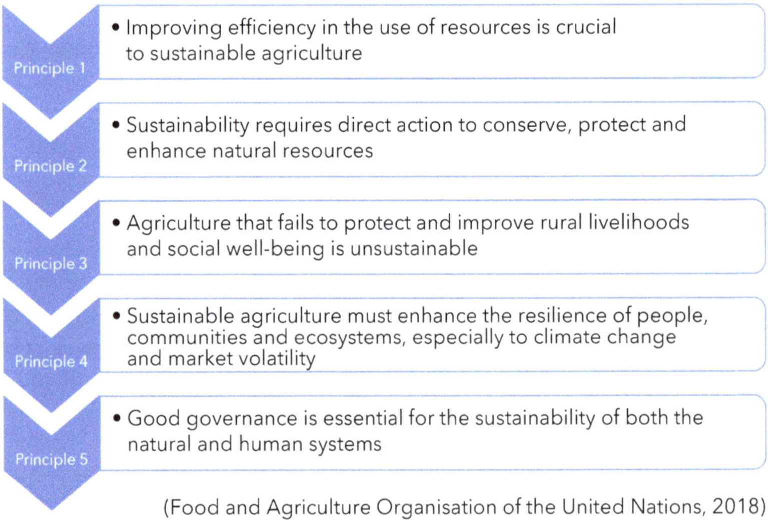

Figure 1 Key principles for sustainability in food and agriculture (FAO UN). Adapted from: Food and Agriculture Organisation of the United Nations (2018).

Table 1 Practices for sustainable agriculture

Practice	Meaning
Agroecology	'Agroecology is the holistic study of agroecosystems including all the environmental and human elements. It focuses on the form, dynamics and functions of their interrelationship and the processes in which they are involved' (Gomiero et al., 2011)
Integrated agriculture	'Integrated agriculture is not governed by specific regulations but its goal is still to reduce both farm management costs and its environmental impact as much as possible, aiming at the long-term sustainability of farming practices' (Gomiero et al., 2011)
Organic agriculture	'...organic agriculture is regulated by laws and needs to strictly follow a specific set of norms. Such norms, among others, forbid the use of agrochemicals and strictly regulate the use of drugs in animal rearing; they also forbid the use of GMOs...Organic agriculture, other than crop productivity, aims at preserving soil fertility, reducing soil erosion, conserving water, biodiversity, landscape, ecological functionality, and reducing global change' (Gomiero et al., 2011)
Sustainable intensification	'Sustainable intensification...seeks to increase agricultural output while keeping the ecological footprint as small as possible' (Rockström et al., 2017)

with social and environmental imperatives are essential for the long-term viability of organisations in the agri-sector and, at national and international level, for achieving the targets laid out in the Sustainable Development Goals. The remainder of this chapter deals with the value of biodiversity and its conservation by the food agri- and agri-processing sectors. This chapter makes a case for protecting biodiversity and provides examples of initiatives being taken to transition from environmentally insensitive to more sustainable forms of agriculture.

2 Overview of the importance of biodiversity and its conservation by the agri sector

'Biological diversity' or 'biodiversity', according to the United Nations Convention on Biological Diversity (CBD and U, 1992), refers to the

> 'variability among living organisms from all sources ,including inter alia, terrestrial, marine and other aquatic ecosystems, and the ecological complexes of which they are part: this includes diversity within species, between species and of ecosystems'.

The ecological case for protecting biodiversity is clear. Nature has an intrinsic value which is independent of economic activity, including farming. As a result, protecting biodiversity is a moral, ethical and social imperative (Naess and Sessions, 1986; Khisty, 2006). From a purely normative and subjective perspective, we consider that the attitudes of those in the agri-sector must change to ensure the continuity of human and non-human life and to give effect to the fact that humanity cannot be seen as separate from biodiversity. If current levels of biodiversity loss continue, key ecosystem services will cease to operate with the very real possibility of 'biological annihilation' (Ceballos et al., 2017).

Even if an apocalyptic outcome is avoided, the loss of plant and animal species will have massive economic implications. With almost half of all suitable land already being cultivated, the agri-sector simply cannot afford to lose additional land to desertification because poor farming practices have already disrupted nutrient recycling, cleared forested areas or over-utilised water reserves (Gomiero et al., 2011). Many plant and animal species are used directly for food, medicines, clothing and the production of basic materials. In other cases, biodiversity forms an integral part of the agricultural process. The honeybee is an excellent example. This humble insect is responsible for pollinating an array of crops, many of which will be unavailable if trends in the declining bee populations continue unchecked. In theory, pollination could be carried out manually or with sophisticated machinery but not without compromising crop quality and yields while adding significantly to the cost of food, which is already becoming unaffordable for most of the world's approximately 8 billion people (Atkins and Atkins, 2016).

Bees are only one example of an essential ecosystem service on which the agricultural sector is dependent. Other species are also responsible for pollinating crops, recycling waste, ensuring water quality or keeping pests and diseases under control. Considered collectively, ecosystem services add anywhere between US$125 and US$145 trillion to the global economy. These estimates are thought to be conservative but already highlight the enormous financial value of biodiversity and the fact that related ecosystem services cannot be replicated by the agri-sector (Atkins et al., 2019).

Biodiversity loss continues to occur at an unprecedented rate. The pressures on biodiversity and ecosystem services are only expected to increase because of rapidly expanding populations and associated demands for more food of a high-quality and at affordable prices (Leibel, 2012). Table 2 summarises the major impacts on biodiversity and ecosystem services resulting from agricultural production.

According to the WWF (2022), the challenge is to:

> 'improve our food production systems without exceeding the carrying capacity of ecosystems and the planet to support the livelihoods, food and nutrition needs of current and future generations and to deliver other essential ecosystem services'.

Agriculture has become an industrial activity where the aim is to increase production and limit costs but this has had a tremendous impact on the environment. The agri-sector needs to transition from a maximisation of

Table 2 Biodiversity impact of agricultural activity

Impact	Explanation
Habitat conversion	'Conversion of forests, grasslands and wetlands to farmland results in loss or fragmentation of habits. This, in turn, contributes to the degradation of ecosystem services such as flood and erosion control'.
Water Usage	'Approximately 70% of all water usage has been attributed to agriculture. Many areas are affected by lowering water tables and water shortages as a result of largescale irrigation. Inefficient irrigation systems not only waste significant amounts of water but can also contribute to increasing salinity levels, resulting in reduced productivity or desertification'.
Pollution	'Excess or inappropriate application of crop inputs such as fertilizers, herbicides and pesticides can result in damage to surrounding areas via run-off or spray drifting into surrounding areas'.
Inappropriate cultivation techniques	'Farming in hilly areas without due regard for contour consideration can lead to extensive erosion during rainy periods while over-cultivation of fragile soils can lead to wind erosion, soil compaction and loss of soil micro-organisms - all factors contributing to reduced productivity of the affected land'.
Introduction of invasive species	'New species introduced for crops, shade or pest control can become invasive and spread much further than originally intended. Invasive species are a major threat to ecosystems because they can lead to overcrowding or the replacement of native species. This can have a significant effect on local and global economies'.
Farming in marginal areas	'Planting and clearing forests in marginal areas can be risky for the wider landscape. Forest trees can be difficult to re-establish and micro-climates once offered by forest canopies are lost, leading to drier conditions and loss of habitats. Furthermore, the abandonment of farms following the depletion of soil fertility and/or lack of access to water can result in accelerated soil erosion and increased number of invasive species'.

(Leibel, 2012, p. 19)

production logic (where the environment is an ancillary concern) to a 'farming with biodiversity' one (where nature is integrated into agricultural processes) (WWF, 2022). Foley et al. (2011) advocate four broad strategies which would greatly increase food production while improving the environment. This includes limited agricultural expansion, improving crop yields on existing farmland, improving and increasing the resource efficiency of agricultural production and shifting behaviours and diets to bypass inefficiencies of rearing livestock. These strategies would reduce the rate of biodiversity loss, lower greenhouse gas emissions and lead to large quantities of crops of better quality using land already under cultivation (Foley et al., 2011; West et al., 2013).

3 Frameworks and guidelines dealing with biodiversity in the context of sustainable agriculture

Numerous frameworks, guidelines or codes of best practice exist or are under development regarding biodiversity and biodiversity protection by the agri-food sector. Various certification systems and other initiatives are also in place to encourage a more proactive approach to the conservation of biodiversity and sustainable agriculture. These are expected to play an increasingly important role in the agri-food sector as consumers, regulators and other stakeholders place more pressure on food producers to mitigate their impact on biodiversity and report on the steps being taken to advance sustainable agriculture and production (West et al., 2013; Maroun et al., 2018).

The remainder of this section deals with a selection of these frameworks, guidelines, codes and certification programmes. The first four examples deal specifically with biodiversity and agriculture. The remaining examples cover guidelines which focus more generally on sustainability but can be adapted for managing the agri-processing sector's biodiversity impact.

3.1 Biological Diversity Protocol

The Biological Diversity Protocol (BDP) provides a standardised approach to account for and consolidate biodiversity information on an organisation's impacts and dependencies on ecosystems and species. It complements existing environmental measurement and financial accounting systems by blending components of the Greenhouse Gas Protocol, the UN System of Environmental-Economic Accounting and double-entry bookkeeping. The BDP also incorporates elements of emerging forms of accounting which are intended to address the risk of loss of species developed mainly by the academic community (for details see Atkins and Maroun, 2018; Atkins and Atkins, 2019; Atkins and Macpherson, 2022)

The BDP aims to enable any organisation to identify, measure, account for and manage its impacts on biodiversity. The protocol covers various business applications, ranging from site management and internal reporting to external mandatory and/or voluntary disclosures. The BDP's suggested process provides organisations with a structure to create ecosystem accounts which are compatible with financial accounting (Endangered Wildlife Trust, 2020). Figure 2 illustrates how the protocol focuses on the biodiversity 'inventory', followed by the accounting for and reporting on biodiversity impact (Sustainable Flows, 2022).

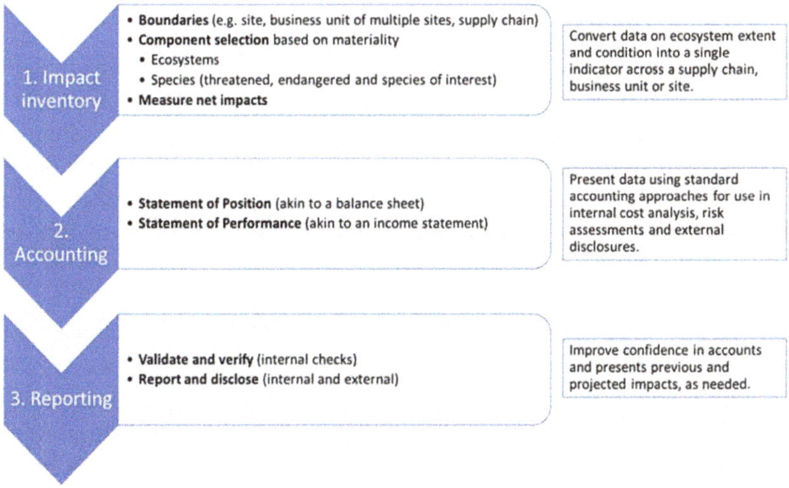

Figure 2 The Protocol's stepwise process to create ecosystem accounts that are compatible with financial accounting. Adapted from: Sustainable Flows (2022).

3.2 The Post-2020 biodiversity framework

The 2011-2020 Strategic Framework for Biodiversity and its Aichi Targets expired in 2020. At the time of writing this chapter, discussions are currently underway on the Post-2020 Global Biodiversity Framework.

The Post-2020 Global Biodiversity Framework, akin to the 'Paris agreement' on climate change, will set 21 targets and 10 milestones for governments to meet by 2030. These include protecting a minimum of 30% of the world's oceans and land, reducing pesticide use by at least two-thirds, eliminating plastic waste and increasing financial resources dedicated to biodiversity to at least US$200 billion annually. The framework has four long-term goals related to the 2050 Vision for Biodiversity. Each goal has corresponding milestones to assess progress in 2030. The long-term goals relate to:

- enhanced integrity of all ecosystems;
- valuing, maintaining or enhancing nature's contributions to people through conservation and sustainable use;
- fair and equitable sharing of the benefits from the utilisation of genetic resources; and
- closing the gap between available financial resources and other means of implementation, and those necessary to achieve the 2050 Vision (Diversity, C. o. B 2022).

3.3 Biodiversity Guidance

The Biodiversity Guidance accompanies the Natural Capital Protocol and is designed to help businesses better incorporate the value of biodiversity into natural capital assessments. The framework has been developed as a stepwise approach to incorporate biodiversity into natural capital assessments. It guides businesses and financial institutions through the process of including the value of biodiversity in their natural capital assessments and how to apply this information in internal decision-making.

Figure 3 demonstrates the stepwise approach, reflecting 4 stages, namely: 'why' (framing guidance), 'what' (scoping guidance), 'how' (measure and value guidance) and 'what next' (application guidance). This stepped approach guides businesses through the process of identifying, measuring and valuing biodiversity in natural capital assessments and how to integrate this information into decision-making processes (Capitals Coalition and Cambridge Conservation Initiative, 2020).

3.4 Framework for action on biodiversity for food and agriculture

This framework addresses biodiversity for the food and agriculture sector as a whole. It aims to improve the sustainable use and conservation of biodiversity for food and agriculture at species and ecosystem levels based on systematic and integrated approaches.

> 'The Framework addresses biodiversity for food and agriculture as a whole, including associated biodiversity and wild foods, as a basis for food security, nutrition and health, sustainable food and agriculture, and poverty reduction and livelihoods. It strengthens the Commission's existing sectoral global plans of action for genetic resources for food and agriculture and their harmonious implementation. It aims to promote coordinated action across all sectors of food and agriculture - and

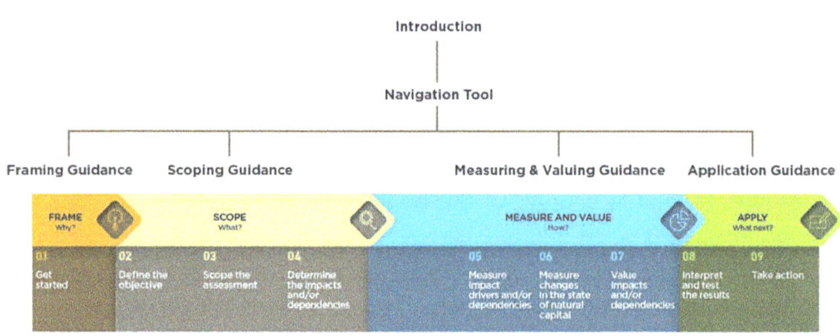

Figure 3 Biodiversity guidance and the corresponding protocol steps. Adapted from: Capitals Coalition and Cambridge Conservation Initiative (2020).

more widely - to improve the sustainable use and conservation of biodiversity for food and agriculture at genetic, species and ecosystem levels, and to contribute to the achievement of the SDGs and the implementation of the post-2020 global biodiversity framework' (Food and Agriculture Organisation of the United Nations, 2018)

3.5 Global reporting initiative

The Global Reporting Initiative (GRI) has an existing standard addressing biodiversity namely, *GRI 304: Biodiversity*. The standard requires a proactive approach to managing biodiversity and is not specific to a particular industry. Topic-specific disclosures are included in sustainability reports (or equivalent). For example, the significant impacts of products and services on biodiversity should be explicated, such as the introduction of invasive species, pests or pathogens in the case of agricultural-related activities[1].

GRI304 can be applied together with the *GRI Biodiversity Reporting Resource*, released in 2007. This was developed to assist organisations in reporting on their biodiversity performance. As the tool explains,

> 'reporting offers organisations an opportunity to explain their relationship with biodiversity. In what way does the organization respond to negative impacts on biodiversity deriving from its activities? What are positive impacts?' (Global Reporting Initiative, 2007).

The GRI Sector Standard Project for Agriculture, Aquaculture and Fishing is a pilot project of the GRI Sector Program. This project aims to identify and describe the significant impacts and stakeholder concerns for the agriculture, aquaculture and fishing sectors from a sustainable development perspective. It documents and provides an authoritative reference for material environmental impacts by broader agri-business. The intention is to increase transparency and promote more consistent reporting by organisations in the respective sectors. Prioritised due to these sectors' significant sustainability impacts, the project calls on companies to disclose 26 likely material topics of which biodiversity is one. The new sector-specific standard for agriculture, fisheries and aquaculture is expected to be published in the first quarter of 2023.

3.6 Sustainable development goals

The United Nations (UN) adopted the 2030 Agenda for Sustainable Development in 2015, which includes the Sustainable Development Goals (SDGs). These goals and targets address a range of global challenges.

[1] The GRI's GSSB appointed a multi-stakeholder Technical Committee in November 2021 to revise *GRI 304: Biodiversity*. This is long overdue give the importance of biodiversity and the fact that the standard was last revised in 2006.

> 'Four of these goals are linked to food production and security: declining biodiversity (SDG 15), loss of ecosystem services and agroecosystem stability caused by increasing stress from food production intensification and climate change (SDG 13), declining soil health caused by agricultural practices (SDGs 2 and 6), and dependence on synthetic fertilisers and pesticides to maintain high productivity (SDG 2). To achieve these SDGs, the agriculture sector must take a leading role in reversing the many negative environmental trends apparent in today's agricultural landscapes to ensure that they will adapt and be resilient to climate change in 2030 and beyond' (Shahmohamadloo et al., 2022).

Biodiversity and ecosystems feature prominently across many of the Sustainable Development Goals (SDGs) and associated targets. They contribute directly to human well-being and development priorities.

3.7 Task force on nature-related disclosures

The Taskforce on Nature-Related Financial Disclosures (TNFD) is an international initiative building on a model developed by the Taskforce on Climate-Related Financial Disclosures (TCFD). Its mission is to provide a framework for how organisations can address environmental risks and opportunities to channel capital flows into positive action. TNFD seeks to provide a framework for organisations to report on risks from biodiversity loss and ecosystem degradation. In turn, improving the availability of data and information will enable organisations to integrate nature-related risks more accurately and reliably into decision making.

On 15 March 2022, the TNFD published a draft version of its proposed framework. The draft framework includes three core sections: key concepts and definitions; draft disclosure recommendations; and guidance on nature-related risk and opportunity assessment. The TNFD plans to test and refine its framework in 2022 and launch it in 2023. Its approach will complement and be aligned with the TCFD. In particular, the TNFD framework will be structured around the same four pillars: governance, strategy, risk management, and metrics and targets (TNFD, 2022).

3.8 Certification or verification programmes

There are numerous labels and certification standards for agricultural products. Some certification programmes, for example, give prominence to biodiversity and ecological criteria. Several agricultural certification systems are working to incorporate biodiversity into their standards.

The following list provides an overview of key pro-biodiversity guidelines, as extracted from the guidance offered by the EU Business @ Biodiversity Platform (2022):

- International Federation of Organic Agricultural Movements (IFOAM) (http://www.ifoam.org);

- Conservation Agriculture (CA) techniques focuses on the protection and sustainable use of natural resources (http://www.fao.org/ag/ca/);
- GACP (Good Agricultural and Collection Practices (for medicinal plants) (http://whqlibdoc.who.int/publications/2003/9241546271.pdf);
- Certification Global GAP Agriculture (GAP = Good Agricultural Practices), focuses mainly on food safety (http://www.globalgap.org/cms/front_content.php?idcat=48);
- Certification FLO, Fairtrade, focuses on social and environmental criteria (http://www.fairtrade.net);
- International Organic Accreditation Service (IOAS) (http://www.ioas.org/);
- Rainforest Alliance, sustainable agriculture program and certification (http://www.rainforest-alliance.org/); and
- Sustainable Agriculture Initiative (SAI) Platform, focuses on principles and practices for the sustainable production of arable and vegetable crops, coffee, dairy, fruit and sustainability in water management (http://www.saiplatform.org).

Certification standards are not the same as an environmental audit which confirms that companies in the agri-sector are operating sustainability. They do, however, provide a benchmark which encourages agri-business to implement and adhere to minimum standards and can be used by stakeholders to evaluate products and hold organisations accountable.

3.9 Voluntary incentives for biodiversity-focused business

There are several types of voluntary incentives to encourage biodiversity conservation by businesses in the agri-sector and other industries. These include mechanisms such as bio trade arrangements and conservation covenants. In 1996, the United Nations Conference on Trade and Development launched the BioTrade Initiative. BioTrade Principles and Criteria have been established to inform businesses and governments wishing to support the conservation and sustainable use of biodiversity in addition to the fair and equitable sharing of biodiversity-related benefits through trade (United Nations Conference on Trade and Development, 2014; EU Business @ Biodiversity Platform, 2022).

The initiatives can include, for example, embedding biodiversity conservation in the title deed of land when transferring it to a new buyer. Stewardship payments can be included as part of the transfer to fund managers responsible for delivering environmental services for public benefit and achieving biodiversity targets. The arrangements promote the collection, production, transformation and commercialisation of goods and services derived from native biodiversity according to criteria developed by the UN.

3.10 Other initiatives and good practices

Companies in the food and agri-sectors are increasingly interested in promoting more sustainable agricultural practices to distinguish themselves from competitors and in response to stakeholder pressure. Examples of these efforts include the Sustainable Agriculture Initiative Platform (SAI Platform) and Business for Social Responsibility's Food and Agriculture Group.

The former is an organisation created by the food industry to support the development of sustainable agricultural and related practices among different stakeholders in the food value chain. This is achieved through a series of activities including, for example, researching improved farming techniques, assisting farmers to implement best practices and developing a performance assessment tool to guide self-evaluations and external reviews of sustainable agricultural practices (SAI Platform, 2022).

The Business for Social Responsibility's Food and Agriculture Group focuses on sustainable water in its corporate members' supply chains. BSR's projects include helping growers with sustainable farming and capacity-building, engaging stakeholders on challenging sustainability issues and setting globally relevant sustainability targets (BSR, 2022). Other examples of efforts to drive sustainable agriculture are discussed in more detail in Section 4.

4 Sustainable agriculture case studies: global initiatives and companies

This section includes short case studies on, or illustrations of, sustainable agricultural practices including biodiversity conservation. Section 4.1 deals with South African-focused initiatives promoting responsible land use and biodiversity protection. Section 4.2 briefly discusses interventions by specific South African businesses.

The focus is on South Africa because the country provides an excellent setting for examining sustainable agriculture because of the significance of the agri-sector to its economy. South Africa is also a developing economy where the importance of managing socio-economic and environmental issues is paramount, especially given the fact that the country is home to significant areas of pristine wilderness and endangered biodiversity. At the same time, the country is regarded as a corporate governance and integrated reporting pioneer (Maroun and Cerbone, 2020) with the result that its agri-sector should have experience addressing issues such as biodiversity loss, food sustainability and responsible supply chains.

The details provided later are based only on public information. Detailed case studies or ethnographies have not been completed with the result that the extent to which different initiatives are being implemented and their impact cannot be discerned accurately. Nevertheless, the details later should provide useful input into any analysis of sustainable agriculture, especially in a developing economy setting.

4.1 South African initiatives

The review of South African initiatives linked directly or indirectly to sustainable agriculture and biodiversity conservation is not exhaustive. The cases dealt with in this section of the chapter include the Sustainability Initiative of South Africa, Badger Friendly Honey and Biodiversity and Wine Initiative (BWI). These cases were selected due to their prominence in the local agri-sector.

4.1.1 Sustainability initiative of South Africa

In 2017, the WWF-SA, in partnership with Blue North, initiated the development of a comprehensive Sustainability Initiative of South Africa (SIZA) Environmental Standard applicable to the South African fruit sector.

> 'The SIZA Environmental Standard and supporting Assurance model has been designed to assist growers in evaluating their current compliance and environmental risks, both at a farm and regional/catchment level. This process is accomplished through the completion of a Self-Assessment Questionnaire (SAQ), the results of which feed into a risk profile report, which can be used to address market requirements, whilst also informing the drafting of site-specific improvement plans and farm environmental management plan and monitoring system. The questions included in the self-assessment questionnaire runs from minimum legal requirements to leading practice across four main topics such as water, soil, energy materials and waste, and farm ecosystems and biodiversity' (Sustainability Initiative of South Africa, 2022).

The eight principles which the standard addresses are summarised in Fig. 4.

4.1.2 Badger friendly honey

Honeybees provide essential pollination services to crops and form the basis of the honey industry, which in South Africa has an estimated turnover of ZAR3.2

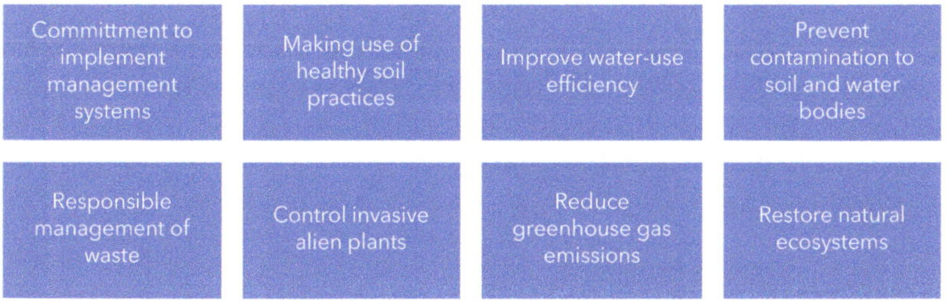

Figure 4 SIZA Environmental Standard Principles. Adapted from: Sustainability Initiative of South Africa (2022).

billion (Atkins and Atkins, 2016). Honey badgers are often seen as a pest by the honey industry because their appetite for bee larvae results in frequent raids on beehives. Badgers were trapped and killed in an effort to protect beehives but without regard for the vital role they play in maintaining a healthy ecosystem.

In 2001, a partnership was formed between the South African Bee Industry Organisation, the Endangered Wildlife Trust, The Green Trust and The Wildlife and Environmental Society of South Africa under the banner of the 'Badger Friendly Honey Initiative'. This initiative trains beekeepers to raise hives out of reach of honey badgers, audits adherence by beekeepers to sustainable practices and accredits participating honey producers.

The 'Badger Friendly Honey Initiative' has halted a decline in badger populations and changed beekeepers' perception of the animal. The initiative is also an excellent example of how consumer awareness, coupled with action by farmers and retailers, can promote biodiversity protection without compromising the quality of produce and adding materially to costs (SANBI, 2006; Conservation South Africa, 2022).

4.1.3 Biodiversity and wine initiative

South Africa is the world's eighth largest producer of wine. Approximately 90% of the country's wine is produced in the Cape Floristic Region, which is one of the world's richest areas for biodiversity. The BWI was formed in 2004 as a collaboration between the South African wine industry and the conservation sector with the objective of minimising loss of an already threatened natural habitat and contribute to the sustainability of local wine production. The BWI has developed biodiversity guidelines for implementation by wine farmers who are assisted with assessing the conservation value of their land in ecological and economic terms.

A major achievement of the BWI has been to embed biodiversity guidelines into the South African wine industry. The initiative promotes transitioning to more sustainable agriculture practices which improves the sustainability of vineyards and promotes the conservation of natural habitats. Consumers can identify BWI members' wine by the prominent label displayed on participating members' wine bottles. The label includes a Cape Sugarbird on a protea with the slogan 'conservation in action'. The initiative is currently funded by the Green Trust (a WWF-Nedbank partnership), CEPF, Wines of South Africa and the RMB Fund (SANBI, 2006; Conservation South Africa, 2022).

4.2 Company case studies

Specific examples of sustainable agricultural practices at the company-level include the *Farming for the Future Project* by Woolworths, several initiatives run by Bayer and steps being taken by Unilever to mitigate the environmental impact of agri-business. The companies' efforts are discussed briefly later.

4.2.1 Woolworths' farming for the future

Woolworths Holdings Limited (Woolworths) is a South African-based retail group. The Group operates a chain of stores offering a selected range of clothing, food, homeware, beauty products and financial services. The Group also has stores in selected African countries and the Middle East. It has a subsidiary in Australia.

Woolworths sells locally grown fresh produce sourced from various suppliers. As part of the Woolworths' 'Good Business Journey', *Farming for the Future* was adopted in 2009 to address the many agricultural challenges that face South Africa, namely water quality and scarcity, poor quality soils, food security, climate change and rising farming costs. Woolworths concluded that a change in approach was needed to produce high-quality food while protecting biodiversity and continuing to support the livelihood of those in the agricultural sector.

King and Thobela (2014) are privy to the details of the *Farming for the Future Project*. The initiative was developed by a multi-stakeholder group including Woolworths' technologists, buyers, produce suppliers and scientific experts under the guidance of a technical advisory committee. The committee helped determine the key areas of priority for farming and packaging/processing. These include

- 'soil management (soil chemical composition, soil nutrient status, fertilisation practices, soil carbon content and soil cover);
- irrigation water management (calculation of irrigation requirement, measurement of soil moisture, water use efficiency, water chemical composition and water health);
- environmental legal requirements (environmental management, water management, agricultural management, heritage management and property zoning);
- biodiversity management (conservation of endangered species, alien invasive plant management, fire management, game and problem animal management and erosion management);
- waste and wastewater management (general farm waste, agri-industry waste, hazardous waste, process wastewater and sewerage waste);
- cooling and energy (measurement of energy use, program to ensure continual improvement and refrigerant used);
- pest and plant management (legal agrochemical usage, integrated pest and disease management and integrated weed management); and

- substrate and run-off management (substrate type, calculation of fertilization requirements, practical implementation of a fertiliser program, run-off water management and structure)'.

(King and Thobela, 2014)

Farming for the Future was adopted by the company's leading suppliers but subsequently extended to other suppliers. A critical success factor for the programme is the use of an annual audit process. King and Thobela explain that:

> '...Each supplier begins with a baseline audit and in line with Woolworths' strategy of continuous improvement, the supplier pass mark for the assessment increases every year. On the first audit, the supplier needs to achieve 50%, second audit 55% and third audit 60%. Effectively, this means that even if a supplier scores the same as he did last year, or even slightly more in a specific aspect, he may still be below the pass mark for the current year'.

Farming for the Future hinges on a partnership between Woolworths and its suppliers cultivated over several years. Sustainability targets are not imposed on producers but incorporated into their business models as part of a process of releasing financial, environmental and social benefits for farmers, Woolworths and consumers. *Farming for the Future* provides an excellent example of a retailer's role in driving positive change in the agricultural supply chain and providing customers with more responsible and sustainable choices (King and Thobela, 2014).

4.2.2 AVI limited

AVI's extensive brand portfolio includes more than 50 brands and spans a range of categories including hot beverages, sweet and savoury snacks, frozen convenience foods, out-of-home ranges, personal care products, cosmetics, shoes, accessories, and fashion apparel.

I&J is one of AVI's frozen food brands and is a leading fishing company and manufacturer of high quality chilled and frozen foods. I&J (as part of AVI's Sustainability Report) states that the company 'strives to lead initiatives to manage fishing effort and protect breeding areas off the South African coast'. A key example of how this is done is the partnership with the WWF-SA and other major South African fishing companies formed in 2009 and called the Responsible Fisheries Alliance (RFA). The alliance is intended to ensure that all stakeholders understand and support the implementation of fishing practices which protect marine ecosystems. I&J is also a signatory to a Participation Agreement with WWF-SA's Sustainable Seafood Initiative. The agreement includes, inter alia, environmental standards specific to each area of operation to which the company and suppliers are expected to adhere.

I&Js' abalone farm is an example of its biodiversity conservation policies in action. The abalone farm is situated immediately adjacent to the sea in an area of high biodiversity. To minimise any potential harm, I&J (in partnership with the WWF and broader abalone farming industry) has adopted the Global Abalone Standards and has been audited by the Aquaculture Stewardship Council. Accreditation was granted in November 2015 and renewed in 2019 (AVI Limited, 2020).

4.2.3 Unilever

In 2021, Unilever adopted 'Regenerative Agriculture Principles' which are a 'new approach to farming that works in harmony with nature to ensure the long-term viability and resilience of land'. The initiative is an international one but is also relevant in a South African setting.

Building on Unilever's Sustainable Agriculture Code adopted in 2010, the 'Regenerative Agriculture Principles' outline five priority areas: (1) regenerating soils, (2) protecting water, (3) increasing biodiversity, (4) developing climate solutions and (4) improving farmer livelihoods. For example, Unilever and its subsidiaries partner with farmers to promote the use of organic fertilisers and improve biodiversity on cultivated land to support essential ecosystem services including pollination and nutrient recycling. This is complemented by partnerships with NGOs and universities to explore alternate farming practices, train farmers and create key performance indicators for sustainable agriculture including the amount of carbon and water captured by soil.

The Regenerative Agriculture Principles programme is based on four key tenets grounded in the focusing on the renewal of natural resources and arable land and supporting local communities:

- 'Have positive impacts from agricultural practices on soil health, water and air quality, carbon capture and biodiversity;
- Enable local communities to protect and improve their environment and wellbeing;
- Produce crops with sufficient yield and nutritional quality to meet existing and future needs, while keeping resource inputs as low as possible; and
- Optimise the use of renewable resources while minimizing the use of non-renewable resources' [Supply Chain (Rhys Thomas), 2021; Unilever, 2021].

5 Conclusion and future trends

Promoting a more sustainable agri-sector is critically important for ensuring global food security without compromising on environmental and other social objectives espoused by the Sustainable Development Goals. While there are several frameworks, guidelines and initiatives in place addressing different aspects of agricultural activity and related biodiversity issues, there is no consolidated approach to assist countries and companies respond to the sustainability challenges faced by the agri-sector. What is needed is an integrated approach to managing the impact of the entire agricultural supply chain with a focus on the different capitals or resources on which the agri-sector is dependent.

The concept of 'integrated thinking' or the proactive management of the economic, environmental and social factors originates in the early environmental accounting literature (Gray, 1992) and has gained momentum following the release of the *International Integrated Reporting Framework* in 2013 (IIRC, 2021). In the context of agri-business, the objective is to ensure that short-term financial goals do not compromise longer-term objectives laid out by the Sustainable Development Goals. Food production and distribution can be profit-orientated but the impact that agricultural activity has on the environment must also be considered. At the same time, biodiversity loss, declining soil quality and land availability are no longer ancillary concerns but key business risks which must be addressed proactively at the strategic and operational levels. The goal is to achieve food security and generate reasonable financial returns in a way that does not compromise the functioning of ecosystem services and give rise to material environmental challenges which must be addressed by future generations.

Different models for integrated thinking have emerged (see De Villiers et al., 2020) but the one outlined by Maroun and Atkins (2021) is presented later because it deals specifically with soil health. As such, it can be easily modified and expanded to address sustainable agriculture (Fig. 5).

Starting with the 'biological' and 'strategic context', the aim is to identify and define the impact of agricultural activity on biodiversity. This can include, for example, habitat conversion, water usage and the introduction of invasive species as per Table 2. The scientific or technical nature of the problem is evaluated and contextualised according to the business model by reflecting on implications for the organisation's strategy and operations, other members of the value chain and consumers. Risks associated with poor agricultural practices are then identified and linked to specific mitigation plans. Opportunities for implementing initiatives which mitigate adverse biodiversity implications, such as those outlined in Section 4, are evaluated.

Figure 5 refers to policies for soil health management. These can be expanded to deal with the other key aspects of sustainable agriculture including those outlined by Pretty (2008) or the Food and Agriculture Organisation of the United Nations (2018) (see also Fig. 1).

Policies are of little value if they are not backed by action. Maroun and Atkins (2021, p. 20) call for performance reporting which includes specific key

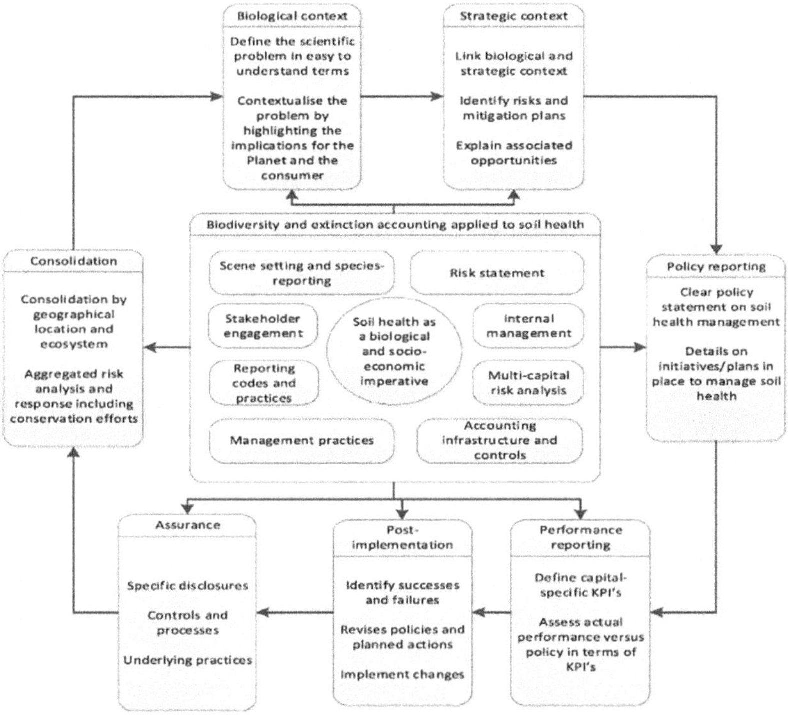

Figure 5 Integrated approach to sustainable agriculture. Adapted from: Maroun and Atkins (2021, p. 20).

performance indicators. Examples could include: an increase in biodiversity on land under cultivation, biological targets for the health/functionality of ecosystem services, waste products recycled or disposed of responsibly and the number of value chain members/business partners adhering to codes on sustainable agricultural practices such as those detailed in Section 3. Performance should be evaluated against targets or objectives. Unfavourable variances should lead to remedial action. Favourable ones should be identified as best practices and replicated in other parts of the business and at other members of the value chain. Table 3 highlights existing and emerging guidelines discussed in the preceding sections which can inform how organisations operationalise the 'components' of the integrated thinking model.

Assurance will play an important role in evaluating operating activities, tracking progress against targets and enabling accountability for environmental performance. Assurance can include internal auditors responsible for overseeing controls designed to reduce biodiversity loss; certification by external parties of goods and services as being sustainably sourced/produced

Table 3 Mapping of integrated thinking elements to existing agriculture/biodiversity guidelines

Guideline	Element of the integrated thinking model
Biological Diversity Protocol	Strategic context, biological context, performance reporting
Post-2020 Biodiversity Framework	Policy reporting, performance reporting
Biodiversity Guidance	Performance reporting
Framework for Action on Biodiversity for Food and Agriculture	Policy reporting, strategic context, biological context
GRI	Accounting principles, performance reporting
SDGs	Policy reporting, strategic context, biological context, performance reporting
TNFD	Performance reporting

and active monitoring by organisations' governance bodies to drive longer-term sustainability objectives.

The integrated thinking model culminates in high quality and transparent reporting on the impact of policies and initiatives on the business model. This reporting takes place internally to assist with performance evaluation and decision-making. External reporting to stakeholders is also essential. Other members of the value chain should be informed of the steps taken to align with generally accepted sustainability standards such as those outlined in Section 3 or 4. Regulators at regional, national and international levels can use external reports to evaluate performance by company, industry and geographical location. Other stakeholders, such as NGOs and consumers, will also find the reports useful for informing their engagement with companies and holding members of the agri-sector accountable for their biodiversity impact.

Ideally, this integrated approach to managing food producers, agri-processors, distributors and retailers can be scaled to deal with the entire value chain rather than a specific entity. Preparing 'consolidated multi-capital accounts' which capture the effect of different parts of the agri-sector on the environment from a geographical or eco-system perspective (see Russell et al., 2017; Atkins and Maroun, 2020) will go a long way to understanding and minimising the global food industry's environmental footprint

Finally, this chapter deals with several core sustainability-related issues specific to the agricultural sector but is not a comprehensive account of existing initiatives designed to improve the long-term viability of agri-business and limit its impact on biodiversity. There is an urgent need for additional research to identify, understand and measure the sector's dependence on biodiversity. For example, there is no generally accepted framework for managing and reporting on the economic, social and environmental factors relevant to members of the agri-sector. The model outlined earlier is a reasonable starting point but it is yet to be tested. For example, the types of key performance indicators which optimise sustainability targets should be examined in more detail including methods for measuring the social and environmental costs of agricultural activity. Estimates of the value of ecosystem services are very broad and developing a more detailed model for gauging changes in the state of these ecosystems and attributing the costs of environmental degradation will go a long way to making the case for sustainable agriculture and holding the industry accountable for how it puts food on the table.

6 Where to look for further information

- Capitals Coalition and Cambridge Conservation Initiative. 2020. Integrating biodiversity into natural capital assessments. Available: https://capitalscoalition.org/wp-content/uploads/2020/10/Biodiversity-Guidance_COMBINED_single-page.pdf.
- Endangered Wildlife Trust. 2020. The Biological Diversity Protocol Available: https://www.nbbnbdp.org/uploads/1/3/1/4/131498886/biological_diversity_protocol__bd_protocol_.pdf.
- EU Business @ Biodiversity Platform 2022. Agriculture Sector and Biodiversity Conservation Best Practice Benchmarking.
- Leibel, N. 2012. Protecting biodiversity in production landscapes: a guide to working with agribusiness supply chains towards conserving biodiversity.

7 References

Atkins, J. and Atkins, B. 2016. *The Business of Bees: An Integrated Approach to Bee Decline and Corporate Responsibility*, Milton Park, Abingdon, Oxon: Routledge.

Atkins, J. and Atkins, B. 2019. *Around the World in 80 Species. Exploring the Business of Extinction*, London: Routledge.

Atkins, J. and Macpherson, M. 2022. *Extinction Governance, Finance and Accounting: Implementing a Species Protection Action Plan for the Financial Markets*, Abingdon, Oxon, United Kingdom: Taylor & Francis.

Atkins, J. and Maroun, W. 2018. Integrated extinction accounting and accountability: building an ark. *Accounting, Auditing and Accountability Journal* 31(3), 750–786.

Atkins, J. and Maroun, W. 2020. The Naturalist's Journals of Gilbert White: exploring the roots of accounting for biodiversity and extinction accounting. *Accounting, Auditing and Accountability Journal* 33(8), 1835–1870.

Atkins, J., Maroun, W. and Atkins, B. 2019. How can accounting, integrated reporting and engagement prevent extinction? In: Atkins, K. and Atkins, B. (Eds). *Around the World in 80 Species. Exploring the Business of Extinction*, London: Routledge.

AVI Limited 2020. *Sustainability Report*. Available at: https://www.avi.co.za/wp-content/uploads/2020/10/Sustainability-Report-2020.pdf [Accessed 19 May 2022].

BSR 2022. Food, beverage, and agriculture. Available at: https://www.bsr.org/en/industry-focus/food-beverage-and-agriculture [Accessed 9 May 2022].

Capitals Coalition and Cambridge Conservation Initiative 2020. *Integrating Biodiversity into Natural Capital Assessments*. Available at: https://capitalscoalition.org/wp-content/uploads/2020/10/Biodiversity-Guidance_COMBINED_single-page.pdf.

CBD and U 1992. Abgerufen am [Online]. United Nations Convention on Biological Diversity (vol. 5).

Ceballos, G., Ehrlich, P. R. and Dirzo, R. 2017. *Biological Annihilation via the Ongoing Sixth Mass Extinction Signaled by Vertebrate Population Losses and Declines*. Proceedings of the National Academy of Sciences.

Conservation South Africa 2022. *Green Choice*. Available at: http://biodiversityadvisor.sanbi.org/wp-content/uploads/2014/07/GreenChoice_brochure-2011_low-res.pdf [Accessed 19 May 2022].

De Villiers, C., Hsiao, P.-C. K. and Maroun, W. 2020. *The Routledge Handbook of Integrated Reporting*. 2 Park Square, Milton Park, Abingdon, Oxon OX14 4RN: Routledge.

Diversity, C. o. B 2022. *Preparations for the Post-2020 Biodiversity Framework*. Available at: https://www.cbd.int/conferences/post2020.

Endangered Wildlife Trust 2020. The biological diversity protocol. Available at: https://www.nbbnbdp.org/uploads/1/3/1/4/131498886/biological_diversity_protocol__bd_protocol_.pdf.

EU Business@Biodiversity Platform 2022. *Agriculture Sector and Biodiversity Conservation - Best Practice Benchmarking*, European Network for Rural Development: Belgium. Retrieved from https://policycommons.net/artifacts/2010376/agriculture-sector-and-biodiversity-conservation/2762819/ on 03 Nov 2023.

Foley, J. A., Ramankutty, N., Brauman, K. A., Cassidy, E. S., Gerber, J. S., Johnston, M., Mueller, N. D., O'Connell, C., Ray, D. K., West, P. C., Balzer, C., Bennett, E. M., Carpenter, S. R., Hill, J., Monfreda, C., Polasky, S., Rockström, J., Sheehan, J., Siebert, S., Tilman, D. and Zaks, D. P. 2011. Solutions for a cultivated planet. *Nature* 478(7369), 337–342.

Food and Agriculture Organisation of the United Nations 2018. Transforming food and agriculture to achieve the SDGs. 20 interconnected actions to guide decision-makers. Available at: http://www.fao.org/3/I9900EN/i9900en.pdf.

Global Reporting Initiative 2007. *Biodiversity: A GRI Reporting Resource*. Available at: https://www.yumpu.com/en/document/read/21483104/biodiversity-a-gri-reporting-resource-global-reporting-initiative [Accessed 25 April 2022].

Gomiero, T., Pimentel, D. and Paoletti, M. G. 2011. Is there a need for a more sustainable agriculture? *Critical Reviews in Plant Sciences* 30(1–2), 6–23.

Gray, R. 1992. Accounting and environmentalism: an exploration of the challenge of gently accounting for accountability, transparency and sustainability. *Accounting, Organizations and Society* 17(5), 399–425.

IIRC 2021. International integrated reporting framework. Available at: https://integratedreportingsa.org/the-international-ir-framework-2021/ [Accessed 19 Jan 2021].

Khisty, C. J. 2006. Meditations on systems thinking, spiritual systems, and deep ecology. *Systemic Practice and Action Research* 19(4), 295–307.

King, L. and Thobela, S. 2014. Woolworths farming for the future. *International Food and Agribusiness Management Review* 17(1030-2016-83014), 161–166.

Leibel, N. 2012. Protecting biodiversity in production landscapes: a guide to working with agribusiness supply chains towards conserving biodiversity.

Maroun, W. and Atkins, J. 2021. A practical application of accounting for biodiversity: the case of soil health. *Social and Environmental Accountability Journal* 41(1–2), 37–65.

Maroun, W. and Cerbone, D. 2020. *Corporate Governance in South Africa*. (Vol. 2). Walter de Gruyter GmbH & Co KG: (Berlin / Boston).
Maroun, W., Usher, K. and Mansoor, H. 2018. Biodiversity reporting and organised hypocrisy: the case of the South African food and retail industry. *Qualitative Research in Accounting and Management* 15(4), 437–464.
Naess, A. and Sessions, G. 1986. The basic principles of deep ecology. *The Trumpeter* 3(4), 3–6.
Paoletti, M. G., Gomiero, T. and Pimentel, D. 2011. Introduction to the special issue: towards a more sustainable agriculture. *Critical Reviews in Plant Sciences* 30(1-2), 2–5.
Pretty, J. 2008. Agricultural sustainability: concepts, principles and evidence. *Philosophical Transactions of the Royal Society of London. Series B, Biological Sciences* 363(1491), 447–465.
Rockström, J., Williams, J., Daily, G., Noble, A., Matthews, N., Gordon, L., Wetterstrand, H., DeClerck, F., Shah, M., Steduto, P., de Fraiture, C., Hatibu, N., Unver, O., Bird, J., Sibanda, L. and Smith, J. 2017. Sustainable intensification of agriculture for human prosperity and global sustainability. *Ambio* 46(1), 4–17.
Russell, S., Milne, M. J. and Dey, C. 2017. Accounts of nature and the nature of accounts: critical reflections on environmental accounting and propositions for ecologically informed accounting. *Accounting, Auditing and Accountability Journal* 30(7), 1426–1458.
SAI Platform 2022. *Sustainable Agriculture for a Better World*. Available at: https://saiplatform.org/ [Accessed 9 May 2022].
SANBI 2006. Biodiversity business. Available at: https://www.sanbi.org/wp-content/uploads/2018/03/biodiversitybusiness.pdf [Accessed 19 May 2022].
Shahmohamadloo, R. S., Febria, C. M., Fraser, E. D. G. and Sibley, P. K. 2022. The sustainable agriculture imperative: A perspective on the need for an agrosystem approach to meet the United Nations Sustainable Development Goals by 2030. *Integrated Environmental Assessment and Management* 18(5), 1199–1205.
Supply Chain (Rhys Thomas) 2021. *What Is Unilever's New Regenerative Agriculture Initiative?* Available at: https://supplychaindigital.com/sustainability/what-unilevers-new-regenerative-agriculture-initiative [Accessed 7 April 2022].
Sustainability Initiative of South Africa 2022. *SIZA*. Available at: https://siza.co.za/ [Accessed].
Sustainable Flows 2022. *Biological Diversity Protocol*. Available at: https://sustainableflows.com/biological-diversity-protocol/ [Accessed 4 May 2022].
TNFD 2022. *TNFD Nature-Related Risk & Opportunity Management and Disclosure Framework*. Available at: https://framework.tnfd.global/.
Unilever 2021. The Unilever regenerative agriculture principles with implementation guidelines 2021. Available at: https://assets.unilever.com/files/92ui5egz/production/489410442380812907bc3d97be02ccda1a44ab4b.pdf [Accessed 7 April 2022].
United Nations Conference on Trade and Development 2014. The Business of Biotrade: using biological resources sustainability and responsibly. Available at: https://unctad.org/system/files/official-document/ditcbcc20094_en.pdf [Accessed 9 May 2022].
West, P. C., Biggs, R., McKenney, B. A. and Monfreda, C. 2013. *Feeding the World and Protecting Biodiversity*. Encyclopedia of Biodiversity (Second Edition), Academic Press. 426–434.
WWF 2022. *Sustainable Agriculture*. Available at: https://www.wwf.org.za/our_work/initiatives/sustainable_agriculture/.

Chapter 3

Avocado cultivation and biodiversity challenges

Yvette Lange and Warren Maroun, University of the Witwatersrand, South Africa

1 Introduction

Avocados, also known as alligator pears or butter fruits, are thought to have originated in what is today, Mexico. Archaeological evidence suggests that avocados have been part of the human diet as long ago as BC10 000. References to the fruit have been found in reports by English and Spanish explorers dating to the sixteenth and seventeenth centuries. Avocados may have been introduced to other parts of the world as early as 1750 (Indonesia) and 1809 (Brazil). By the beginning of the 1900s, the plants had made their way to numerous countries such as South Africa, Kenya and Australia (Ingram et al., 2013).

It was only in the 1950s, however, that the food became popular, initially as an addition to salads (Ingram et al., 2013). An increase in diseases related to poor diet, coupled with the growth of a health-conscious middle class, has contributed to a more than 300% increase in avocado cultivation and consumption since the 1970s (Ingram et al., 2013). Regarded by some as a 'superfood', the green pear-shaped fruit is high in nutrients and an excellent source of mono-unsaturated fats, vitamins, minerals and fibre. Regular consumption of avocados is thought to reduce the risk of heart disease, lower cholesterol, regulate blood sugar levels and contribute to better overall health (Ingram et al., 2013; Krosofsky, 2021). In addition to being a 'nutritional powerhouse', avocados make an excellent meal by themselves or as an addition to different dishes. Their growing popularity throughout the world means that avocados are a high-value crop for farmers in both developed and

http://dx.doi.org/10.19103/AS.2023.0128.21

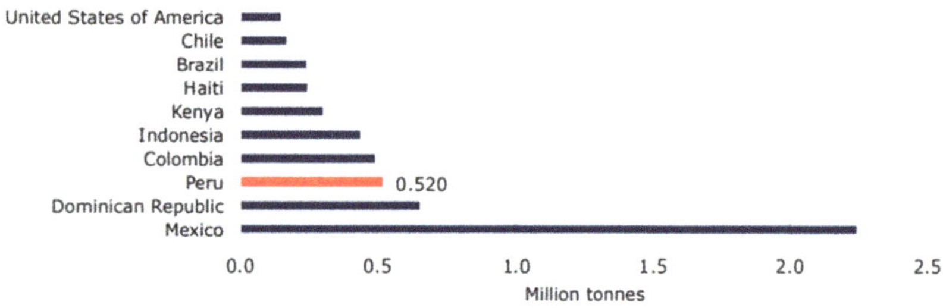

Figure 1 Total production of avocados by country, 2019. Adapted from: European Commission (2021).

developing economies where they are referred to as 'green gold' (Krosofsky, 2021).

As shown in Fig. 1, Mexico produces the most avocados followed by the Dominican Republic and Peru (World Economic Forum, 2020; European Commission, 2021).

Each year, 11 billion pounds of avocado are consumed worldwide with Europe, North America and Asia accounting for most of this consumption. The USA alone eats its way through an estimated US$5 billion worth of avocados annually. The global market for avocados is valued at just over US$14 billion and, with an estimated annual growth rate of 9%, is one of the fastest growing markets (World Economic Forum, 2020).

Unfortunately, the massive increase in avocado production comes with serious environmental impacts. For example, avocado production is water-intensive (Esteve-Llorens et al., 2022). The demand for 'green gold' is seeing significant amounts of land being converted from pristine wilderness to mono-agriculture use. Biodiversity loss, soil degradation and declining water quality are the outcomes (World Economic Forum, 2020). The adverse impact of avocado farming is especially severe, given that much of the cultivation takes place in wetter regions where landscapes can support greater varieties of biodiversity. Direct and indirect greenhouse gas emissions (Krosofsky, 2021), additional waste from the use of non-recyclable packaging (Kedzierski et al., 2020) and complex socioeconomic problems (Magrach and Sanz, 2020) add to the total cost of avocado cultivation and challenge the belief that the fruit is worth its weight in gold.

This section responds to calls for additional research on the link between the demand for 'superfoods' and environmental impact in more detail with a specific focus on avocado-biodiversity (see Magrach and Sanz, 2020). The biodiversity issues associated with avocado production are outlined (Section 2) followed by case studies dealing with efforts to cultivate avocados more sustainably (Section 3). A summary of key trends and suggested areas for future research follow in Section 4.

2 Biodiversity challenges within avocado cultivation

Avocado cultivation comes with significant biodiversity-related challenges including, for example, deforestation, over-exploitation of freshwater supplies and incorrect use of agrochemicals which contribute to destroying soil quality. Pertinent biodiversity problems are summarised in Fig. 2 and are outlined in the following sections.

2.1 Monoculture

Avocados are often a mass-scale monoculture crop. This type of agriculture contributes to lower direct costs of production with short-term benefits for farmers, retailers and consumers but has adverse implications for the environment. Most notably, large areas of land are cleared to make way for single plant species with an immediate impact on biodiversity and the healthy functioning of ecosystems (Bianchi et al., 2006). For example, essential pollination by different insects and mammals which depend on biodiversity is compromised with implications for the remaining flora and, paradoxically, the productivity of crops cultivated on converted land (Aizen and Harder, 2009; Atkins and Atkins, 2016).

The concentration of a species for agricultural use increases its susceptibility to pests and disease, especially because naturally occurring species which would otherwise have kept pests and pathogens under control may no longer be supported by reduced habitats in highly cultivated areas (see Bianchi et al., 2006; Panizzi, 2013). Perhaps most problematic is that soil quality is adversely affected by industrialised agricultural practices leading to reduced organic content, reduced water retention, reduced diversity of microbial life in the soil and reduced nutrient recycling (Schalk et al., 2014). Faced with a reduction of productive capacity, the quality and quantity of crop yields decline and farmers must compensate with more artificial fertilisers and pesticides which

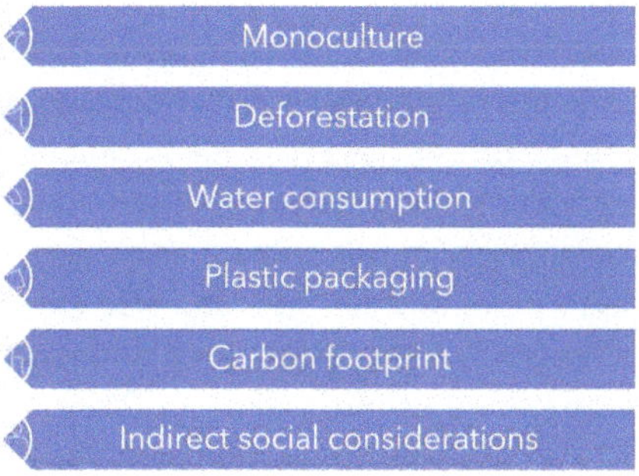

Figure 2 Summary of the biodiversity-related considerations.

compromises biodiversity further while adding to the costs of production (World Economic Forum, 2020; Okafor, 2022).

2.2 Deforestation

Mono-agriculture and deforestation go hand-in-hand. Before the boom in the international avocado market, small-scale farmers integrated with the local environment and contributed to a type of agro biodiversity. Deforestation could be limited if farms were dispersed and did not require industrialised levels of cultivation. Cleared land would also be used to farm different species such as maize, legumes and fruits with avocados being only one example. Different types of avocados would also be cultivated leading to genetic diversity and reduced environmental impact (De la Vega-Rivera and Merino-Pérez, 2021).

In contrast, agribusiness requires large tracts of land to satisfy the world's growing appetite for avocados. To maximise space and access to light, virgin land must be stripped to allow for large-scale planting. In most cases, the large and well-established trees being removed are keystone species (an organism that helps hold the surrounding eco-system together) which are critical for maintaining a healthy ecosystem (De la Vega-Rivera and Merino-Pérez, 2021; Esteve-Llorens et al., 2022). Although not only attributable to avocado cultivation, a report by Global Forest Watch states that 98% of deforestation in Mexico occurs because of agricultural expansion. Forest lands with diverse wildlife have been destroyed to produce avocados and other crops. Large areas have been intentionally burned to bypass a Mexican Law allowing producers to change the land use to commercial agriculture. Deforestation intensifies the effects of climate change and leads to other serious environmental problems (Magrach and Sanz, 202; World Economic Forum, 2020; Okafor, 2022).

As an illustration, wide-scale deforestation in Michoacán, an area in Mexico used intensively for avocado farming, has likely contributed to a change in micro-climates. The region has uncharacteristically hot and dry spells with irregular rainfall. Cold seasons are less intense (Avocado Institute of Mexico, 2022). Reduced forest cover is thought to play a key role in Michoacán's changing weather and has led to other changes in flora and fauna. Monarch butterflies arrive late to the region and in fewer numbers (Avocado Institute of Mexico, 2022). Temperate forests have become fragmented and the removal of well-established trees (which also produce seeds) means that genetic diversity is reduced, forests take longer to re-generate and the plants which do remain are more vulnerable to pests and disease (see also Panizzi, 2013). The ability of forests to enable nutrient recycling, store much-needed water and prevent soil erosion is reduced (World Economic Forum, 2020; De la Vega-Rivera and Merino-Pérez, 2021) with adverse implications for the animals living in these spaces.

The consequences of deforestation for people are also significant. On the economic front, often vulnerable populations are deprived of access to high-quality sources of food, water, medicine and other consumables. Eco-tourism becomes impossible and, because deforestation typically contributes to a decline in soil quality (see Sahani and Behera, 2001), the costs of production increase for the agribusiness sector which, paradoxically, drives deforestation. This, in turn, leads to higher prices for end-consumers compounding the problem of inequitable access to nutritious food and declining levels of health among poor populations. On the cultural front, areas being cleared for farming are often pristine wilderness with a significant intrinsic value. The forests are more than a respite from the trappings of modern life; they are an integral part of the identities of the people living in and close to them. The loss of these areas is a moral defeat, especially when considering the time and effort which will be required to restore deforested areas to their original state (Maroun and Atkins, 2018; Buchling and Atkins, 2020).

2.3 Water consumption

An estimated 9.5 billion litres of water are used daily to produce avocados. Stated differently, one kilogram of avocados requires 0.75 tons of water. When comparing this water consumption to other fruits, like bananas, avocados are seven times more water intensive. As a result, regions used for avocado plantations often experience water scarcity. Despite the attempts at regulating the amount of water consumed by avocado farmers, the impact on groundwater levels or fresh water supplied by rivers and streams is significant (Sommaruga and Eldridge, 2021; Okafor, 2022).

For example, the world's largest avocado producer relies on the extraction of water from local aquifers to supplement other supplies of fresh water to sustain its avocado farms. Unintended consequences have resulted. The Mexican Government has reported a significant number of earthquakes of different magnitudes with many of these occurring in the regions where most intensive cultivation of avocados occurs. While the link between avocados and earthquakes has not been established definitively, it is thought that the over-exploitation of groundwater leads to subterranean caverns forming which, when they collapse into the space left by extracted water, cause movement in the land above (World Economic Forum, 2020).

Chile, which is already classified as a water-stressed country, offers a second example. The country has increased its avocado crop by more than tenfold from 1961 when just 3000 hectares were being cultivated (Magrach and Sanz, 2020). The result has been a massive increase in water consumption by the agriculture sector. The combined effects of traditionally low rainfall, soils which have low water retention and the added stress from climate change lead to an increase in the risk of desertification and further biodiversity loss (Magrach and Sanz, 2020). Similar challenges occur in Peru where much of the avocado farming is concentrated in coastal areas which are already classified as hyper-arid climates. Inadequate water management strategies, even when farmers make use of sophisticated irrigation systems, mean that naturally occurring water sources are placed under extreme pressure. Over-exploitation of water is compounded by the effects of fertiliser run-off and contamination from pesticides. The precise long-term impact on biodiversity and the sustainability of local avocado producers is difficult to determine (Esteve-Llorens et al., 2022).

2.4 Packaging

How food items are packaged is often overlooked by consumers. Many producers use non-biodegradable packaging to protect avocados during transportation, to store them or to sell them to customers. Examples include plastic films, containers and linings which are often single-use items which consumers dispose of in landfills (see Kedzierski et al., 2020). Even when packaging is recyclable, not all consumers will take the time and effort to dispose of it correctly. The 'throw-away culture' leads to irresponsible use of plastics, unsightly waste and a mounting pollution problem.

Discarded plastic litters urban areas, blocks drainpipes and poses a health risk to communities. Over time, plastic bags, containers and linings break down into smaller pieces which contaminate the environment. It is estimated that thousands of tonnes of plastic are floating in the world's oceans and rivers posing a material threat to animal life. The long-term consequences for people who ingest the plastic found in seafood and drinking water are still to be determined (for details see Eriksen et al., 2014; Okafor, 2022).

2.5 Carbon footprint

Carbon footprint refers to an activity's greenhouse gas emissions. As a whole, agriculture contributes 24% to global greenhouse gas emissions (Khan et al., 2021). Avocado cultivation has an especially high carbon footprint. Large amounts of fertiliser are used by the sector leading to direct and indirect emissions. Electricity and diesel are required to operate the machinery used to cultivate, process and store avocados before they are supplied to the market.

As discussed in Section 2.3, large volumes of water are required to grow avocados. Much of this must be pumped to farms and then distributed by irrigation systems leading to additional energy consumption and greenhouse gas emissions (Esteve-Llorens et al., 2022). Finally, most of the world's avocados are exported from South America to the USA, Europe and parts of Asia along extended distribution networks. Cold storage is required during processing and shipping, leading to further carbon emissions. It is estimated that two small avocados in a single packet have a carbon footprint of 846.36 g, almost twice the amount of a kilogram of bananas (Sustainable Food Trust, 2020; Okafor, 2022).

2.6 Socioeconomic impact

Avocados have become an important source of income and foreign currency, especially for the world's poorest countries which happened to cultivate and export the fruit in quantity. While 'green gold' has benefited many commercial operations, it comes with a material environmental cost (as outlined previously) and social implications.

Branding avocados as a 'superfood' has turned them into a commodity in demand. Crop production is dominated by large organisations that do not always prioritise social factors when faced with an opportunity for short-term financial returns. Consumers have few guarantees that the avocados they are eating have been sourced from farms which, for example, pay fair wages, refrain from child labour, adhere to adequate health and safety standards and invest in the advancement of local communities which supply essential human capital (Magrach and Sanz, 2020). Other socioeconomic concerns include the following:

- Price inflation which makes it difficult for locals to purchase fruits grown in their homeland;
- Unfair distribution of profits in the value chain, especially where small-scale farmers are concerned;
- Land dispossession, especially by rural and indigenous communities, to make way for commercial cultivation;
- Food insecurity and health-related problems when small-scale farmers elect to grow only avocados instead of different food types; and
- The cost and complexity of complying with industry and international standards which impose additional burdens on the farmers who, paradoxically, have the least impact on how the sector functions.

Perhaps most notable is the link between avocados and organised crime. In many South American countries, the main hub for avocado farming globally, the lucrative avocado business has led to criminal groups targeting producers and distributors in a bid to control the cash flows from avocado exports (De la Vega-Rivera and Merino-Pérez, 2021). For example, farmers in Michoacán are at the mercy of dangerous cartels which demand payment in exchange for being allowed to operate unmolested (Magrach and Sanz, 2020; Okafor, 2022).

3 Sustainable avocado cultivation: global initiatives and case studies

Several initiatives are being explored to reduce the environmental impact of avocado farming. Examples are provided in the following sections.

3.1 Using water more effectively

To promote sustainable agricultural practices and prevent long-term damage to the environment, governments (and other stakeholders) are encouraging farmers to change how they work the land. This includes the introduction of agroecological systems of cultivation and forestation, including the diversification of crops.

With a focus on water, modern best practice is emerging making use of under-utilised water resources and providing subsidies for sustainable water technology solutions and efficient irrigation methods (Alva Tech, 2020; Magrach and Sanz, 2020). Examples include:

- Micro-sprinklers and drip irrigation can increase avocado yield substantially. These systems use up to half the amount of water compared to traditional sprinkler systems and prevent soil erosion (Alva Tech, 2020).
- Local government can assist by providing water storage facilities and investing in the infrastructure required to capture rainwater. This allows groundwater tables and aquifer levels to recover while creating 'water buffers' during dry spells (Alva Tech, 2020).
- Treating wastewater is another solution. According to the 2021 UN Worldwide Water Development Report, 'an estimated 380 km^3 of nutrient-rich wastewater is generated annually. This totals 15% of all global agricultural water withdrawals and could provide enough water to irrigate 42 million hectares'. For example, in Chile and Spain, the increased planting of avocado trees combined with low rainfall left water reservoirs depleted and resulted in the over-utilisation of groundwater. Local governments responded with wastewater treatment plants which supplied much-needed water to farmers while offering an environmentally friendly solution for dealing with human effluent (Alva Tech, 2020).

3.2 Fairtrade and trade agreements

Environmentally aware consumers are voicing their concerns about the negative biodiversity and other environmental impacts associated with avocado farming (Alva Tech, 2020). The goal is for consumers globally to support eco-minded cultivation by only purchasing 'sustainability-certified' avocados.

Under a sustainable certification system, avocado growers adhere to required economic, environmental and social practices verified by third parties. A similar approach is being followed for coffee, cocoa and banana producers which are certified by organisations such as Fairtrade and Rainforest Alliance. Avocado certification will reduce environmental impact by compelling producers to operate according to minimum environmental standards while also contributing to a fairer distribution of income throughout the value chain (O'Connell, 2021). This is the case if standards are applied consistently by major and other producers of 'green gold' (Häberli, 2019).

Global G.A.P., for instance, is an international organisation promoting safe and sustainable agriculture. It sets voluntary standards for the certification of agricultural products based on best practices which cover cultivation, distribution and packaging. Standards outlined by Global G.A.P. (or equivalent) could be adapted for the avocado industry to drive minimum levels of environmental responsibility and awareness and to provide a basis for holding growers, regional industries and governments accountable for the biodiversity-related costs of avocado production. 'Sustainable avocado standards' would address, at a minimum, three core areas:

- Social sustainability including issues such as liveable wages for farmworkers and fair profits shares for land-owners;
- Environmental sustainability, for example, water allocation in water-stressed areas, responsible use of pesticides and the substitution of oil-based fertilisers with organic ones; and
- Economic sustainability to address, for example, competitive production methods, the effective use of state subsidies and the optimisation of distribution networks.

(Häberli, 2019)

3.3 Policy changes

While consumer choices are a compelling way to shift crop production and consumption, this alone is unlikely to be sufficient to halt and then reverse the adverse effects of avocado farming on biodiversity. Policy changes will also be required with a focus on both the demand and supply side of the market.

The coordinated governance of land allocation and use is an example. The Rio Conventions address issues such as desertification and land degradation, biodiversity loss and the need to tackle climate change. The 2014 Paris Agreement includes commitments to promote a holistic and coordinated approach to land use in support of sustainable food production and global food security (UNFCCC, 2012; Magrach and Sanz, 2020). A number of the Sustainable Development Goals (SDGs) are relevant.

- Achieving zero hunger (Goal 2) is only possible if agricultural practices are carefully coordinated and managed to limit biodiversity impact while optimising production. Policies to address how avocados are cultivated and supplied to markets are required. These need to be easy to implement in major growing regions to ensure the long-term viability of the sector and its role in contributing to food security.
- Access to clean water and sanitation (Goal 6) is undermined if avocado growers use water resources irresponsibly. Cooperatives, industry bodies and local governments have a role to play in ensuring that avocados are cultivated to the extent that farms can be sustainably supplied with fresh water. As discussed in Section 3.1, steps to reduce water usage, recycle water with waste treatment plants and protect against chemical contamination of water supplies must be taken and actively monitored.
- Affordable and clean energy (Goal 7) is relevant given the energy-intensive nature of avocado cultivation. Renewable energy solutions should form a key part of policy development aimed at improving the sector's long-term sustainability. Considerations include grants to assist small-scale farmers with solar and wind-power generation, biofuel production using waste material and optimisation of delivery networks to reduce the distance between farms and marketplaces.
- Life on land (Goal 15) covers the need to use existing areas under cultivation more effectively to limit further deforestation. The variety of crops grown in different areas, including the types of avocado species being cultivated, should form part of an integrated growing strategy which promotes agro biodiversity rather than the cultivation of a single type of avocado over large areas. Policies for the reforestation of areas which are no longer under cultivation, including minimum standards for reintroducing biodiversity, are also essential.
- Finally, taking steps to protect biodiversity and mitigate the effects of climate change (Goals 13–15) provides an overarching objective which can only be achieved if responsible production is promoted and enforced (Goal 12). Sustainability certifications discussed in Section 3.2 are a relevant policy consideration. Equally important are governance frameworks which stress the responsible sourcing of agricultural goods by retailers and efficient mechanisms for holding unsustainable suppliers accountable.

3.4 Consumer behaviours

In addition to focusing on producers, SDG12 focuses on consumer behaviour, something which is critical for promoting sustainable food production (Magrach and Sanz, 2020). Individuals should be more conscious of the environmental impact of the food they are consuming and alter their buying decisions to include more locally grown foods and those in season to lower the demand for imported goods with a higher carbon footprint (Okafor, 2022).

Research confirms that plant-based diets offer important health benefits while such foods can also lower greenhouse gas emissions. However, in the case of avocados, plant-based diets have environmental challenges. A balance must be struck between lowering our dependence on animal products and ensuring that plant-based substitutes are sourced responsibly (Esteve-Llorens et al., 2022). Consumer behaviours, and the influence of these, will be an important consideration in the future cultivation of avocados and the relative demand for the supply of avocados and its subsequent impact on the environment and affected biodiversity.

3.5 Monitoring and measurement

Monitoring and measurement of performance and related issues within the agri-food sector is explored in Chapter 2 of this book. In this chapter, the recently released relevant Sector Standard by the Global Reporting Initiative (GRI, 2022) is examined briefly.

'GRI 13: Agriculture, Aquaculture and Fishing Sectors' (GRI-13) was approved by the Global Sustainability Standards Board on 19 May 2022 and released on 28 June 2022. The GRI has released this Sector Standard in acknowledgement of the fact that agriculture, aquaculture and fishing rely heavily on biodiversity and ecosystems. As a result, understanding and managing the impact these sectors have on the environment is a critical part of the broader sustainable development project.

GRI 13 describes 26 topics which are likely to be material for all organisations involved in crop cultivation, animal production, aquaculture and fishing. These are summarised in Fig. 3.

The Standard details and sets expectations for what companies in these sectors need to report on regarding their shared and specific impacts on the economy, environment and people. While the GRI Sector Standard for agriculture, aquaculture and fishing is effective for company reports or other materials published on or after 1 January 2024, early adoption is encouraged (GRI, 2022).

3.6 Socioecological frameworks

According to socioecological frameworks, humans and nature are interconnected, with humans drawing benefits from nature and impacting ecosystems (Gray and Milne, 2018). Khan et al. (2021) develop a 'cosmopolitanism framework'

Figure 3 GRI-13 reporting topics.

which calls for a multi-dimensional approach to avocado production. The framework is illustrated using experiences in Mexico but could be adapted to other regions where large-scale avocado cultivation takes place. The framework is reproduced in Fig. 4.

At the planetary level, key risks such as climate change, loss of species, over-utilisation of water and conversion of forested land to agricultural use are highlighted. Specific targets (which are not indicated in Fig. 4) are set to provide a frame of reference (Khan et al., 2021). Consistent with earlier research in environmental accounting and reporting, setting targets assists with the conceptualisation of problems, the development of remedial actions and the advancement of accountability for financial and extra-financial performance (Atkins and Maroun, 2018). The responsibility for achieving targets rests with individuals, firms and governments.

Governments have an important role to play in working to achieve collective action being taken by the avocado industry in realising sustainable development objectives. To this end, country-level targets are set and linked to cultural, moral and governance-related 'performance measures'. These include the following:

- Setting appropriate export strategies and monitoring priorities (cultural);

Planetary Boundaries	Country targets	COSMOPOLITANISM targets (soft or hard targets as performance measures)			
Environmental context (limits)		Avocado	Cultural	Moral	Governance
Climate change Atmospheric CO_2 : 350ppm ***Rate of terrestrial and marine biodiversity loss*** Extinction rate: >10 species per 1m species / year ***Interference of nitrogen and phosphorous cycles*** Concentration of Nitrogen removed from atmosphere : 35m tonnes / year Concentration of phosphorus flowing into oceans 11m tonnes / year	MEXICO TBD	**Individual**	Local farmers; Avocado and community representatives with equitable stakeholder say into policy formation. Monitoring of plant and animal species within their colonies and populations.	Rights and codes of conduct for workers on farms. Protection of avocado, its biome and associated animal species.	Setting of minimum salaries for labourers on farms. Setting minimum standards for working conditions. Longer term impact assessment of pesticides and fertilisers.
Ozone depletion in stratosphere Stratospheric ozone concentration <5% reduction, 276 Dobson units ***Ocean acidification*** Surface ocean average arogonite saturation arogonite > 80% of pre industrialisation ***Global fresh water consumption*** Consumptive blue fresh water use 4000km³/year consumptive use ***Land use planning*** Area of global forested land as % of original forest cover 75%; cropland 15%; Area of biome as % of original forest 85% (tropical) 50% temperate 85% boreal		**Firm**	Open local forums for local communities - corporate engagement. Local environmental and human species health accountability. Longer term well being responsibility and accountability. Investment into biome and human factors in decision making.	Re-balancing of the distributive rights of small farmers vs large players. Rights of local communities impacted by business. Promotion of broadening ethical standards as management actions and board accountability.	Breaking powerful cartels, gangs and illegal trade. Staged and level appropriate licensing and certification. Monitoring and reporting biodiversity issues across the industry - from tree to table. Embedding ecological resource considerations within longer term strategies.
Aerosol loading Global anthropogenic aerosol optical depth : TBD ***Chemical loss (novel entities)*** TBD		**Government**	Distributive monitoring and control e.g. Export strategies for different countries. Educational and scientific programmes supported by corporate investment.	Mandatory ecological wellbeing targets and annual reporting by organisations. Establishing better biosphere farming opportunities. Biodiversity research programmes that promote respect to local communities and individuals as a priority.	Independent regulation of land usage; environmental wellbeing and water usage. Preservation and protection of biome species and environment. Better planning and control of irrigation and land adaptation - empowering the local level.

Figure 4 'Cosmopolitan resilience opportunities'. Adapted from: Khan et al. (2021).

- Setting ecological targets for the sector which must be reported on by individual growers (moral); and
- Specific regulatory interventions over land usage, water consumption and biodiversity conservation (governance).

(Khan et al., 2021)

Country-level targets can be applied to ecosystems, geographical areas and regions. This allows them to provide a type of consolidated assessment of environmental impact and overcome the problem of linking issues like biodiversity losses to specific firms or requiring only some of the industry members to take remedial actions (Cuckston, 2017; Atkins and Maroun, 2020).

At the firm or industry level, targets are set for factors such as the health of species in areas under the organisation's control, investment in natural capital development, fair allocation of income across the value chain and the incorporation of environmental issues into strategy development and risk assessment (Khan et al., 2021). The firm-level targets contextualise broader country-level ones.

Between businesses and individuals, the focus should be on incentivising actions which contribute to the 'well-being' of farmers. Examples include improving working conditions for labourers, animal well-being, responsible use of pesticides and the adoption of effective irrigation techniques (Khan et al., 2021). Together with targets set by individual farmers, the firm/industry initiatives provide a practical means for avocado producers to become more sustainable and for their collective actions to mitigate the impacts of their activities on biodiversity. The model follows an integrated approach to improving environmental performance and accountability by co-opting the key stakeholders in the value chain, focusing on specific objectives and encouraging actions which address economic, environmental and social factors concurrently.

3.7 Case study examples

This section outlines selected case studies highlighting either instances where biodiversity conservation is under threat as a direct result of avocado cultivation or where initiatives are being successfully implemented to conserve biodiversity.

3.7.1 The Kenyan elephants

The Amboseli ecosystem in Kenya is one of Africa's most familiar landscapes. In 2019, 192 000 people visited the national park bringing in millions of dollars. Amboseli is under threat from population growth and the accompanying expansion of crop farming. The conservation solution has been to create wildlife corridors allowing animals to move in and out of the ecosystem. These corridors are, however, under threat by avocado farmers.

Kenya's exports of avocados have increased substantially in recent years. Part of this rise is attributed to a trade deal with China signed in 2019, where avocado consumption is also on the increase. Kenyan agribusinesses, KiliAvo Fresh Ltd in particular, responded to the growing demand for 'green gold' by announcing plans for the development of new avocado plantations. Their decision has angered conservationists and local community leaders because of its proximity to important wildlife corridors. Stakeholders are also concerned that if avocado producers are allowed to encroach on the corridors, a dangerous precedent will be set encouraging further demands on wilderness areas by avocado and other farmers.

The effect of a blocked corridor will mean that wildlife will not be able to migrate and disperse to access grazing and water, with catastrophic effects. KiliAvo was initially granted a licence to proceed but the decision was suspended following objections from local communities and conservation organisations. The company insists that it has followed the correct processes to expand its operations and that new farms are located outside of wildlife corridors. They have also been quick to highlight synergies between avocado farming and socioeconomic development:

> 'There are lots of benefits for locals and even the country as a whole. Farms like this bring in employment, increase land value and show the locals how to better manage their land' (Clark, 2020).

Whether or not the immediate economic benefits will outweigh longer-term environmental considerations remains to be seen. The key point is that the tension between avocado farming and biodiversity is not limited to South America. It is equally relevant in Africa where conflict between industry and the environment is becoming more serious as populations grow and Africa attempts to address its economic woes.

(Clark, 2020)

3.7.2 Association of Avocado Exporting Producers and Packers of Mexico (APEAM)

Promoting responsible farming practices is a keystone of the Association of Avocado Exporting Producers and Packers of Mexico (APEAM). APEAM has confirmed its dedication to accountable environmental practices and its attention to industry labour issues by creating the 'Green Agenda'.

The Green Agenda is a comprehensive research report which provides a diagnosis of the avocado industry's sustainability in Michoacán. The goal of the project is to achieve a greater level of sustainability for the industry and the communities of Michoacán, while increasing positive competition and responsible farming practices. The Mexican avocado industry is aligning its sustainability ambitions with the goals of the UN's 2030 Agenda for Sustainable Development. An official agreement was signed between APEAM and the UN on September 9, 2020.

To protect sustainability, the avocado orchards in Michoacán which seek to export their fruits must follow sustainable avocado farming guidelines under the supervision of APEAM:

- 'No hunting: this protects the natural biodiversity of the area and creates a balance within the ecosystem;
- No loose dogs: untethered canines help to prevent pollution brought in from their waste, while also protecting the territory of the natural fauna;
- No cattle grazing: this safeguards the cleanliness of the soil, the natural water table, and local flora in the groves;
- Providing workers with services and education on sanitation secures an environment in line with the health and well-being of the orchard staff; and
- Regular blood tests: conducting blood tests on the crew responsible for applying fertilizers ensures that agrochemicals used within the orchards are not contaminating the surrounding environment or the farmers themselves'.

(Avocado Institute of Mexico, 2022)

Another example of APEAM's work in the biodiversity conservation space is its reforestation initiative (launched in 2011). Since then, more than 2200 acres have been replanted with various species native to the area. Sustainable farming consists of preserving the forests of the region, which in turn leads to better soil and air quality, less erosion and cleaner water for all the inhabitants of Michoacán (Avocado Institute of Mexico, 2022).

3.7.3 Greencell

Greencell, part of the Westfalia Fruit Group, is an established fresh fruit and vegetable supplier in England. The company manages over 100 000 tons of fruit per year, sourcing from over 30 countries. Greencell has implemented standards and commitments to ensure the sustainable and ethical production of its goods. This entails, *inter alia*, requiring growers to have SEDEX registration. SEDEX is a community for responsible businesses which covers the sourcing of foods from suppliers which adhere to strict environmental and ethical standards and are subject to third-party audits to ensure compliance with policy requirements. The initiative extends to the sourcing and procurement of avocados which Greencell acknowledges are grown in areas of significant biodiversity importance.

(Westfalia and Fruit, 2022).

4 Conclusion and future trends

The environmental impact of avocado cultivation and, in particular, the negative effect on biodiversity, cannot be ignored. Avocado farming is contributing significantly to deforestation, irresponsible use of water, soil degradation and climate change. The avocado supply chain needs international standards to drive sustainable farming and distribution practices while enabling organisations, industries and governments to be held accountable for direct and indirect environmental impacts (World Economic Forum, 2020; Okafor, 2022).

With the demand for 'green gold' set to grow in the coming years, strategies for sustainable cultivation of the fruit need to be developed. These have to balance environmental objectives with social and economic imperatives, especially given that much of the world's supply of avocados comes from developing economies. Initiatives need to be tailored according to the biodiversity risks, the needs of local communities and the capacity of policymakers to develop and enforce regulations (González-Estudillo et al., 2017). A concerted effort by environmental bodies, international institutions (such as the United Nations and World Bank) and consumer groups is needed to drive much-needed change at producers and promote a culture of considering environmental factors when making decisions about the food being put on families' tables.

Important questions remain on the environmental and social impacts of superfood production and consumption. For example, the effect of land use changes as the demand for avocado production rapidly increases must be quantified. This is important for understanding the extent to which avocado farming is impacting biodiversity and predicting the long-term implications of industrial food production. Feeding the diverse diets of a growing population without destroying biodiversity is a related avenue for future research. This line of enquiry should go hand-in-hand with exploring how consumers can be made more aware of the environmental costs of the food they consume, how assurance providers can be deployed to accredit sustainable farming practices and how monitoring bodies can ensure that the avocado industry adheres to applicable social and environmental standards.

Equally relevant is considering the role of industry regulators, governments and supra-national bodies in developing laws, codes of best practice and international guidelines which can be used to promote more sustainable agricultural practices and enable accountability for biodiversity impact. That the world is facing a serious environmental problem is axiomatic. Practical solutions for balancing human needs with the importance of protecting biodiversity are less obvious.

5 References

Aizen, M. A. and Harder, L. D. 2009. The global stock of domesticated honey bees is growing slower than agricultural demand for pollination. *Current Biology* 19(11), 915–918.

Atkins, J. and Atkins, B. 2016. *The Business of Bees: an Integrated Approach to Bee Decline and Corporate Responsibility*. Abingdon, Oxon: Routledge.

Atkins, J. and Maroun, W. 2018. Integrated extinction accounting and accountability: building an ark. *Accounting, Auditing and Accountability Journal* 31(3), 750–786.

Atkins, J. and Maroun, W. 2020. The naturalist's journals of Gilbert White: exploring the roots of accounting for biodiversity and extinction accounting. *Accounting, Auditing and Accountability Journal* 33(8), 1835–1870.

Avocado Institute of Mexico 2022. The Green agenda: A cycle of sustainability. Available at: https://avocadoinstitute.org/the-green-agenda-a-cycle-of-sustainability/.

Bianchi, F. J., Booij, C. J. and Tscharntke, T. 2006. Sustainable pest regulation in agricultural landscapes: a review on landscape composition, biodiversity and natural pest control. *Proceedings. Biological Sciences* 273(1595), 1715–1727.

Buchling, M. and Atkins, J. 2020. Chapter 29. Reporting on more than just natural capital. In: De Villiers, C., Hsiao, P.-C. K. and Maroun, W. (Eds). *The Routledge Handbook of Integrated Reporting*. Abingdon, Oxon: Routledge.

Clark, N. 2020. *Avocados v Elephants: Is the Green Gold Putting Wildlife at Risk?*

Cuckston, T. 2017. Ecology-centred accounting for biodiversity in the production of a blanket bog. *Accounting, Auditing and Accountability Journal* 30(7), 1537–1567.

De La Vega-Rivera, A. and Merino-Pérez, L. 2021. Socio-environmental impacts of the avocado boom in the meseta purépecha, Michoacán, Mexico. *Sustainability* 13(13), 7247.

Eriksen, M., Lebreton, L. C., Carson, H. S., Thiel, M., Moore, C. J., Borerro, J. C., Galgani, F., Ryan, P. G. and Reisser, J. 2014. Plastic pollution in the world's oceans: more than 5 trillion plastic pieces weighing over 250,000 tons afloat at sea. *PLoS ONE* 9(12), e111913.

Esteve-Llorens, X., Ita-Nagy, D., Parodi, E., González-García, S., Moreira, M. T., Feijoo, G. and Vázquez-Rowe, I. 2022. Environmental footprint of critical agro-export products in the Peruvian hyper-arid coast: A case study for green asparagus and avocado. *Science of the Total Environment* 818, 151686.

European Commission 2021. Impact of the agreement on biodiversity: the case of avocado production in Peru. Available at: https://trade.ec.europa.eu/doclib/docs/2021/july/tradoc_159724.pdf.

González-Estudillo, J. C., González-Campos, J. B., Nápoles-Rivera, F., Ponce-Ortega, J. M. and El-Halwagi, M. M. 2017. Optimal planning for sustainable production of Avocado in Mexico. *Process Integration and Optimization for Sustainability* 1(2), 109–120.

Gray, R. and Milne, M. J. 2018. Perhaps the Dodo should have accounted for human beings? Accounts of humanity and (its) extinction. *Accounting, Auditing and Accountability Journal* 31(3), 826–848.

GRI 2022. *GRI 13: Agriculture, Aquaculture and Fishing Sectors 2022 Frequently Asked Questions (FAQs).*

Häberli, C. 2019. Please make avocados sustainable again! In: International Affairs Forum. Center for International Relations, Washington, DC, 44–47.

Ingram, D. K., Elks, C. M., Davenport, G. M. and Roth, G. S. 2013. 14 potential health benefits of avocados. *Bioactives in Fruit* 337), 337-352.

Kedzierski, M., Frère, D., Le Maguer, G. and Bruzaud, S. 2020. Why is there plastic packaging in the natural environment? Understanding the roots of our individual plastic waste management behaviours. *Science of the Total Environment* 740, 139985.

Khan, N., Kakabadse, N. K. and Skouloudis, A. 2021. Socio-ecological resilience and environmental sustainability: case of avocado from Mexico. *International Journal of Sustainable Development and World Ecology* 28(8), 744–758.

Krosofsky, A. 2021. Lean, mean, but not so green: how avocados impact the environment. Available at: https://www.greenmatters.com/p/avocado-environmental-impact.

Magrach, A. and Sanz, M. J. 2020 Environmental and social consequences of the increase in the demand for 'superfoods' world-wide. *People and Nature* 2(2), 267–278.

Maroun, W. and Atkins, J. 2018. The emancipatory potential of extinction accounting: exploring current practice in integrated reports. *Accounting Forum* 42(1), 102–118.

O'Connell, S. 2021. The avocado in your Super Bowl guacamole is bad for the environment. You can make it better. Available at: https://www.washingtonpost.com /climate-solutions/2021/02/04/climate-solutions-avocados/.

Okafor, J. 2022. *Environmental Impact of Avocados.*

Panizzi, A. R. 2013. History and contemporary perspectives of the integrated pest management of soybean in Brazil. *Neotropical Entomology* 42(2), 119–127.

Sahani, U. and Behera, N. 2001. Impact of deforestation on soil physicochemical characteristics, microbial biomass and microbial activity of tropical soil. *Land Degradation and Development* 12(2), 93–105.

Schalk, V. D. M. L., John, R. U. W., Charlene, J., Ruan, V., Sarah, J. D. and Matthew, A. 2014. The unknown underworld: understanding soil health in South Africa. *South African Journal of Science* 110, 1–4.

Sommaruga, R. and Eldridge, H. M. 2021. Avocado production: water footprint and socio-economic implications. *EuroChoices* 20(2), 48–53.

Sustainable Food Trust 2020. Why our love for avocados is not sustainable. Available at: https://sustainablefoodtrust.org/news-views/avocados-sustainability/.

Tech, A. 2020. Can sustainable farming technologies save the avocado? Available at: https://www.alva-water.com/post/are-avocado-s-green-is-water-intensive-agriculture-under-threat.

UNFCCC 2012. *The Rio Conventions Action on ADAPTATion.*

Westfalia and Fruit 2022. Our sustainability story. Available at: https://www.hmh.co.za/wp -content/uploads/2019/05/WestfaliaSustainabilityStoryMay2019.pdf.

World Economic Forum 2020. Avocado: the 'green gold' causing environment havoc. Available at: https://www.weforum.org/agenda/2020/02/avocado-environment-cost -food-mexico/.

Chapter 4

The circular economy and sustainable soybean farming in Brazil: an integrated approach

Wayne van Zijl, University of the Witwatersrand, South Africa

1 Introduction

Farming has been in existence for millennia. The earliest evidence of a cultivated land is found in Israel. The Ohalo II people, residing off the shore of the Sea of Galilee, used the trial-and-error approach 23 000 years ago to cultivate cereal crops (Snir et al., 2015). With the advanced technological state of modern society, current farms do not need to rely on trial and error. Instead, farms can use the latest scientific research and incorporate more macro-level considerations into their techniques, product choices and structural setups to ensure food security, affordability and environmental sustainability (Van Vliet et al., 2013; Lin, 2018; Pivoto et al., 2018).

Activists and vocal stakeholders promoting sustainability are often perceived as being emotional, reducing their credibility in the eyes of big business. There are fears that activists are trying to turn businesses into not-for-profit companies. This chapter approaches sustainable agriculture from a neutral stance, illustrating the need for change and resulting synergistic benefits for not only the environment but also businesses and other stakeholders. In doing so, the business imperative is maintained, enhancing the acceptability of change

http://dx.doi.org/10.19103/AS.2023.0128.23

in the name of sustainability. The chapter also provides activists and businesses with a framework that each party can understand. The framework encourages productive and meaningful engagements (Maroun and Atkins, 2018) by coupling the International Integrated Reporting Council's (IIRC) six capitals with the broad concept of a circular economy (see IIRC, 2013, 2021; Tedesco et al., 2022).

To illustrate the proposed framework's usefulness, the chapter provides an overview of Brazil's soybean farming and how the sector can operate to improve farming and the country's sustainability. Brazil has been selected because of its iconic standing and global importance for biodiversity and influence on climate and weather patterns (Shukla et al., 1990; Doughty et al., 2015; Amaral et al., 2021).

The remainder of this chapter is structured as follows: Section 2 provides a brief overview of the circular economy, the IIRC's six capitals and their amalgamation to form a blended framework. Sections 3 and 4 discuss farming, with a focus on soybean farming in Brazil. The application of the proposed framework to aspects of Brazil's soybean farming is provided in Section 5, followed by an overview of the importance of accounting and accountants.

2 The circular economy, the integrated reporting framework and agriculture

The Industrial Revolution ushered in a production system characterized by raw material consumption and 'unbridled generation of waste' (Tedesco et al., 2022, p. 397). Products are not designed to be indefinitely repaired or disassembled, leading to the exponential growth of waste and rapid depletion of non-renewable resources (Jabbour et al., 2013). The impact of this 'take-make-use-dispose' system on the environment was largely ignored (Tedesco et al., 2022, p. 398). Years later, land degradation, volatile resource prices, non-renewable resource scarcity, biodiversity loss and rapidly changing weather patterns are examples of the negative impacts of a linear production model (Jabbour et al., 2013). These devastating effects have led to a strong focus on environmental protection and responsibility (Lai and Stacchezzini, 2021). Research indicates that the linear production system has reached its limits in terms of providing production efficiency and aiding in the development of new technologies (Tedesco et al., 2022). A different approach must become the new norm and focus not only on organizational sustainability but also on social and environmental sustainability.

A circular economy may be part of the solution. Geissdoerfer et al. (2017) state that a circular economy is

> '...a regenerative system in which resource input and waste, emission, and energy leakage are minimised by slowing, closing, and narrowing material and energy loops. This can be achieved through long-lasting design, maintenance, repair, reuse, remanufacturing, refurbishing, and recycling' (p. 766).

The circular economy draws inspiration from nature where fauna and flora are reliant on the products or death of each other for nutrients. As a simple illustration, grass is eaten by deer that are eaten to nourish predators. When both deer and predators die, they nourish the grass. The key differentiator between linear and circular systems is a purposeful design to be repairable or to allow for the easy dismantlement of end-of-life products so they can be used as inputs for new goods and services. This framework is well suited to agricultural operations (Secco et al., 2020; Tedesco et al., 2022).

The IIRC Integrated Reporting (IR) framework's six capitals can be integrated into the circular economy logic using integrated thinking. The six capitals are:

- Human;
- Social and relationship;
- Natural;
- Intellectual;
- Manufactured; and
- Financial.

The IR framework emphasizes the interrelatedness of all capitals, where the enhancement of one often comes at the expense of another (IIRC, 2021). This requires business leadership to practice integrated thinking. Maroun et al. (2022, p. 21) believe that integrated thinking can be gauged by a company's 'integrated awareness and understanding; integrated leadership commitment and capability; integrated structures; integrated organisational performance management; and integrated external communication'. A verbal commitment to defining 'value' by all six capitals must be accompanied by substantive action to create the organizational structures to embed and adopt integrated thinking so as to enable integrated decision-making. This also necessitates the creation of systems to collect, analyse and report on all capitals (IIRC, 2021; Maroun et al., 2022).

In this sense, the IR framework encourages companies to view their operations as a closed-loop system (as does a circular economy). Both the IR framework and the circular economy also require a more long-term consideration of 'value' than is the case with a linear production system. For the latter, short-term profit taking is crucial. Efficiencies and minimization of waste production are understood only in terms of immediate financial capital gains. With shareholders as the primary focus, capitalism and linear production systems eagerly traded decreases in natural, human and social and relational capital for increases in financial and manufactured capital (Jabbour et al., 2013; Flower, 2015; IIRC, 2021; Tedesco et al., 2022). A circular economy and IR framework (CE-IR) approach shifts the priority from enhancing financial capital to balancing all six capitals or maximizing the number of capitals that can be positively impacted. To illustrate, Table 1

Table 1 Linear production and CE-IR approaches

Capital	Linear production	CE-IR
Financial	+40[a]	+10
Human	−25	−2
Social and relational	−15	+5
Natural	−25	−2
Intellectual	+5	+5
Manufactured	+15	+5
Net effect*	**−5**	**+21**

[a] Note: Units are used to illustrate the concept. They do not mean anything beyond their illustration value in portraying relative changes.

presents a linear production and CE-IR response to the question of mechanizing a farm operation to stay competitive in the market.

Under a linear production system, the solution is to mechanize the farm operations as fast as possible. Rapidly mechanizing the farm operation results in significant contributions to financial and manufactured capital but at great cost to human, social and relational and natural capital. There is also a net 'loss' (see Table 1) as mechanization, product packaging and transport strategies readily trade natural, human and social and relational capital for financial capital (see also Fearnside, 2001).

The CE-IR approach would consider the significant toll the change would have on human, natural and social and relational capitals. The company must do something to address its competitiveness in the market and help to keep food costs down. Allowing itself to be liquidated is not in the best interests of anyone. Workers rely on the company for money, and society relies on farms to produce sufficient food at reasonable prices. CE-IR attempts to balance the impact on all six capitals as follows:

- A transitional plan is developed to up-skill employees for different jobs over a set period (Lin, 2018). This protects human and social and relational capital.
- Mechanization is implemented over time to allow for the re-deployment of workers elsewhere whilst introducing faster and cheaper forms of sowing and harvesting. Intellectual and manufactured capitals are enhanced slightly over time.
- Financial capital is enhanced but not nearly as much as under a linear production approach.
- Finally, natural capital is significantly less negatively impacted, as the combination of human and mechanical farming is less damaging to the environment than a purely mechanical method.

Table 1 reflects that extracting maximum financial capital value comes at the high cost of many other capitals. It illustrates that 'a decline in natural capital can never be off-set by an increase in financial capital. Future generations cannot live off the profits of today's companies' (Flower, 2015, p. 8). This concept is key to creating a sustainable future where the current generation leaves the planet in no worse condition than how they received it.

Adopting a more balanced strategy allows more capitals to be enhanced, albeit only slightly, and has a reduced negative effect on others. It is unlikely that a perfectly balanced system will be created in the short term (Geissdoerfer et al., 2017), but this is at least a step in the right direction. In addition, an overall positive outcome may result where the focus is balance as opposed to the maximization of one or two capitals whilst treating the others as subordinate (De Villiers and Maroun, 2017). Farming's significance to all humanity provides a compelling case to apply a CE-IR framework to create a sustainable future.

3 Farming in Brazil: an overview

Farming is a vital activity for many countries, especially developing countries, in terms of economic, environmental and social (EES) development. Agriculture can provide a springboard to develop economies into industrial, manufacturing and then high-income service economies. In addition, a strong agricultural sector contributes to a country's food security and reduces malnutrition. These are crucial foundation blocks for a developing economy but require agricultural sectors to be modernized (Lin, 2018).

By improving farming processes, agricultural productivity increases, leading to eventual agricultural surpluses. These surpluses help countries accumulate financial capital (foreign currency) via exports. The accumulated wealth can be used to further positive structural changes. Socioeconomic transformation helps increase wages, alleviate poverty and ensure appropriate nourishment among the youth of a country. Without proper nourishment, children may have disabilities or their growth may be stunted. Both cases reduce children's ability to become productive members of society. Agriculture can lead to a positive reinforcing cycle where increasing wages and better nutrition lead to decreasing malnutrition and increasing numbers of healthy, productive adults (Lin, 2018). As the number of productive adults exceeds the number of those required to work directly or indirectly in the agriculture sector, the number of adults available to engage in industrial, manufacturing and high-income services increases, reinforcing positive socioeconomic transformation.

But is all modernization good? Too much modernization and mechanization can lead to reduced labour requirements, leading to negative effects on the planned structural transformation (Lin, 2018). Kamali et al. (2017) investigated genetically modified (GM), non-GM and organic soybean farming in Brazil to determine which of the three had better EES outcomes. Their results were inconclusive, and the authors attribute this to the sensitivity of each factor to various inputs and processes in the agricultural sector. They demonstrate how important, careful and proactive government policy-making is to the agricultural sector and which systems will benefit the country the most (see also Sabourin et al., 2020).

Using GM seeds is one form of modernization that allows more indiscriminate use of herbicides[1] as well as sowing different climates with different soil conditions (Fearnside, 2001; Proterra Foundation, 2021). Other recent developments include the use of unmanned aerial vehicles (UAVs), drones and microprocessors to monitor crops with unprecedented efficiency, accuracy and continuity. Drones are fitted with infrared cameras and GPS technology to improve crop risk management and decision-making. For example, early signs of failing crop health can be detected timeously and addressed accurately to minimize losses and wastage (see Pivoto et al., 2018). Other technological developments include smart farming (SF) where small-scale microprocesses and sensors are used to aid crop management and decision-making. When SF is coupled with cloud computing and artificial intelligence, the possibilities to enhance both the productivity and the sustainability of farming are significant (Kamali et al., 2017; Lin, 2018; Pivoto et al., 2018; da Silva et al., 2021). SF also has implications for the employment requirements of agriculture and the skill levels of farmers. This iterates the importance of careful management to ensure a balance between reducing reliance on human labour and creating alternate jobs in the industrial, manufacturing, and high-income service sectors of a country's economy (Lin, 2018).

Two major stumbling blocks to modernizing agriculture are the extent of public and private investment into relevant and continuous research and development (R&D) and convincing farmers to utilize new technology. The latter is particularly problematic in developing countries. Modern farming techniques and systems typically require specific know-how and carefully controlled and planned systems. For instance, high-yielding seeds may need stringent routines and structures for water and other inputs (Lin, 2018). Government policies and planning must take these factors into account to ensure the successful implementation of their policies in the agricultural sector.

Brazil is situated along the eastern coast of South America. It has approximately 7500 km of coastline that extends deep into the interior of the continent. Brazil covers 8.5 million km^2 between the equator and Tropic

[1] Some GM soybeans are modified to make them resistant to herbicides, such as Roundup-Ready™ (Fearnside, 2001; Proterra Foundation, 2021).

of Capricorn. It has significant environmental value because of its vast tropical rainforest. Brazil boasts a variety of climatic regions, including desert, equatorial, tropical, semiarid, oceanic and subtropical areas (Food and Agriculture Organization of the United Nations (FAO) 2022). Many tribal nations inhabited Brazil before Portugal's Pedro Alvares Cabral landing there in 1500. The word Brazil is believed to have been derived from the Portuguese word *pau-brasil*, meaning brazilwood. This was a common trade item used in the textile industry. *Brasa* means 'red like an ember' and refers to the deep red dye derived from brazilwood.

Family farming - defined as those with 'limited areas; predominance of family labour; family management; gross income primarily from agricultural production; and residence on or near the farm' - came under threat in the late 2010s (Sabourin et al., 2020, p. 46; Ferreira Costa, 2021). This was after 20 years of pro-family farming policy support since 'the election of left-wing governments' in 1995 (Sabourin et al., 2020, p. 46). After 2015, government policies supporting family farming in the form of financial and technical assistance and organizational support lost popularity. The 'fragilization of the progressive governments in place' and political and economic crises played a role in dismantling family farming operations (Sabourin et al., 2020, p. 60).

Brazil's major agricultural products include coffee, sugar, soybeans, manioc, rice, maize, cotton, wheat, milk and meat (FAO, 2022). Their main exports are soybeans and meat. Soybeans have performed well, strengthened by the coronavirus disease 2019 pandemic and demand from China (Biz Latin Hub, 2022; Brazilian Farmers, 2022). As a key agricultural product of Brazil, soybeans are the focus of this chapter.

4 Soybean farming in Brazil

Soybean farming dates back to 2500 BC in China (Alves et al., 2003) and has become a worldwide staple. Soybean is a legume, and its popularity stems from its high protein content, reaching almost 40%. Soybean plants can grow up to 2 m in height. They are branching, self-fertilizing plants with white or purple flowers that bear pods. The plants' seeds can have a variety of colours but are commonly brown and contain one to four seeds per pod (Alves et al., 2003; Encyclopaedia Britannica, 2022).

Soybean performs best in warm climates with fertile, well-draining sandy loam. They provide benefits to the soil by promoting the growth of dinitrogen-to-ammonia-processing bacteria in their root systems, making soybean an excellent crop in multicrop farming systems (Alves et al., 2003; Encyclopaedia Britannica, 2022). Soybeans are used as ingredients for both animal and human food products owing to their high protein value and relatively cheap production costs. Soybean is also used as an ingredient for non-food products such as paints, fire extinguishers and fertilizers, demonstrating its versatility and value (Encyclopaedia Britannica, 2022).

Soybean farming is a key agricultural product of Brazil and is exported as soybeans, soybean meal and soybean oil. Soybean farming is primarily carried out in the southern and central-western regions of Brazil using either conventional or organic farming (Kamali et al., 2017; Cattelan and Dall'Agnol, 2018). Soybean had humble beginnings in Brazil as an experimental crop between its introduction in 1882 by the United States until 1940 (Alves et al., 2003; Cattelan and Dall'Agnol, 2018). Soybean entered Brazil via Salvador and remained in Brazil's Bahia State where the climate is tropical. Because the soybean variety introduced to Brazil needed a tropical climate, its production was focused in the Bahia State. It was not until 1941 that Brazil first reported a significant production of soybean in the State of Rio Grande do Sul, with approximately 7651 ha of soybean and an approximate production of 9146 t per annum (Alves et al., 2003; Cattelan and Dall'Agnol, 2018).

Before 1941, soybean was primarily used as biomass in cattle and pig feed. After 1941, soybean increasingly became used as a grain-producing crop. Soybean production increased from 457 t in 1941 to 25881 t in 1949. Soybean experienced even greater growth from the 1970s after governmental interventions were implemented to enhance wheat production. Soybean benefitted as a good summer crop to rotate with wheat as a winter crop. Soybean cultivars more suited to different regions also allowed soybean to be used more widely in Brazil (Cattelan and Dall'Agnol, 2018).

The 1970s–1990s saw great success implementing Integrated Pest Management (IPM) (Panizzi, 2013). IPM emphasized that the use of a combination of pest control methods should be both cost-effective and environmentally sustainable. Identifying economic injury levels (EILs) was key to the program. EILs refer to the minimum amount of insects that will cause crop losses equal to the cost of pest management. The program grew from strength to strength by conducting surveys to identify the populations and spatial spread of both pests and natural enemies. By following a combined approach of using natural enemies and targeted pesticides based on surveys and seasons, a 50% reduction in pesticide use was seen (Panizzi, 2013).

In the 1980s, viruses that affect specific pests such as velvet bean caterpillar, which causes defoliation of soybean plants, were also used effectively. For example, a nuclear polyhedrosis virus (AgNPV) that was naturally occurring in Brazil was improved to be highly effective. AgNPV began as homemade solutions from collected diseased caterpillars. Later, a wettable powder formulation was used. In the 1990s, egg parasitoids were developed, where a parasite lays eggs inside stink bugs' eggs, thus destroying them. With IPM, careful planning is required. Because growers used AgNPV and parasitoids, only targeted pesticides that would not negate their other pest management measures could be used (Panizzi, 2013).

From the 2000s onwards, soybean farming experienced another rise accompanied by no-tillage farming, the use of multiple crops, a lack of updates to ensure IPM remained effective and rapidly decreasing pesticide, herbicide and fungicide costs (Panizzi, 2013; Amaral et al., 2021). GM seeds were also introduced in the early 2000s that were primarily modified for herbicide resistance (Proterra Foundation, 2021). These developments led to a significant reduction in the use of IPM in Brazilian soybean farming. Despite the proven effectiveness of IPM, the use of AgNPV and natural enemies was replaced with the 'easiness' and low cost of pesticides and herbicides (Oda and Soares, 2000; Panizzi, 2013, p. 123; Proterra Foundation, 2021).

Unfortunately, the new acceleration of soybean production meant that important biomes also came under threat. Brazil's Amazon and grassland biomes were affected and are of significant importance because of their concentration of biodiversity, biomass productivity and carbon trapping (Fearnside, 2001; Amaral et al., 2021). While soybean production can reduce unemployment and poverty (Panizzi, 2013; Lin, 2018), the new, mainly mechanized, soybean plantations being developed required less human involvement, resulting in net increases in unemployment and increasing displacement of subsistence farmers. These subsistence farmers relocated and went on to deforest virgin forest that would often, itself, become part of new soybean productions, creating a cycle of destruction (Mamede and De Araújo, 2008; Van Vliet et al., 2013). Soybean production also brings with it major habitat destruction from the cleared land for planting coupled with the 'dragging effect' (Fearnside, 2001, p. 24). This is where additional habitat destruction occurs due to logging to build the required infrastructure, clearing land to develop transport routes and use of agrochemicals (Fearnside, 2001).

Because of Amazonian deforestation, pressure from international groups, such as Greenpeace, saw the incorporation of SoyM (Soy Moratorium). It was implemented in 2006 by the Brazilian Association of Vegetable Oil Industries (ABIOVE) and the National Association of Grain Exporters (ANEC). Under the programme, the ABIOVE and ANEC would not buy soy from, or fund crops grown in, deforested areas in the Amazon biome. SoyM ran until 2016 and relied on remote sensor imaging techniques to identify soybean farming municipalities. SoyM's objectives are now permanent and are believed to have contributed to reports that only 1% of Amazon deforested areas were used for soybean production. This resulted in the recognition and approval of Brazilian-exported soybean by European customers (Amaral et al., 2021).

5 Improving soybean farming in Brazil: using a circular economy and integrated reporting framework

The CE-IR framework is most effective when utilized on a national scale as opposed to on a case-by-case basis. Because the ideal situation is a closed-loop system, multiple companies, institutions and the government enhance the effectiveness of the CE-IR framework's potential where the output and waste of one system constitute the inputs for another (Secco et al., 2020; Tedesco et al., 2022). It is also more efficient to consider the six capitals at a national level as opposed to only at a business-by-business level, as illustrated in more detail in the following sections.

Society and business constitute a mutually beneficial relationship where each is dependent on the other for its survival. Society needs goods and services, while businesses need customers. In addition, attempting to make all businesses not-for-profit companies is foolhardy. Individuals will only risk their livelihoods (and financial capital) in exchange for the potential of a greater return (see Jensen and Meckling, 1976). But demanding super-profits at the cost of future generations' ability to live is unacceptable (Flower, 2015; Atkins et al., 2022). A balance must be struck.

Governments are tasked to manage the socioeconomic welfare of their citizens. It is in a government's best interests to safeguard food security and affordability, promote exports to generate foreign income and manage unemployment. Farming, coupled with tertiary education initiatives and a CE-IR framework, can provide synergistic solutions for many government mandates. As noted in Section 3, a lack of up-to-date data contributed to IPM strategies being disregarded (Panizzi, 2013). The section also discusses how SF can enhance efficiencies and reduce pesticide reliance and the risk of crop failure (Pivoto et al., 2018). Lin (2018) reflected on the role of mechanization in ensuring efficient yields and affordability but cautioned that planning is vital to ensure an equitable transition. Lastly, Kamali et al. (2017) discussed the use of GM, non-GM and organic farming.

Starting with funding, approaching the right financiers is vital, as financiers determine the farm's targets. Green finance and government subsidies are useful first steps. Green finance includes investors and lenders who wish to balance economic and environmental outcomes (Wang and Zhi, 2016; van Zijl et al., 2022). These financiers may be willing to accept lower financial capital returns in lieu of increased natural and social and relational capital returns, making them a good strategic partner in a CE-IR framework.

Utilizing a CE-IR framework, the government can partner with green financiers to promote more sustainable soybean farming. To ensure soybean farming is profitable and sustainable, the government can institute policies to link tertiary education subsidies to specific soybean-farming topics.

These include creating and improving SF technology[2] (Pivoto et al., 2018), ensuring IPM metrics[3] are kept up to date (Panizzi, 2013), identifying new IPM technologies,[4] developing sustainable packaging for soybean products and developing technologies for using pests and natural enemies as alternative protein sources for both human and animal foodstuff ingredients (see Nesic and Zagon, 2019; Chavez and Uchanski, 2021).

Some of SF's biggest challenges are upskilling people to operate SF technologies (for example, drone pilots and IT engineers) and educating farmers to improve the uptake of SF. Promoting SF at universities and Technikons promotes human and social and relational capitals by equipping locals with the tools and knowledge needed to implement SF technologies at soybean farms. Workers who will be made redundant by mechanization can also be funded by green finance and the government to attend SF courses, aiding the inevitable transition to a balance of labour and mechanization in farming. As workers and tertiary education graduates are well educated in SF, its uptake should be considerably enhanced. The result is enhanced intellectual, human, social and relational, natural and financial capitals (see Maroun and Atkins, 2021).

With regard to higher degrees, students need relevant and impactful topics that will also provide them with opportunities in the marketplace. Providing master's and PhD students with topics in IPM; virology; biochemistry and electrical, mechanical and bio-engineering ensures that important R&D is undertaken that has benefits to local farms and the economy in general. Access to data is often a barrier to research, but with government-farmer-university partnerships, this barrier is removed. Intellectual capital is created that benefits both the local farmers and the economy when new (and excess) technologies and skills are exported. Some of these financial benefits can be used to finance further topics to close the education financing loop. Local R&D also reduces reliance on overseas intellectual capital, thus reducing costs and improving the country's socioeconomic stability by uplifting the country to have a better balance between labour-intensive agricultural and service activities (Lin, 2018).

To illustrate a component of the closed loop, IPM research can specifically focus on the integration of pests, natural enemies and alternative protein sources. Some of the research findings have warned that maintaining a perfect pest-natural enemy balance in crops and the surrounding habitats is extremely challenging. It may not be feasible and responsible for the surrounding environments (see Bianchi et al., 2006; Karp et al., 2018). It may, therefore, be useful to create insect farms near soybean farms. Pest-specific insects can be leased to farmers to control pests, while re-harvested insects can be used as protein sources. Excess insects can be used as a fertilizer for crops (see

[2] Drones, microprocessors and UAVs.
[3] EIL and natural enemies.
[4] Such as new viruses, parasitoids, herbicides, pesticides and fungicides.

Chavez and Uchanski, 2021), while plant waste from crops can be used as food for insect farms. This is one example of looking to create closed-loop systems where the waste and end-of-life products of one component serve as the input for the same or different systems (see Secco et al., 2020; Tedesco et al., 2022).

Under the 'take-make-use-dispose' linear production system, financial capital was prioritized at all costs. Land not already used for another enterprise was cheaply turned into agricultural land, often through 'slash and burn' techniques (Van Vliet et al., 2013, p. 1455). Important biomes are destroyed by the 'dragging' effect to supply materials for infrastructure and make way for transport routes (Fearnside, 2001). A CE-IR framework acknowledges the non-transferability of key biodiversity biomes. To protect and restore biomes, tourism is advanced as a component of a CE-IR framework. Because of rapid soybean expansion, many important biomes were cleared. While the SoyM has prevented further deforestation, reforestation is needed. Active tourism can be used to target environmentally conscious consumers who wish to play a part in reforestation initiatives (Andersson, 2007). Active tourism is particularly attractive in Brazil because the Amazon is already synonymous with biodiversity and is viewed as a key area requiring protection and rehabilitation (Shukla et al., 1990; Fearnside, 2001; Mamede and De Araújo, 2008; Van Vliet et al., 2013; Assis et al., 2020; Amaral et al., 2021).

Drawing on the Amazon's inherent value and natural beauty, active tourism can be used to fund the reforestation efforts of previous soybean farms. In this tourism experience, tourists would travel to see the devastating effects of deforestation and physically plant new trees or tend to newly replanted areas. This provides employment to soybean farmers who must cease operations (human capital), rehabilitates important biomes (natural capital), contributes funding to the local economy and reforestation efforts (financial capital) and increases environmental awareness through tourists' word of mouth and social media. Environmental education can also form part of this initiative, especially to drive the use of sustainable product packaging, linking back to university research in this area. Active tourism can also extend to insect farming and harvesting. If accompanied by education to create awareness about this alternative protein source, its benefits and its place in ecosystems, then its future acceptability and success are strengthened.

Spin-off industries are important to the CE-IR framework. Research should actively seek uses for waste products from soybean farming as well as create industries for those products. This may take the form of insect farms, IPM and drone/UAV pilot training facilities, sustainable packaging production, chemical production plants and retailers and even seed modification facilities.

To realize a CE-IR framework at the national level, planning and monitoring technologies must be available. Because of the inherent diversity of the capitals, comparing them is a complex task that requires accountants to develop means of identifying, recognizing and measuring the different capitals.

6 The role of accounting in developing a circular economy

The tools to identify, recognize and measure items for financial reporting have undergone many significant changes over time (Hopwood, 1987). Accounting has developed from mere counting in primarily agricultural periods to single-entry then double-entry accounting. Entirely different sub-disciplines have also been developed with management accounting and control, social accounting and, more recently, sustainability and integrated reporting (see Bryer, 1993; Bryer, 2000; Ravenscroft and Williams, 2009; Atkins and Maroun, 2015; Flower, 2015). Once again, the current needs of society require an update to existing accounting technologies or the development of entirely new ones.

To date, researchers have experimented with several alternative accounts to provide useful information and enable broader accountability. These include 'monetising' nature (see Atkins et al., 2015), pictorial accounts (see Gallhofer and Haslam, 1996; Atkins and Maroun, 2020), 'nature diaries'/'nature signaling' (see Sullivan and Hannis, 2017; Atkins and Maroun, 2020; Büchling and Atkins, 2020; Büchling and Maroun, 2023; Büchling and Maroun, 2021), emancipatory (see Gallhofer and Haslam, 1996; Atkins et al., 2023; Atkins and Macpherson, 2022) and shadow accounts (see Dey et al., 2010). The point is that accountants should not be quick to judge what the future of accounting may hold. They should be encouraged to experiment widely. So long as the reporting is useful and improves each successive generation's ability to live sustainably, it should be given full consideration and support.

The need to develop the ability to account for all six capitals is important because *something not measured is something not done*. For companies to feel the pressure to adopt new systems and change their modus operandi, society needs the ability to review environmental performance and make judgements (see Ravenscroft and Williams, 2009; Pandya et al., 2021). In addition, companies that wish to accelerate their sustainability require accounting to reflect accurately their performance in all six capitals. This will allow companies to explain why, for example, they have lower financial returns but are still better performers overall and should be rewarded and supported. New reporting technologies should also address so-called 'greenwashing' that undermines the progress of substantive and credible environmental accounting (see de Freitas Netto et al., 2020; van Zijl et al., 2022). Lastly, as with all accounts, new technologies must be capable of being assured to enhance their credibility.

7 Conclusion

This chapter briefly discussed the circular economy and IIRC's IR framework. It provided a means to combine principles from both sources to create a CE-IR framework that attempts to create sustainable closed-loop systems while using the interrelatedness of the six capitals to plan and articulate the creation or erosion of the value of a business over the short, medium and long term.

This chapter also provided an overview of Brazil's soybean farming, and then, through the CE-IR framework, it demonstrated trade-offs and key considerations. Specifically, the need for planning at the national level and to integrate many components of different systems to close the loop was illustrated. Finally, it highlighted the importance of accounting and accountants in the sustainability journey, along with examples of alternative means of reporting useful information.

With multiple competing needs, a CE-IR framework may be more capable of enhancing more capitals and provide an overall greater benefit compared to allowing financial capital to hold every other capital hostage or be treated as subordinate. Without compromising the business imperative or the need to create returns for shareholders and creditors, this chapter demonstrated how all capitals can be equally considered to create value for stakeholders.

Future research is needed to expand on the CE-IR framework idea and provide a detailed roadmap for its implementation. Case studies can be conducted over relevant aspects of soybean farming raised in this chapter under a CE-IR framework. Further research can also help analyse existing cases of government-university-farming partnerships to identify the advantages, disadvantages and areas for improvement.

8 References

Alves, B. J. R., Boddey, R. M. and Urquiaga, S. 2003. The success of BNF in soybean in Brazil. *Plant and Soil* 252(1), 1-9.

Amaral, D. F., De Souza Ferreira Filho, J. B., Chagas, A. L. S. and Adami, M. 2021. Expansion of soybean farming into deforested areas in the amazon biome: the role and impact of the soy moratorium. *Sustainability Science* 16(4), 1295-1312.

Andersson, T. D. 2007. The tourist in the experience economy. *Scandinavian Journal of Hospitality and Tourism* 7(1), 46-58.

Assis, T. O., De Aguiar, A. P. D., Von Randow, C., De Paula Gomes, D. M., Kury, J. N., Ometto, J. P. H. B. and Nobre, C. A. 2020. CO2 emissions from forest degradation in Brazilian Amazon. *Environmental Research Letters* 15(10), 104035.

Atkins, J. and Macpherson, M. 2022. *Extinction Governance, Finance and Accounting: Implementing a Species Protection Action Plan for the Financial Markets*. Taylor & Francis.

Atkins, J. and Maroun, W. 2015. Integrated reporting in South Africa in 2012: perspectives from South African institutional investors. *Meditari Accountancy Research* 23(2), 197-221.

Atkins, J. and Maroun, W. 2020. The Naturalist's Journals of Gilbert White: exploring the roots of accounting for biodiversity and extinction accounting. *Accounting, Auditing and Accountability Journal* 33(8), 1835–1870.

Atkins, J., Atkins, B. C., Thomson, I. and Maroun, W. 2015. "Good" news from nowhere: imagining utopian sustainable accounting. *Accounting, Auditing and Accountability Journal* 28(5), 651–670.

Atkins, J., Doni, F., Hassan, A. and Maroun, W. 2022. Revealing Plato's 'Shadow Kingdom': rendering pandemic risk explicit in extinction engagement. In Atkins, J. and Macpherson, M. (Eds). *Extinction Governance, Finance and Accounting. Implementing a Species Protection Action Plan for the Financial Markets* (vol. 26) (1st edn.). Routledge.

Atkins, J. F., Atkins, B., Maroun, W., Barone, E. and Gozman, D. 2023. Conservation through conversation? Therapeutic engagement on biodiversity and extinction between NGOs and companies. *Business Strategy and the Environment* 32(5), 2631–2647.

Bianchi, F. J., Booij, C. J. and Tscharntke, T. 2006. Sustainable pest regulation in agricultural landscapes: a review on landscape composition, biodiversity and natural pest control. *Proceedings. Biological Sciences* 273(1595), 1715–1727.

Biz Latin Hub 2022. Brazilian agriculture exports rise 33.7%, soy hits record. In *Biz Latin Hub*. Biz Latin Hub Group.

Brazilian farmers 2022. Brazilian agricultural exports grow in July 2021 and surpass 2020 levels. *Brazilian Farmers*. Brazilian Confederation of Agriculture and Livestock.

Bryer, R. A. 1993. Double-entry bookkeeping and the birth of capitalism: accounting for the commercial revolution in medieval northern Italy. *Critical Perspectives on Accounting* 4(2), 113–140.

Bryer, R. A. 2000. The history of accounting and the transition to capitalism in England. Part one: theory. *Accounting, Organizations and Society* 25(2), 131–162.

Büchling, M. and Atkins, J. 2020. Reporting on more than just natural capital. *The Routledge Handbook of Integrated Reporting*. Routledge.

Büchling, M. and Maroun, W. 2021. Accounting for biodiversity and extinction: the case of South African national parks. *Social and Environmental Accountability Journal* 41(1–2), 66–97.

Büchling, M. and Maroun, W. 2023. Biodiversity reporting practices of the South African national parks. *Social Responsibility Journal* 19(1), 138–165.

Cattelan, A. J. and Dall'agnol, A. 2018. The rapid soybean growth in Brazil. *OCL* 25(1), 1–12.

Chavez, M. and Uchanski, M. 2021. Insect left-over substrate as plant fertiliser. *Journal of Insects as Food and Feed* 7(5), 683–694.

Da Silva, K. A., Nicola, V. B., Dudas, R. T., Demetrio, W. C., Maia, L. D. S., Cunha, L., Bartz, M. L. C., Brown, G. G., Pasini, A. and Kille, P. 2021. Pesticides in a case study on no-tillage farming systems and surrounding forest patches in Brazil. *Scientific Reports* 11, 1–14.

De Freitas Netto, S. V., Sobral, M. F. F., Ribeiro, A. R. B. and Da Luz Soares, G. R. 2020. Concepts and forms of greenwashing: a systematic review. *Environmental Sciences Europe* 32, 1–12.

De Villiers, C. and Maroun, W. 2017. *Sustainability Accounting and Integrated Reporting*. Routledge.

Dey, C., Russell, S. and Thomson, I. 2010. Exploring the potential of shadow accounts in problematising institutional conduct. *Social Accounting and Public Management*. Routledge.

Doughty, C. E., Metcalfe, D. B., Girardin, C. A. J., Amezquita, F. F., Durand, L., Huaraca Huasco, W., Silva-Espejo, J. E., Araujo-Murakami, A., Da Costa, M. C., Da Costa, A. C. L., Rocha, W., Meir, P., Galbraith, D. and Malhi, Y. 2015. Source and sink carbon dynamics and carbon allocation in the Amazon Basin. *Global Biogeochemical Cycles* 29(5), 645–655.

Encyclopaedia Britannica 2022. *Soybean. Encyclopaedia Britannica*. Encyclopaedia Britannica, Inc.

FAO 2022. Brazil [Online]. Available at: https://www.fao.org/3/y4632e/y4632e09.htm [Accessed 24 May 2022].

Fearnside, P. M. 2001. Soybean cultivation as a threat to the environment in Brazil. *Environmental Conservation* 28(1), 23–38.

Ferreira Costa, C. G. 2021. Increased vulnerability of family farming in the context of a weakened FNS agenda due to austerity measures in Brazil. *Economía Agraria y Recursos Naturales* 21(1), 103–134.

Flower, J. 2015. The international integrated reporting council: a story of failure. *Critical Perspectives on Accounting* 27, 1–17.

Gallhofer, S. and Haslam, J. 1996. Accounting/art and the emancipatory project: some reflections. *Accounting, Auditing and Accountability Journal* 9(5), 23–44.

Geissdoerfer, M., Savaget, P., Bocken, N. M. P. and Hultink, E. J. 2017. The Circular Economy–A new sustainability paradigm? *Journal of Cleaner Production* 143, 757–768.

Hopwood, A. G. 1987. The archaeology of accounting systems. *Accounting, Organizations and Society* 12(3), 207–234.

IIRC 2013. The international framework: integrated reporting. Available at: http://www.theiirc.org/wp-content/uploads/2013/12/13-12-08-THE-INTERNATIONAL-IR-FRAMEWORK-2-1.pdf [Accessed 1 October 2013].

IIRC 2021. *International <IR> Framework*. International Integrated Reporting Council.

Jabbour, C. J. C., Santos, F. C. A., Fonseca, S. A. and Nagano, M. S. 2013. Green teams: understanding their roles in the environmental management of companies located in Brazil. *Journal of Cleaner Production* 46, 58–66.

Jensen, M. C. and Meckling, W. H. 1976. Theory of the firm: managerial behavior, agency costs and ownership structure. *Journal of Financial Economics* 3(4), 305–360.

Kamali, F. P., Meuwissen, M. P. M., De Boer, I. J. M., Van Middelaar, C. E., Moreira, A. and Oude Lansink, A. G. J. M. 2017. Evaluation of the environmental, economic, and social performance of soybean farming systems in southern Brazil. *Journal of Cleaner Production* 142, 385–394.

Karp, D. S., Chaplin-Kramer, R., Meehan, T. D., Martin, E. A., Declerck, F., Grab, H., Gratton, C., Hunt, L., Larsen, A. E., Martínez-Salinas, A., O'rourke, M. E., Rusch, A., Poveda, K., Jonsson, M., Rosenheim, J. A., Schellhorn, N. A., Tscharntke, T., Wratten, S. D., Zhang, W., Iverson, A. L., Adler, L. S., Albrecht, M., Alignier, A., Angelella, G. M., Zubair Anjum, M., Avelino, J., Batáry, P., Baveco, J. M., Bianchi, F. J. J. A., Birkhofer, K., Bohnenblust, E. W., Bommarco, R., Brewer, M. J., Caballero-López, B., Carrière, Y., Carvalheiro, L. G., Cayuela, L., Centrella, M., Ćetković, A., Henri, D. C., Chabert, A., Costamagna, A. C., De La Mora, A., De Kraker, J., Desneux, N., Diehl, E., Diekötter, T., Dormann, C. F., Eckberg, J. O., Entling, M. H., Fiedler, D., Franck, P., Frank Van Veen, F. J., Frank, T., Gagic, V., Garratt, M. P. D., Getachew, A., Gonthier, D. J., Goodell, P. B., Graziosi, I., Groves, R. L., Gurr, G. M., Hajian-Forooshani, Z., Heimpel, G. E., Herrmann, J. D., Huseth, A. S., Inclán, D. J., Ingrao, A. J., Iv, P., Jacot, K., Johnson, G. A., Jones, L.,

Kaiser, M., Kaser, J. M., Keasar, T., Kim, T. N., Kishinevsky, M., Landis, D. A., Lavandero, B., Lavigne, C., Le Ralec, A., Lemessa, D., Letourneau, D. K., Liere, H., Lu, Y., Lubin, Y., Luttermoser, T., Maas, B., Mace, K., Madeira, F., Mader, V., Cortesero, A. M., Marini, L., Martinez, E., Martinson, H. M., Menozzi, P., Mitchell, M. G. E., Miyashita, T., Molina, G. A. R., Molina-Montenegro, M. A., O'neal, M. E., Opatovsky, I., Ortiz-Martinez, S., Nash, M., Östman, Ö., Ouin, A., Pak, D., Paredes, D., Parsa, S., Parry, H., Perez-Alvarez, R., Perović, D. J., Peterson, J. A., Petit, S., Philpott, S. M., Plantegenest, M., Plećaš, M., Pluess, T., Pons, X., Potts, S. G., Pywell, R. F., Ragsdale, D. W., Rand, T. A., Raymond, L., Ricci, B., Sargent, C., Sarthou, J. P., Saulais, J., Schäckermann, J., Schmidt, N. P., Schneider, G., Schüepp, C., Sivakoff, F. S., Smith, H. G., Stack Whitney, K., Stutz, S., Szendrei, Z., Takada, M. B., Taki, H., Tamburini, G., Thomson, L. J., Tricault, Y., Tsafack, N., Tschumi, M., Valantin-Morison, M., Van Trinh, M., Van Der Werf, W., Vierling, K. T., Werling, B. P., Wickens, J. B., Wickens, V. J., Woodcock, B. A., Wyckhuys, K., Xiao, H., Yasuda, M., Yoshioka, A. and Zou, Y. 2018. Crop pests and predators exhibit inconsistent responses to surrounding landscape composition. *Proceedings of the National Academy of Sciences of the United States of America* 115(33), E7863–E7870.

Lai, A. and Stacchezzini, R. 2021. Organisational and professional challenges amid the evolution of sustainability reporting: a theoretical framework and an agenda for future research. *Meditari Accountancy Research* 29(3), 405–429.

Lin, J. Y. 2018. Agriculture is key for economic transformation, food security, and nutrition [Online]. Available at: https://www.ifpri.org/blog/agriculture-key-economic-transformation-food-security-and-nutrition [Accessed 17 June 2022]. Washington, DC: International Food Policy Research Institute.

Mamede, M. D. A. and De Araújo, F. S. 2008. Effects of slash and burn practices on a soil seed bank of caatinga vegetation in Northeastern Brazil. *Journal of Arid Environments* 72(4), 458–470.

Maroun, W. and Atkins, J. 2018. The emancipatory potential of extinction accounting: exploring current practice in integrated reports. *Accounting Forum* 42(1), 102–118.

Maroun, W. and Atkins, J. 2021. A practical application of accounting for biodiversity: the case of soil health. *Social and Environmental Accountability Journal* 41(1–2), 37–65.

Maroun, W., Ecim, D. and Cerbone, D. 2022. Refining integrated thinking. *Sustainability Accounting, Management and Policy Journal* 14(7), 1–25.

Nesic, K. and Zagon, J. 2019. Insects-a promising feed and food protein source? *Scientific Journal. Meat Technology* 60(1), 56–67.

Oda, L. M. and Soares, B. E. 2000. Genetically modified foods: economic aspects and public acceptance in Brazil. *Trends in Biotechnology* 18(5), 188–190.

Pandya, A., Van Zijl, W. and Maroun, W. 2021. Fair value accounting implementation challenges in South Africa. *Journal of Accounting in Emerging Economies* 11(2), 216–246.

Panizzi, A. R. 2013. History and contemporary perspectives of the integrated pest management of soybean in Brazil. *Neotropical Entomology* 42(2), 119–127.

Pivoto, D., Waquil, P. D., Talamini, E., Finocchio, C. P. S., Dalla Corte, V. F. and De Vargas Mores, G. 2018. Scientific development of smart farming technologies and their application in Brazil. *Information Processing in Agriculture* 5(1), 21–32.

Proterra Foundation 2021. The non-GM production: how volatility is killing the supply chain in Brazil [Online]. Available at: https://www.proterrafoundation.org/news/the-non-gm-production-how-volatility-is-killing-the-supply-chain-in-brazil/#_ftn1 [Accessed 26 September 2022]. Netherlands: ProTerra Foundation 2022.

Ravenscroft, S. and Williams, P. F. 2009. Making imaginary worlds real: the case of expensing employee stock options. *Accounting, Organizations and Society* 34(6-7), 770-786.

Sabourin, E., Craviotti, C. and Milhorance, C. 2020. The dismantling of family farming policies in Brazil and Argentina. *International Review of Public Policy* 2(1), 45-67.

Secco, C., Da Luz, L. M., Pinheiro, E., De Francisco, A. C., Puglieri, F. N., Piekarski, C. M. and Freire, F. M. C. S. 2020. Circular economy in the pig farming chain: proposing a model for measurement. *Journal of Cleaner Production* 260, 121003.

Shukla, J., Nobre, C. and Sellers, P. 1990. Amazon deforestation and climate change. *Science* 247(4948), 1322-1325.

Snir, A., Nadel, D., Groman-Yaroslavski, I., Melamed, Y., Sternberg, M., Bar-Yosef, O. and Weiss, E. 2015. The origin of cultivation and proto-weeds, long before Neolithic farming. *PLoS ONE* 10(7), e0131422.

Sullivan, S. and Hannis, M. 2017. "Mathematics maybe, but not money": on balance sheets, numbers and nature in ecological accounting. *Accounting, Auditing and Accountability Journal* 30(7), 1459-1480.

Tedesco, M., Simioni, F. J., Sehnem, S., Soares, J. F. and Junior, L. M. C. 2022. Assessment of the circular economy in the Brazilian planted tree sector using the ReSOLVE framework. *Sustainable Production and Consumption* 31, 397-406.

Van Vliet, N., Adams, C., Vieira, I. C. G. and Mertz, O. 2013. "Slash and burn" and "shifting" cultivation systems in forest agriculture frontiers from the Brazilian Amazon. *Society and Natural Resources* 26(12), 1454-1467.

Van Zijl, W., Cerbone, D. and Maroun, W. 2022. Green bonds from an integrated thinking perspective. *Extinction Governance, Finance and Accounting*. Routledge.

Wang, Y. and Zhi, Q. 2016. The role of green finance in environmental protection: two aspects of market mechanism and policies. *Energy Procedia* 104, 311-316.

Chapter 5

Biodiversity considerations in the marine aquaculture and fisheries industries

Dusan Ecim and Warren Maroun, University of the Witwatersrand, South Africa

1 Introduction

'Aquaculture' is the farming of aquatic organisms such as fish, molluscs, crustaceans and aquatic plants (such as seaweed) by way of intervention in the rearing process to enhance production[1]. This involves individual or corporate ownership of aquatic stock and production processes, including the final harvest for consumption (FAO, 2022). When aquatic organisms can be exploited by the public as a common property resource, this is referred to as the 'harvest of fisheries' (FAO, 2022).

The aquaculture and fisheries sectors play an important role in global food security and nutrition (FAO, 2022). Aquatic foods such as fish and other seafood products are an important source of protein, fatty acids, vitamins, minerals and bioavailable micronutrients for more than three billion people (The Nature Conservancy, 2023; FAO, 2022). Both industries supply an income to approximately 12% of the global population and have an estimated joint value of US$422.5 billion as of 2020 (The Nature Conservancy, 2023; FAO, 2022).

Unfortunately, the aquaculture and fisheries industries have material adverse impacts on marine ecosystems[2]. There is increasing exploitation and a

[1] This includes, *inter alia*, stocking, feeding and protection from predators.
[2] The aquatic environment segment includes marine water, fresh water and brackish water. The fish-type market is divided into carps, molluscs, crustaceans, mackerels, sea bream and others including, for example, pompano, snappers, groupers, salmon, milkfish, tuna, tilapia, catfish and sea bass (FAO, 2022).

http://dx.doi.org/10.19103/AS.2023.0128.24

loss of biodiversity, which is impairing the ocean's capacity to provide aquatic foods, maintain water quality and recover from ecological damage (Smith et al., 2020; Zhang and Kendall, 2019). Given the social and economic importance of this agrifood system, the environmental impacts need to be prioritised to ensure the sustainability of both the marine biodiversity and the economic viability of the sector (Maroun et al., 2018).

There are studies dealing with biodiversity impacts on aquaculture (see, for example, Campbell et al., 2021; Davies et al., 2019; Puszkarski and Śniadach, 2022; Ghauri et al., 2022) and, to a larger extent, the biodiversity impacts on fisheries (see, for example, Hadjimichael, 2018; The Nature Conservancy, 2023; Gaillet et al., 2022; Rees et al., 2020), but few scholars deal directly with the impact of fisheries and aquaculture on biodiversity and how firms in these sectors report on their biodiversity management[3]. In this context, the current chapter outlines biodiversity-related risks for companies in the aquaculture and fisheries industries and potential solutions to these challenges per earlier research. How some of the industry's leaders are reporting on their efforts to manage biodiversity-related challenges is considered.

2 Background to the aquaculture and fisheries industries

The global aquaculture market, valued at US$281.5 billion (production = 122.6 million tonnes) in 2020, is expected to reach an estimated US$378 billion by 2027[4]. Asia leads in the sector, contributing to 91.6% of the total global production (Sumesh and Roshan, 2020; FAO, 2022). Global capture fisheries production, valued at US$141 billion (production = 90.3 million tonnes) in 2020, was severely impacted by the COVID-19 pandemic during 2020, and the situation was further exacerbated by China reporting an ongoing reduction in annual catches at the time of writing (FAO, 2022). Despite this, it is expected that aquatic food consumption will increase by 15% by 2030, resulting in an average of 21.4 kg seafood per person (FAO, 2022). Figure 1 illustrates the top ten countries ranked by aquaculture production and includes the related marine capture production quantity and US$ monetary value.

China contributes to 57% of the total global aquaculture production (70.48 million tonnes valued at US$167.87 billion) and to 15% of the global marine capture (13.45 million tonnes). This places China as the global leader in the sector. China is followed by Indonesia (14.85 million tonnes aquaculture; 6.99 million tonnes marine capture) and India (8.64 million tonnes aquaculture; 5.52

[3] For the purpose of this chapter, reports on biodiversity-related matters include sustainability reports, integrated reports and other environmental, social and governance-related reports.

[4] This is before considering the economic impacts of the COVID-19 pandemic, which has impacted growth prospects; however, the marine economy has sufficiently recovered after the pandemic.

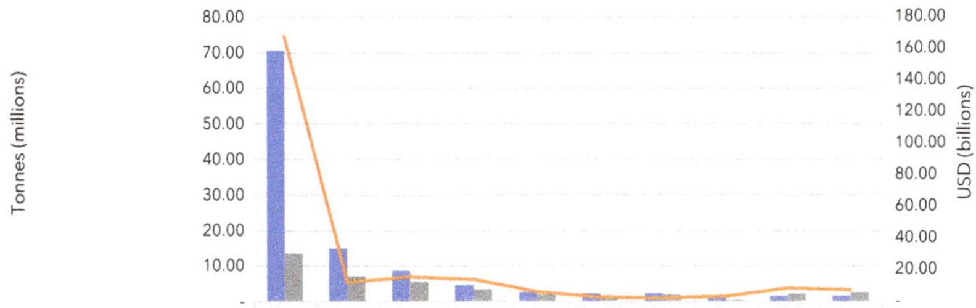

	China	Indonesia	India	Vietnam	Bangladesh	Republic of Korea	Philippines	Egypt	Chile	Norway
Aquaculture production (tonnes 'millions)	70.48	14.85	8.64	4.61	2.58	2.33	2.32	1.59	1.51	1.49
Marine capture production (tonnes 'millions)	13.45	6.99	5.52	3.42	1.92	1.38	1.91	0.42	2.18	2.60
Monetary value (USD 'billions)	167.87	12.09	15.56	14.00	6.30	2.76	2.30	3.26	8.46	7.30

Figure 1 Aquaculture and marine capture production statistics for 2020. Adapted from: FAO (2022).

million tonnes marine capture) in terms of the production quantities. Egypt produces the highest output in Africa, followed by Chile in South America. Norway is the largest producer in Europe.

Figure 1 illustrates the top aquaculture producers, as this is a significant driver of the revenue in the marine sector. Untabulated data show that Peru (5.7 million tonnes), Russia (5.1 million tonnes) and the United States of America (4.3 million tonnes) also contribute significantly to the global marine capture.

Marine aquaculture and farming can have both positive and negative impacts on biodiversity and natural capital. This is dependent on the type of species farmed, as well as the farming methods employed (Ghauri et al., 2022). For example, the farming of mussels, oysters and seaweed can have positive impacts on biodiversity and natural capital, as these species can improve water quality by absorbing excess nutrients and pollutants from the water. They also provide habitats for other species, which can help restore wild fish populations (FAO, 2022; Cabral et al., 2016). Oyster farming, for example, has synergies for marine conservation and is a low-cost tool to improve water quality while creating jobs in coastal communities and growing healthy and sustainable seafood (Cartier and Carpenter, 2014; NOAA Fisheries, 2021). The oysters efficiently remove nitrogen from the aquatic ecosystem and incorporate it into their cells and tissue (NOAA Fisheries, 2021). Nitrogen is then recycled into the aquatic environment by oysters during excretion and consumed by phytoplankton, which, in turn, decreases the toxicity of water, resulting in a highly productive ecosystem that does not require external inputs such as fertilisers (Chapelle et al., 2000). An oyster farm with '100 000 oysters per acre has the capability to filter up to 5 000 000 gallons of water per day, per acre', efficiently reducing nitrogen levels and limiting the growth of harmful algae (NOAA Fisheries, 2021). The cultivation of bivalve shellfish (oysters) and seaweed facilitates new habitats for fish and mobile invertebrate species that,

Table 1 Summary of impacts on major aquaculture and fisheries species

Species group	Examples	Production quantity (thousand tonnes)	Key biodiversity impacts
Finfish in inland aquaculture	Carp, trout, black bass	49 120.5	An integrated management approach must be applied to balance the ecosystem and fisheries requirements. Due to the infrastructure required for inland aquaculture, mitigation measures need to be implemented, for example, habitat rehabilitation, water management programmes and fish-passage systems (to protect species that depend on longitudinal and lateral movements to complete their lifecycle). Agriculture and aquaculture can form integrated farming systems to cycle nutrients, for example, fishponds can provide a source of water for irrigation and irrigation systems can be fished
Finfish in marine and coastal aquaculture	Salmon, seabass	6418.2	Salmon has a strong competitive position in the world market with demand exceeding most fish categories. As a 'blue food', finfish is environmentally more friendly than land-based food, there are negative biodiversity impacts. These include wastes from farms, changes in the nutrient content of water, exceeding carrying capacity, nutrient and wastewater release, escaped finfish impacting the biodiversity habitat, disease outbreaks and parasite abundance on farms. The impacts need to be managed using sustainable technology and farming practices given the direct impact on surrounding aquatic environments
Crustaceans	Shrimp, crawfish, prawn	11 237	Shrimp and prawns have historically been highly traded aquatic commodities that are currently produced primarily in Latin America and East and Southeast Asia. Similar to finfish, farming may result in the destruction of the mangrove ecosystem, a decline in biodiversity due to water and nutrient pollution, sedimentation, saltwater intrusion and diseases that need to be managed with appropriate farming techniques

(*Continued*)

Table 1 (*Continued*)

Species group	Examples	Production quantity (thousand tonnes)	Key biodiversity impacts
Molluscs	Oysters, scallops	17 742.2	The most important bivalve mollusc species for international trade are scallops, clams, oysters and mussels. Molluscs are an important component of global ocean health with oyster reefs also serving as barriers to storms and tides, preventing erosion and protecting productive estuary waters. When oyster reefs are primarily used to harvest commercial oyster meat, they can become degraded. Managing these resources sustainably is important to conserve the marine ecosystem, the food source as well as the economy. Currently, oyster populations are at historic lows with erosion from development, wetland loss and excessive nutrient pollution impacting the viability of shellfish. Overfishing and outdated harvest methods have destroyed or damaged reef structures and reduced oyster populations that require rehabilitation

Developed from FAO (2022); Ford et al. (2012); NOAA Fisheries (2021) and European Commission (2023).

for example, can hide from predators inside an oyster cage (NOAA Fisheries, 2021).

However, should the carrying capacity of the ecosystem be exceeded, the ecological productivity may be reversed. Similarly, a sea-based oyster infrastructure may also adversely impact other aquatic animals' habitats (FAO, 2022). If too many fish or shellfish are grown in a small area, then the water quality can decrease because of waste matter deposited and oxygen depletion. Non-native species can have adverse effects on the local species and disrupt the ecosystem (Davies et al., 2019; FAO, 2022). Appropriate farming practices, education and licences are important to sustainably manage aquaculture (FAO, 2022). Table 1 summarises biodiversity impacts of material groups of species. Biodiversity losses give rise to business risks, particularly for sectors dependent on natural resources, such as aquaculture and fisheries (Smith et al., 2020). They can result in production delays and disruption to supply chains in addition to impacting the quantity, quality and access to essential goods and services by consumers (Atkins et al., 2022). As stakeholders become more environmentally conscious, companies associated with a poor biodiversity track record may also find it difficult to preserve market share, attract or retain key employees and secure reliable supply arrangements. Nevertheless, many organisations - including those in aquaculture and fisheries - have been slow to implement significant reforms aimed at curtailing biodiversity loss. The continued focus on short-term profits and concerns about the cost of re-engineering business models are the primary impediments (Smith et al., 2020; Atkins et al., 2015).

3 Core biodiversity themes established in aquaculture and fisheries academic research

To understand better the link between the aquaculture and fisheries sectors and biodiversity, a Scopus database search was conducted to identify research dealing specifically with biodiversity in these sectors. Scopus is the largest database of peer-reviewed academic literature and is recognised for the quality of its filtering criteria and the fact that it includes journals of good standing and with robust peer-review processes in place (Rinaldi et al., 2018; Ecim and Maroun, 2023).

'Aquaculture and Biodiversity' and 'Fisheries and Biodiversity' were used as terms to search for in the title, abstract and keywords of academic papers. These were subject to the following filters:

- Only academic papers published from 1993 to 2023 were considered.
- Papers dealing with business management and accounting-related topics were selected.
- The papers had to deal with aquaculture and fisheries as a core topic.

There were 82 and 296 papers dealing with biodiversity-related matters in the aquaculture and fisheries sectors, respectively. These papers were analysed to identify key biodiversity themes and areas of research using VosViewer software (see Van Eck and Waltman, 2017). The results are illustrated in Fig. 2. Only core themes were included for the ease of analysis.

The core themes were:

- environmental impact and solutions (green cluster);
- governance approach, policies and stakeholder considerations (red and yellow clusters); and
- sustainable development imperatives (blue cluster).

Each theme is explored in more detail in the following sections.

3.1 Environmental impact and solutions

Managing natural capital plays an important role in maintaining food supply in the aquaculture and fisheries sectors (Ghauri et al., 2022), but there are many challenges.

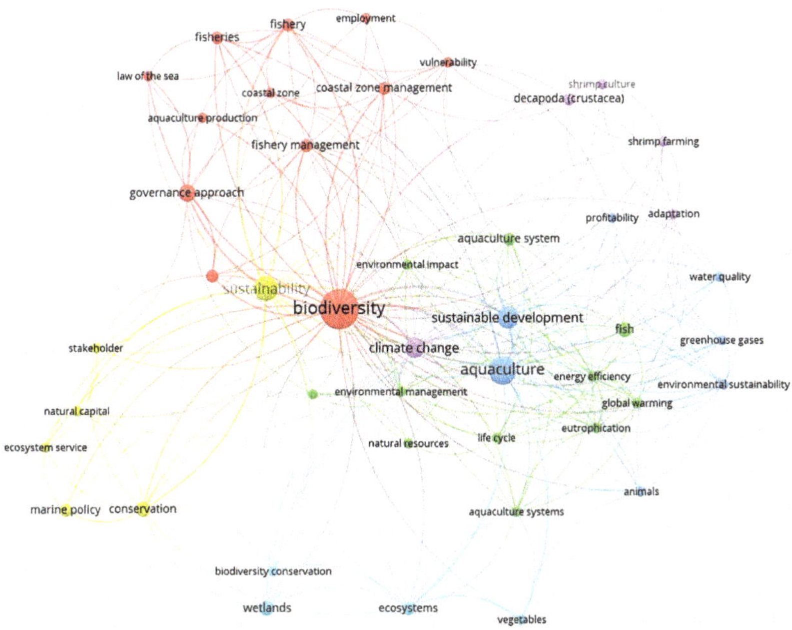

Figure 2 Visualisation of connected keywords for aquaculture, fisheries and biodiversity.

3.1.1 Challenges

Marine ecosystems can be damaged by, for example, habitat disruption, unsustainable farming practices and biodiversity depletion (FAO, 2022). Examples include eutrophication, which takes place because the environment is enriched with nutrients (such as nitrogen and phosphorous) from fertilisers used in aquatic farming, domestic and industrial sewage and wastewater (European Commission, 2023). Eutrophication increases the number of algae species (including toxic species) in the ecosystem, resulting in oxygen deficiencies in the bottom water and overall water quality degradation (European Commission, 2023; FAO, 2022). The water degradation impacts the viability of fish species and the overall balance of the ecosystem. In addition to reduced stocks, eating fish or shellfish harvested from affected areas can be dangerous as a result of toxicity buildup (Smith et al., 2020; European Commission, 2023).

Further habitat disruption can result from mangrove destruction. Mangroves are ecologically and economically important carbon-rich forests that provide a wide range of ecosystem services including a defence against erosion (Ahmed and Glaser, 2016). The destruction of mangroves results in different climatic disturbances, including coastal flooding, sea-level rise, salinity changes and an increase in sea surface temperature (Ahmed and Glaser, 2016; FAO, 2022).

These changes have adverse effects on coastal aquaculture and result in a loss of biodiversity and fewer productive farming areas (ibid).

Unsustainable farming practices may be caused by a lack of sustainability-related knowledge, economic disparity resulting in cheaper, less sustainable practices being used and a prioritisation of economic imperatives over environmental considerations (Smith et al., 2020; FAO, 2022). For example, aquatic farming (such as salmon farming) can result in fish waste and parasites being released into the surrounding water if the correct practices and procedures are not in place (FAO, 2022). This damages the marine habitat and results in disease outbreaks (de Lima Vieira et al., 2021). The escape of farmed or captive fish can lead to genetic 'pollution' of wild fish populations, impacting the native species and resulting in biotic depletion (Ahmed and Glaser, 2016; de Lima Vieira et al., 2021). Aquatic farming (such as the growth of mussels and oysters), which takes place on ropes or nets suspended in water, can impact the movement of species, disrupting and reducing the biodiversity of the ecosystem (FAO, 2022).

Marine capture and fishing have adverse environmental impacts, with the fishing industry experiencing major setbacks in the past 10 years, mainly as a result of overfishing and unsustainable fishing practices (Holm et al., 2017). This includes large quantities of discarded bycatch, which, ecologically, changes the availability of prey, degrades marine ecosystems and lowers the productivity of fisheries (Holm et al., 2017; FAO, 2022). Damage to the ecosystem also results in a loss or shift of breeding grounds of aquatic animals, salinisation[5] of freshwater aquaculture facilities, algal blooms and toxins in seafood organisms (Kibria and Haroon, 2017). These issues impact the aquaculture and fisheries biodiversity and ecological welfare, on the one hand, and the economic viability of the sector, on the other hand.

3.1.2 Solutions

Sustainable farming practices can be achieved by improving access to more efficient technology and financial capital, increasing engagement with stakeholders on biodiversity-related topics, ensuring suitable policies for managing fishing quotas and reducing the amount of nutrients released into marine ecosystems (Marín Del Valle et al., 2022). Technological innovations such as 'Recirculating Aquaculture Systems (RAS)' (see Ahmed and Turchini, 2021) are examples of eco-friendly, water-efficient and highly productive intensive farming systems that reduce water demands and re-use waste waters

[5] Salinisation is the process of increasing salt content in soil and results in productivity changes, an increased risk for less tolerant species and an increase in plant pests and diseases, resulting in the loss of biodiversity (FAO, 2022).

to avoid adverse environmental impacts. Other sustainable farming practices include, for example,

- biofloc technology to recycle waste nutrients as fish food to improve the water quality;
- commercial probiotics to increase animal resistance to pathogens[6] and improve the productivity of aquaculture production processes;
- improving agricultural practices such as restrictions on the use of fertilisers, optimising crop cultivation, using low- and zero-water exchange systems and limiting fertiliser washout from land;
- digital platforms such as automatic water quality sensors, artificial intelligence and biosecurity sensors, which can lower food conversion ratios, improve aquatic stock health and reduce stress in aquaculture production systems;
- low-density culture methods, which can allow for the growth of endangered species in controlled environments before reintroducing the species into the wild;
- multi-species co-culture, which includes crops and aquatic animals to promote ecological intensification;
- rotating the location of farms to avoid overusing the ecosystem and zoning production areas;
- using a closed-contaminant system, which is a type of barrier restricting and controlling interactions between farmed fish and the external aquatic environment; and
- moving the location of aquatic farming either further offshore towards ocean currents, which can efficiently dispose of waste or farming entirely on the 'land' in enclosed bio-secure environments.

(developed from de Lima Vieira et al., 2021; European Commission, 2023; Wan et al., 2020; Zhang and Kendall, 2019; Kibria and Haroon, 2017; Marín Del Valle et al., 2022; FAO, 2022; Kaiser et al., 2010)

The responsibility of marine ecosystems rests not only with the organisations directly involved in the sector but also on stakeholders who impact the marine ecosystems indirectly (Holm et al., 2017). For example, organisations not operating in the marine aquaculture or fisheries industries may still contribute to general wastage, sewage discharge and water pollution with adverse consequences for marine ecosystems (Smith et al., 2020). As a result, these organisations will need to be co-opted as part of an integrated strategy to reduce biodiversity loss in rivers and seas. For example, communities, organisations and local governments in high-risk biodiversity areas can

[6] Stocking of post-larval shrimp sourced from specific pathogen-free broodstock also reduces disease impacts.

preserve biodiversity by protecting the wetlands (Kibria and Haroon, 2017). Wetlands act as nutrient sinks and prevent erosion, filter and purify water and are an important habitat for many species and fish that depend on the wetland for breeding, nesting and feeding (Kibria and Haroon, 2017). Stakeholders can preserve the existing vegetation, including mangroves, by encouraging the introduction of additional wetlands, supporting community efforts to clean up wetland, promoting conservation projects, curbing urban development, setting up compost piles away from the wetland, using proper recycling and waste disposal methods, using dry wells for drainage and reporting any environmental violations to the appropriate authorities (Kibria and Haroon, 2017; FAO, 2022; European Commission, 2023).

Conservation efforts need to be collaborative. Organisations that operate in other agrifood sectors can collaborate with aquaculture producers to manage forestry, food and fish production in coastal areas simultaneously (Kibria and Haroon, 2017; Wan et al., 2020). Ocean resource depletion can be mitigated by collaborating with other stakeholders in the agrifood sector to use biofuel production (such as microalgae oil and its co-product, defatted microalgal biomass) to replace fishmeal (including cultured tilapia, salmon, shrimp, carp, flounder, yellowtail and cod) from wild fish catch and reduce the impact of overfishing (Zhang and Kendall, 2019).

General waste and sewage discharge need to be managed through repurposing, reusing and recycling waste products and the appropriate filtration, biological treatment and disinfection of sewage and wastewater before discharge (Ghauri et al., 2022; FAO, 2022). Conservation groups can launch campaigns to increase general public knowledge and awareness of biodiversity conservation and preservation (FAO, 2022).

Governments need to promote the implementation of frameworks, legislation and policies aimed at biodiversity conservation, sustainable development and the granting of fishing and other licences to individuals and organisations prioritising conservation and ecosystem welfare (Marín Del Valle et al., 2022; FAO, 2022). It is also necessary to approve social protection and insurance schemes for farmers, which includes financing, to ensure that farms adopt environmentally conscious technologies (Ghauri et al., 2022). Education, training programmes and technology subsidies need to be implemented to promote sustainable practices, particularly in developing economies that may lack the resources to implement sustainable initiatives (FAO, 2022).

3.2 Governance approach, policies and stakeholder considerations

A key term used in the context of aquaculture and fisheries is 'blue growth' where the sea is recognised as a driver of the economy. The approach can lead to overexploitation if the growth is not sustainable (Hadjimichael, 2018). In response are efforts to transition to 'blue communities' characterised by multidimensional management of social, cultural, environmental and economic factors (Campbell et al., 2021). Blue communities promote sustainable aquaculture expansion, effective management of fisheries and a clear understanding of the impacts of aquaculture and fisheries across aquatic value chains (FAO, 2022). Appropriate governance and legal frameworks can facilitate this holistic approach to managing the sector.

Good aquaculture governance is important for maximising the industry's contribution to sustainable development and biodiversity management. For example, sound governance and expert training leads to the production of more nutritious aquatic food, additional employment opportunities and increased revenues for organisations and governments (by way of taxes) (FAO, 2022). Good governance practices also support better environmental stewardship by reducing pressure on wild fishery stocks and promoting the responsible use and protection of natural capital, including the land, water, coastal habitats and aquatic resources (FAO, 2022).

Biodiversity protection and ecological issues need to be incorporated into a stakeholder-inclusive approach to managing marine-dependent businesses (King et al., 2022). The organisation needs to act in a socially responsible manner across operations, which involves implementing biodiversity-conscious policies to manage aquaculture farming, harvesting, fishing activities, production activities and distribution. When evaluating environmental performance, it is necessary to discharge accountability to all stakeholders and establish effective internal and external communication channels with them (Rinaldi, 2020). This includes a designated committee (such as a risk or sustainability committee) and a 'Chief Value Officer' to oversee and monitor the biodiversity risks, opportunities, targets and outcomes (King and Atkins, 2016). The governance structures enable transformative environmental accounting and reporting, the establishment of an ethical culture, improved business performance, effective control and legitimacy with the goal of preventing biodiversity losses (King et al., 2022; IOD, 2016). Governance needs to be embedded in a transparent, equitable and enforceable framework that covers the aquaculture production and marine capture process. However, problems arise in implementing a sustainable good governance approach. These include (FAO, 2022; European Commission, 2023):

- decreased accountability by the public and private sectors for the biodiversity impact of direct and indirect actions impacting marine ecosystems;
- inadequate implementation of regulations and policies governing aquaculture and fisheries; and

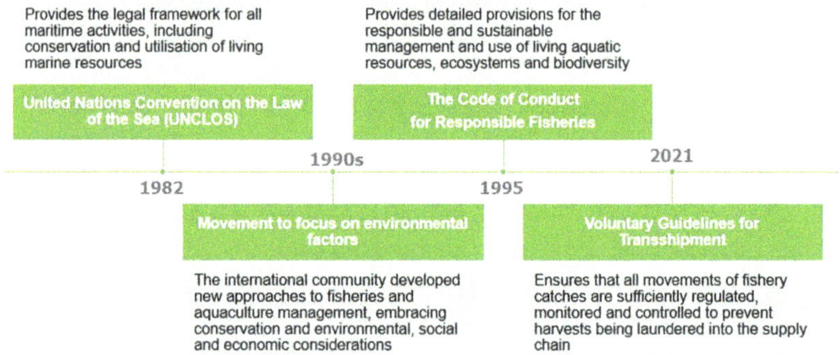

Figure 3 Key frameworks for the governance of aquaculture and fisheries. Adapted from: FAO (2022).

- poor planning over farming sites resulting in disease outbreaks and habitat deterioration.

Although there are limited marine-specific governance frameworks, sustainability initiatives have been in place since the 1980s. For example, in the European Union, legislation to reduce phosphorus and nitrogen loads include The Nitrates Directive (1991), The Urban Waste-Water Treatment Directive (1991) and the Water Framework Directive (2000). The policies are intended to promote the long-term sustainability of the biodiversity and industry, including aquaculture and fisheries (European Commission, 2023). Figure 3 summarises the core frameworks developed to manage the governance of aquaculture and fisheries.

The legal environment includes regulations over, for example, aquaculture permits, coastal zone management, fisheries management, marine spatial planning[7] and adaptive environmental assessment and management (Craig, 2019). Public participation and accountability are promoted and leads to increased knowledge and adaptive governance over marine-related considerations (Craig, 2019). Using multi-stakeholder groups to advise on marine aquaculture policy is advised to promote collaborative problem-solving across the value chain (Siddiki and Goel, 2015). Examples of potential collaborations that can enhance the support of biodiversity-related matters include the Convention on Biological Diversity (CBD) Secretariat, the United Nations Environment Programme (UNEP), the International Labour Organization (ILO) and the International Maritime Organization (IMO) (FAO, 2022).

[7] Marine spatial planning involves multiple stakeholders (including energy, industry, government, conservation and recreation) making informed and coordinated decisions concerning the sustainable use of marine resources.

3.3 Sustainable development imperatives

Several Sustainable Development Goals are relevant to aquaculture and fisheries.

- The achievement of no poverty (Goal 1) and zero hunger (Goal 2) is supported by aquaculture and fishing, given the role that the industry plays in both feeding populations and providing a source of income (Sections 1 and 2). Aquatic foods, as a source of essential fats and protein, also promote health and well-being (Goal 3).
- Responsible consumption and production (Goal 12) is relevant for aquaculture and fisheries in light of the concerns over exploitation of the natural resources. Sustainable farming practices (Section 3.1), supported by suitable education initiatives, are important to maintain ecosystems and ensure a sustainable source of aquatic resources. This assists with managing climate change (Goal 13). Healthy marine ecosystems are an important carbon sink.
- Life below water (Goal 14) involves the conservation and sustainable use of oceans, seas and marine resources. The aquaculture and fisheries industries need to do more to balance economic considerations, growing food demands and the importance of conserving marine ecosystems. The solutions outlined in Section 3.1 offer an immediate way forward and will need to be implemented without delay.

(Ghauri et al., 2022; EIT Food, 2023; FAO, 2022)

Organisations need to accept accountability for their ecological footprint. They should measure, evaluate and analyse their impact on the environment and create policies to rehabilitate ecosystems. This includes integrating the Sustainable Development Goals into business models, performance evaluations and reporting to stakeholders (EIT Food, 2023). Sustainability and integrated reporting are two examples of communication tools that can deal with both financial and extra-financial considerations.

Frameworks focusing on long-term value creation, sustainability and an integrated approach to managing economic and environmental imperatives can be applied by organisations in the aquaculture and fisheries sectors. Examples include the International Integrated Reporting Framework (see IIRC, 2021), which promotes the 'cycle of integrated thinking and reporting, resulting in efficient and productive capital allocation, [acting] as a force for financial stability and sustainability' (IIRC, 2021, p. 2). The Global Reporting Initiative standards can also be used to guide management's approach to and disclosure of biodiversity-related impacts (GRI, 2019) while the Carbon Disclosure Project guides the organisation's engagement with climate change metrics (CDP, 2020).

Regular biodiversity reporting and monitoring can help ensure that organisations have the management control systems in place to monitor impacts on the ecosystem and develop plans to improve or protect biodiversity in the areas under their management. Effective reporting lowers information asymmetry, promotes an awareness of the importance of conserving biodiversity and is becoming increasingly important for addressing regulatory and other demands for transparent accounts on how companies manage natural resources (Atkins and Atkins, 2019; Maroun et al., 2018). The next section analyses the biodiversity reports of various aquaculture and fisheries organisations to understand better the type of information being reported, the issues faced and the solutions being implemented by a sample of industry leaders.

4 Examples of biodiversity reporting

Five companies' integrated/sustainability reports were assessed using a qualitative content analysis to provide further details on biodiversity challenges and solutions and illustrate reporting conventions in the aquaculture and fisheries industries. The selected companies are sector leaders operating in multiple locations. Table 2 provides details on the selected companies.

From each organisation's latest integrated reports, the following information was obtained:

1. The biodiversity issues being reported;
2. Solutions to manage the biodiversity concerns; and
3. Metrics used to evaluate the organisation's performance in terms of long-term sustainability.

In addition, the top ten keywords based on earlier research (refer to Fig. 2) were used to establish the frequency with which the organisation's referred to core biodiversity themes in their reporting. Although this is not a scientific measure of the biodiversity content in the reports, it provides a general sense of how often an organisation discusses important concepts and provides direction in navigating the key issues. The keywords used are summarised in Table 3 with the associated occurrence and link strength. The occurrence measures the number of times the keyword appears in conjunction with other biodiversity-related themes in prior literature. A higher link strength demonstrates the number of cited references two items have in common, showing a stronger link between the keyword and biodiversity-related research.

4.1 The biodiversity issues that were reported and integration into the business model

The frequency of biodiversity themes noted across the reports is included in Fig. 4.

Table 2 Companies analysed for the case study

Company	Market capitalisation (US$ billion)[a]	Company headquarters and listing	Primary operations and selection criteria
Mowi ASA[b]	91.35	Norway; Oslo Stock Exchange	World's largest producer of Atlantic salmon and producer of whitefish and other seafood products. Primary interest is fish farming, mainly salmon, the operations of which are focused on Norway, Scotland, Canada, the Faroe Islands, Ireland and Chile
Leroy Seafood Group[c]	33.5	Norway; Oslo Stock Exchange	Sells farmed and wild-caught seafood products and has made a wide range of investments, including in land-based farming, adding new whitefish fishing vessels, and in its mature salmon operations
Atlantic Sapphire ASA[d]	1.13	Norway; Oslo Stock Exchange OTC Markets Group USA	World's largest land-based aquaculture company, with operations in Denmark and the United States of America. The entity represents operations in the North American jurisdiction. Atlantic Sapphire breeds Bluehouse (land-raised) salmon farming and transforms protein production globally
Tassal Group[e]	0.77	Australia; Australian Stock Exchange	Tassal is the largest producer of Tasmanian grown Atlantic salmon, supplying salmon to both domestic and international markets. The entity represents a jurisdiction in the Australasian jurisdiction
Oceana Group[f]	0.51	South Africa; Johannesburg Stock Exchange	Africa's largest fishing company and one of the most empowered food companies on the JSE. This entity has been selected to incorporate an organisation operating in a developing economy that does still have strong governance and integrated reporting structures

[a] As of January 2023 per Google Finance.
[b] https://mowi.com/investors/reports/.
[c] https://www.leroyseafood.com/en/sustainability/sustainability-library-2021/.
[d] https://atlanticsapphire.com/sust/.
[e] https://tassalgroup.com.au/information/reports/year-2022/.
[f] https://oceana.co.za/sustainability/sustainable-development-reports/.

Mowi reports the highest frequency of core biodiversity-related information. Mowi has achieved sustainability awards, including being ranked

Table 3 Core themes associated with biodiversity research in the aquaculture and fisheries industries

Keyword	Occurrences	Total link strength
Biodiversity	24	362
Sustainability	10	134
Sustainable development	8	144
Climate change	7	103
Governance approach	5	72
Conservation	4	52
Coastal zone management	3	33
Ecosystems	3	53
Fishery management	3	52
Marine policy	3	34

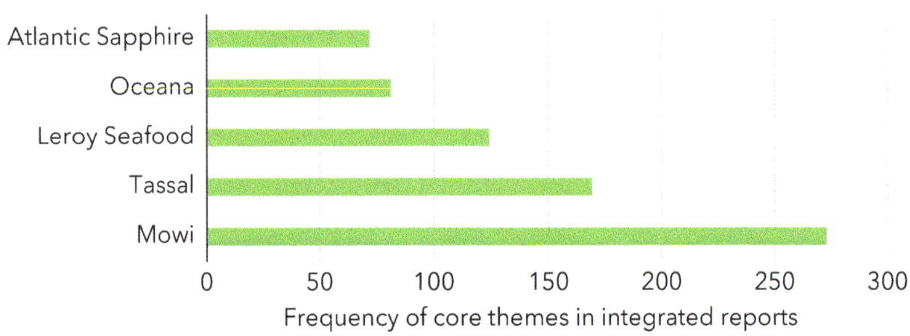

Figure 4 Frequency of biodiversity-related themes reported.

as the most sustainable animal protein producer globally by the FAIRR Initiative, which focuses on environmental, social and governance risks in the global food sector. Mowi is ranked as the second most sustainable seafood company (World Benchmarking Alliance) and has achieved positive sustainability ratings from its CDP and MSCI scores[8]. Mowi's integrated report details material biodiversity threats, planned responses and performance indicators used to evaluate biodiversity impact.

Mowi's integrated report details that salmon has lower greenhouse gas emissions, water use and land use than land-based food such as chicken. The carbon footprint of farm-raised salmon is lower than that of land-based food, and the edible yield (68%) is higher than that of chicken (46%), pork (52%) or lamb (38%). Seafood also avoids food waste as the by-products such as offcuts and trimmings are fully reusable and Mowi clearly details how these by-products are used to highlight production efficiencies.

[8] Morgan Stanley Capital International – rating designed to measure a company's resilience to long-term,industry material environmental, social and governance risks.

Mowi details further contributions to sustainability, including:

- increasing transparency and traceability at site locations;
- working with governments to improve aquaculture and fishing regulations;
- working towards eliminating illegal, unreported and unregulated (IUU) fishing;
- reducing the use of plastic pollution;
- developing plastic packaging that is fully reusable, recyclable or compostable; and
- ensuring good animal welfare.

The details listed earlier are completed by the steps taken to ensure food safety with seafood traceability tests performed to identify any food safety incidents and improve the approach to food safety[9].

Tassal (169 occurrences), Leroy Seafood (124 occurrences) and Oceana (81 occurrences) follow Mowi. Despite operating in a developing country, Oceana Group's integrated reporting is comparable with those prepared by its international peers. This is probably because South Africa is an integrated reporting and thinking pioneer, being the first jurisdiction to outline these concepts in codes of corporate governance (De Villiers et al., 2014). South African-listed companies are also required to prepare an integrated report or to explain why they have not done so (see IOD, 2016). As a result, Oceana has experience in preparing sustainability-related extra-financial information.

The five organisations disclose similar sustainability-related information dealing with governance and stakeholders, environmental impacts, protecting oceans, addressing climate change, incorporating the Sustainable Development Goals into business models and integrating sustainability into key performance indicators (KPIs).

Examples include preserving biodiversity, efficient water use, innovative ways to deal with environmental impact, climate-friendly food production, preventing fish escapes, resource availability, responsible use of medicine and chemicals, responsible waste management and fish health and welfare. Circular economy is also receiving increased attention. This production technology is concerned with the reduction of raw material usage, re-using components and recycling wherever possible. In the context of the seafood sector, Tassal, for example, discloses that waste products from salmon include the heads and frames of the fish, viscera (guts), trims, skins and fish mortalities. Waste products are recovered and recycled through other food production systems in the form of fishmeal and fish oil. Disclosures on the positive and negative impacts on biodiversity across the value chains are reported.

[9] Refer to Mowi's annual report available at https://mowi.com/investors/reports/.

Importantly, biodiversity issues cover only material issues with materiality based on:

- discussions with key stakeholders including employees, shareholders and investors, government and regulatory authorities, consumers, local communities, small-scale fishers, non-profit organisations, suppliers, aquaculture and fisheries organisations and benchmark developers;
- likelihood and magnitude assessments of issues impacting planet, product, people and profit; and
- alignment with the objectives set out in the Sustainable Development Goals.

(see also Edgley et al., 2015; Cerbone and Maroun, 2020)

Independent third-party sustainability and food safety certification enhance the reliability of the information reported and ensure that suitable seafood practices are adhered to. This also ensures that biodiversity challenges are identified by the organisation timeously. Certification can be received from, for example:

- Global Seafood Sustainability Initiative (GSSI), which is a global benchmark tool providing a formal recognition of seafood certification schemes that successfully complete a rigorous and transparent benchmark process, underpinned by the Food and Agriculture Organization Guidelines;
- Global Food Safety Initiative/BRC Global Standard for Food Safety, which aims to improve food safety and business efficiency and is based on internationally recognised standards such as ISO and Codex Alimentarius;
- The Aquaculture Stewardship Council, which manages the global standards for responsible aquaculture (developed by the WWF Aquaculture Dialogues) and certifies responsibly farmed seafood;
- Marine Stewardship Council, which recognises well-managed and sustainable fisheries through a certification programme;
- GLOBAL G.A.P. is a brand of smart farm assurance solutions, which include a range of standards for safe, socially and environmentally responsible farming practices covering the entire production chain; and
- The Global Seafood Alliance is an association dedicated to advancing responsible seafood practices through education, advocacy and third-party assurances.

Each organisation explains its core values and position on sustainability and conserving biodiversity, using their mission and value statements. Inputs into the business model are identified such as, for example, marine biomass, water used and electricity used. Policies set environmental objectives, which are given substance by formal processes and management controls designed to reduce adverse environmental impacts throughout the value chain. Formal risk and opportunity assessments are performed, which span multiple time frames and focus on the interconnection between economic, environmental and social issues.

The Oceana Group, for example, illustrates the controls in place to collect, analyse and report on the inputs and outputs of biodiversity-related data. There are defined committees in place to manage biodiversity risks and opportunities and processes to manage the governance of key issues. The organisation's management controls include, for example, compliance processes, quality control, impact assessments and internal audits designed to minimise the impact of adverse biodiversity.

4.2 Solutions to manage the biodiversity concerns

All of the organisations have addressed the common solutions noted in Section 3.1 to, at least, some extent. Mowi, for example, states that the company operates in a landscape conservation area and needs to perform regular environmental surveys to monitor the impact on farming sites. Atlantic Sapphire includes extensive details on their 'Bluehouse' bio-secure environment. The salmon farming process is kept completely on land to eliminate the threats to wild fish stock as well as the company's fish from sea lice, parasites and other diseases being transferred. This also avoids untreated fish waste being disposed into coastal areas and removes the risk of microplastics, mercury and other contaminants present in the water. The water source comes straight from aquifers, which have not been exposed to any man-made contamination. Tassal has a similar concept of reducing the impact on the aquatic habitats and ecosystems in the form of Ocean Sanctuary Enclosures.

All five organisations disclose how they limit water consumption by using, for example, 'RAS' technology to recirculate and filter water. Detailed Life Cycle Assessments of the supply chain are conducted to understand better the environmental impacts of producing products and to highlight areas for improvement by quantifying the environmental impacts that occur over the life cycle of a product.

4.3 Metrics used to evaluate the organisation's performance in terms of long-term sustainability

Key performance indicators form an important part of managing the sector's environmental impact. Mowi, for example, incorporates the Sustainable Development Goals into the KPI structure where life below water (Goal 14) is evaluated according to the percentage of sites with minimum aquatic environmental impact, the number of biodiversity projects implemented, the number of fish escape incidents and compliance with sustainable feeding policies, including the use of antibiotics and medicine. These targets are aligned with the GRI and Sustainability Accounting Standards Board (SASB) index and provide a quantifiable measure of performance that allows the organisation to report more accurately on how sustainability is being achieved.

Based on the KPIs used by the companies under review, the following metrics can be incorporated into a performance evaluation/structure that mitigates environmental impact:

- Energy consumption and the percentage of renewable energy;
- Greenhouse gas emissions;
- Sustainability certifications including the percentage of the harvested volume certified by GSSI recognised standards;
- Percentage of reusable, recyclable or compostable plastic packaging and farming plastic equipment;
- Percentage of waste volume sent to landfills;
- Reports on the number of escaped fish;
- Fish welfare statistics including the percentage of average monthly survival, average monthly stocking density and the Operational Welfare Indicator rating;
- Antimicrobial use per tonne biomass produced;
- Percentage of fish treated with non-medicinal treatment systems;
- Percentage of freshwater withdrawal from water-stress areas;
- Percentage of sites with minimal benthic[10] impact;
- Accidental and intentional mortalities of birds and mammals;
- Food waste reduction in fish unsold products;
- Number of biodiversity-related projects;
- Sustainable feed ratios including Fish in Fish Out (FIFO), recapture FIFO, feed conversion ratio, forage fish dependency ratio, percentage soy from deforestation-free areas and the percentage of inclusion of emerging feed raw materials;
- Training and further education initiatives;
- Whistleblowing statistics related to negative biodiversity practices; and
- Rankings using international metrics such as CDP scores.

Comparative data and both positive and negative outcomes from performance reviews need to be reported to provide a balanced and transparent report. Accurate and complete reporting on KPIs is also essential for holding organisations accountable for their biodiversity impact.

[10] Benthic refers to anything associated with, or occurring on the bottom of, a body of water.

5 Conclusion and future trends

As a significant contributor to economic activity, global food security and nutrition, aquaculture and fisheries are economically and socially important. However, the industry can over-exploit the aquatic natural resources. It faces significant biodiversity threats that impact the long-term viability and sustainability of the sector and the related ecosystems.

This chapter highlights the biodiversity issues faced by the aquaculture and fisheries industries and includes relevant solutions to manage these risks and promote sustainable practices. Figure 5 summarises the process of applying an integrated thinking logic to the biodiversity assessment in aquaculture and fisheries. Ten steps are illustrated to manage the biodiversity assessment. These are not intended to be applied sequentially but to form part of an iterative review of, and response to, biodiversity impacts.

Figure 5 emphasises that an organisation needs to identify its inputs, processes, outputs and outcomes related to natural capital and biodiversity impacts. The organisation can then assess material direct and indirect environmental impacts and develop solutions to manage its impact on ecosystems. The governance approach needs to maximise the organisation's contribution to sustainable development and biodiversity management and provide the systems and processes for the organisation to be held accountable. Governance includes management control systems to monitor impacts on the ecosystem and develop plans to improve or protect biodiversity. It incorporates stakeholder engagement to support risk identification, receive feedback on the operation of other components of the governance system and identify emerging problems.

The organisation needs to monitor and adhere to relevant policies and regulations and consider how sustainability initiatives are implemented and aligned with the industry's best practices, such as the active commitment to the Sustainable Development Goals. Specific sustainability initiatives need to be implemented mindful of the feedback received from stakeholders, policy and regulatory developments, internal assessments of direct and indirect impacts and the capacity of governance structures to co-ordinate and control interventions. Post-implementation reviews, including the use of suitable key performance indicators to gauge progress on how biodiversity management is being integrated into business models will be paramount.

Steps 1 to 8 in Fig. 5 correspond to the core biodiversity themes developed from the prior literature. The final two steps deal with assurance and reporting to stakeholders. Relevant sustainability certifications, including food safety certification, are essential to enhance reliability and promote accountability. Finally, reporting on biodiversity-related matters ensures that organisations' successes and failures are reported to stakeholders. This plays a key educative

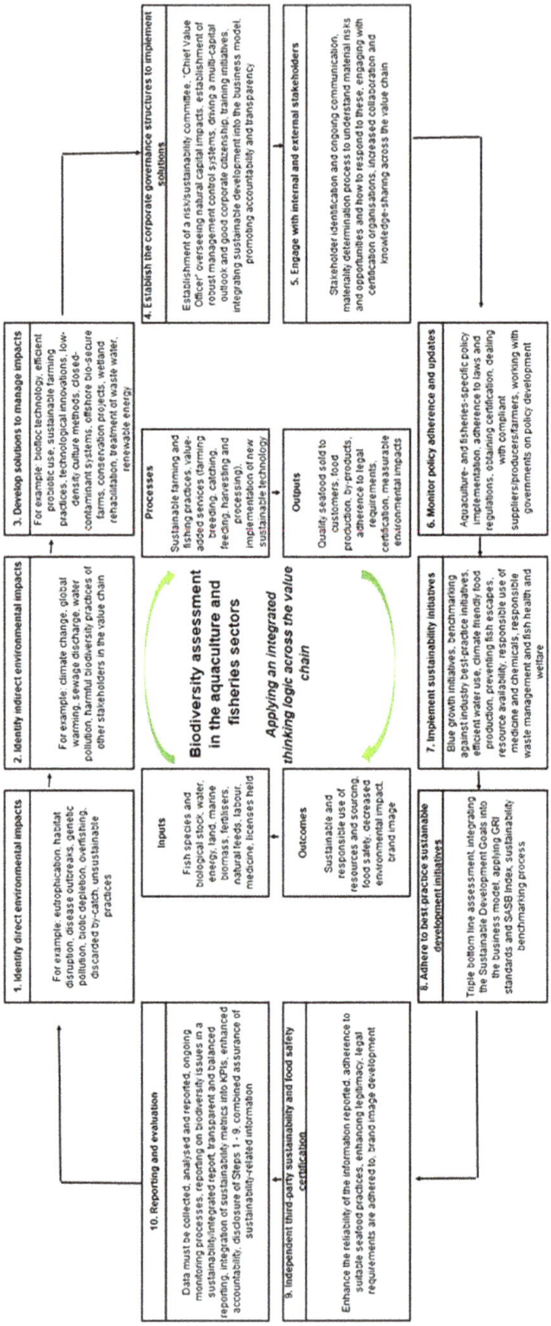

Figure 5 Biodiversity assessment in the aquaculture and fisheries sectors.

role and also promotes transparency, accountability and the efficient allocation of financial and extra-financial resources.

Despite the progress made on technological innovations and prioritising sustainability objectives, research needs to be conducted into how operations can be changed to lower environmental impacts and the type of information that needs to be considered for internal management purposes and external reporting. This can be supported by more detailed case studies illustrating how the integrated thinking logic summarised in Fig. 5 can be operationalised by aquaculture and fisheries firms.

The academic literature tends to focus on developed economies such as the United States of America and European countries (see Puszkarski and Śniadach, 2022; Campbell et al., 2021; Garza-Gil et al., 2016; Siddiki and Goel, 2015). From a developing economy perspective, research focuses on China as the leading aquaculture production jurisdiction (see Wan et al., 2020; Marín Del Valle et al., 2022; Davies et al., 2019). There is limited research on how organisations in other developing economies can navigate biodiversity challenges and use financially viable solutions to protect and conserve the environment while still benefitting economically and contributing to food security in often vulnerable regions. How developed economies can assist in promoting biodiversity solutions in developing regions needs to be investigated through increased knowledge-sharing collaborations.

Failure to address the biodiversity concerns will have economic and environmental ramifications. Aquatic foods have the potential to feed and nourish the world's growing population, but the sector's growth must be sustainable. Key industry players need to reflect on the long-term sustainability of the sector and on more biodiversity-friendly practices to ensure that there will always be fish in the sea and food on people's plates.

6 Where to look for further information

Further information on biodiversity impacts, including specific issues encountered in the fisheries and aquaculture industry, can be found using the following sources.

The following articles provide a good overview of biodiversity and sustainability reporting:

- Atkins, J. and Maroun, W. 2018. Integrated extinction accounting and accountability: building an ark. *Accounting, Auditing & Accountability Journal*, 31(3), 750-786.
- Baboukardos, D., Mangena, M. and Ishola, A. 2021. Integrated thinking and sustainability reporting assurance: International evidence. *Business Strategy and the Environment*, 30, 1580-1597.

The following academic journals provide a range of papers dealing with aquaculture, fisheries and biodiversity implications:

- Marine Policy https://www.sciencedirect.com/journal/marine-policy.
- Journal of Cleaner Production https://www.sciencedirect.com/journal/journal-of-cleaner-production.

Finally, the following practitioner guidance and websites provide an overview of sustainability and biodiversity frameworks/guidelines:

- FAO https://www.fao.org/3/cc0461en/cc0461en.pdf.
- The European Commission environmental policies department which is responsible for policies aiming to protect the environment and biodiversity, minimise risks to human health, and promote the transition to a circular economy which is available at: https://commission.europa.eu/index_en.
- Oceana Group Limited: A White Paper: The Fishing Industry's Role in Adapting to Climate Change available at: https://oceana.co.za/wp-content/uploads/2016/09/20121206i-Publication-Oceana-Climate-Change-White-Paper-update-v2.pdf.

7 References

Ahmed, N. and Glaser, M. 2016. Coastal aquaculture, mangrove deforestation and blue carbon emissions: is REDD+ a solution? *Marine Policy* 66, 58–66. DOI: 10.1016/j.marpol.2016.01.011.

Ahmed, N. and Turchini, G. M. 2021. Recirculating aquaculture systems (RAS): environmental solution and climate change adaptation. *Journal of Cleaner Production* 297. DOI: 10.1016/j.jclepro.2021.126604.

Atkins, J. and Atkins, B. 2019. *Around the World in 80 Species. Exploring the Business of Extinction*. London: Routledge.

Atkins, J., Atkins, B. C., Thomson, I. and Maroun, W. 2015. 'Good' news from nowhere: imagining utopian sustainable accounting. *Accounting, Auditing and Accountability Journal* 28(5), 651–670. DOI: 10.1108/AAAJ-09-2013-1485.

Atkins, J., Doni, F., Hassan, A. and Maroun, W. 2022. Revealing Plato's 'Shadow Kingdom': rendering pandemic risk explicit in extinction engagement. In: Atkins, J. and Macpherson, M. (Eds). *Extinction Governance, Finance and Accounting. Implementing a Species Protection Action Plan for the Financial Markets* (1st edn.) (vol. 26). Milton Park, Abingdon, Oxon: Routledge.

Cabral, P., Levrel, H., Viard, F., Frangoudes, K., Girard, S. and Scemama, P. 2016. Ecosystem services assessment and compensation costs for installing seaweed farms. *Marine Policy* 71, 157–165. DOI: 10.1016/j.marpol.2016.05.031.

Campbell, L. M., Fairbanks, L., Murray, G., Stoll, J. S., D'Anna, L. and Bingham, J. 2021. From blue economy to blue communities: Reorienting aquaculture expansion for community wellbeing. *Marine Policy* 124, 104361–104387. DOI: 10.1016/j.marpol.2020.104361.

Cartier, L. E. and Carpenter, K. E. 2014. The influence of pearl oyster farming on reef fish abundance and diversity in Ahe, French Polynesia. *Marine Pollution Bulletin* 78(1-2), 43-50. DOI: 10.1016/j.marpolbul.2013.11.027.

CDP 2020. CDP insight action [Online]. Available at: https://www.cdp.net/en [Accessed June 2021].

Cerbone, D. and Maroun, W. 2020. Materiality in an integrated reporting setting: insights using an institutional logics framework. *The British Accounting Review* 52(3), 2-18. DOI: 10.1016/j.bar.2019.100876.

Chapelle, A., Ménesguen, A., Deslous-Paoli, J.-M., Souchu, P., Mazouni, N., Vaquer, A. and Millet, B. 2000. Modelling nitrogen, primary production and oxygen in a Mediterranean lagoon. Impact of oysters farming and inputs from the watershed. *Ecological Modelling* 127(2-3), 161-181. DOI: 10.1016/S0304-3800(99)00206-9.

Craig, R. K. 2019. Fostering adaptive marine aquaculture through procedural innovation in marine spatial planning. *Marine Policy* 110. DOI: 10.1016/j.marpol.2019.103555.

Davies, I. P., Carranza, V., Froehlich, H. E., Gentry, R. R., Kareiva, P. and Halpern, B. S. 2019. Governance of marine aquaculture: pitfalls, potential, and pathways forward. *Marine Policy* 104, 29-36. DOI: 10.1016/j.marpol.2019.02.054.

de Lima Vieira, J., dos Santos Nunes, L., Rodrigues de Menezes, F. G., Vieira de Mendonça, K. and Viana de Sousa, O. 2021. An integrated approach to analyzing the effect of biofloc and probiotic technologies on sustainability and food safety in shrimp farming systems. *Journal of Cleaner Production* 318, 1-8. DOI: 10.1016/j.jclepro.2021.128618.

De Villiers, C., Rinaldi, L. and Unerman, J. 2014. Integrated Reporting: insights, gaps and an agenda for future research. *Accounting, Auditing and Accountability Journal* 27(7), 1042-1067. DOI: 10.1108/AAAJ-06-2014-1736.

Ecim, D. and Maroun, W. 2023. A review of integrated thinking research in developed and developing economies. *Journal of Accounting in Emerging Economies* 13(3), 589-612. DOI: 10.1108/JAEE-02-2022-0046.

Edgley, C., Jones, M. J. and Atkins, J. 2015. The adoption of the materiality concept in social and environmental reporting assurance: A field study approach. *The British Accounting Review* 47(1), 1-18. DOI: 10.1016/j.bar.2014.11.001.

Food, E. I. T. 2023. Can sustainable aquaculture help to achieve the un SDGs? [Online]. Available at: https://www.eitfood.eu/blog/can-sustainable-aquaculture-help-to-achieve-the-un-sdgs#:~:text=of%20aquatic%20species.-,Aquaculture%20has%20a%20key%20role%20to%20play%20in%20helping%20to,%2314%3A%20Life%20below%20water. [Accessed January 2023].

European Commission 2023. Our oceans, seas and coasts [Online]. Available at: https://ec.europa.eu/environment/marine/good-environmental-status/descriptor-5/index_en.htm [Accessed January 2023].

FAO 2022. Towards blue transformation. *The State of World Fisheries and Aquaculture*. DOI: 10.4060/cc0461en.

Ford, J. S., Pelletier, N. L., Ziegler, F., Scholz, A. J., Tyedmers, P. H., Sonesson, U., Kruse, S. A. and Silverman, H. 2012. Proposed local ecological impact categories and indicators for life cycle assessment of aquaculture. *Journal of Industrial Ecology* 16(2), 254-265. DOI: 10.1111/j.1530-9290.2011.00410.x.

Gaillet, G., Asselin, A.-C. and Wermeille, A. 2022. Sustainable fisheries: towards operationalization of decision making accounting for biodiversity. *Journal of Cleaner Production* 362, 132103-132133. DOI: 10.1016/j.jclepro.2022.132103.

Garza-Gil, M. D., Vázquez-Rodríguez, M. X. and Varela-Lafuente, M. M. 2016. Marine aquaculture and environment quality as perceived by Spanish consumers. The case of shellfish demand. *Marine Policy* 74, 1-5. DOI: 10.1016/j.marpol.2016.09.011.

Ghauri, S., Jackson, E. L., Marinova, D. and Mohammadi, H. 2022. Agricultural co-operatives for managing natural capital to achieve UN Sustainable Development Goals 12-15: A conceptual framework. *Journal of Co-Operative Organization and Management* 10(2), 100188-100198. DOI: 10.1016/j.jcom.2022.100188.

GRI 2019. Consolidated set of GRI sustainability reporting standards available. Available at: https://www.globalreporting.org/standards/gri-standards-download-center/?g=ae2e23b8- 4958-455c-a9df-ac372d6ed9a8.

Hadjimichael, M. 2018. A call for a blue degrowth: unravelling the European Union's fisheries and maritime policies. *Marine Policy* 94, 158-164. DOI: 10.1016/j.marpol.2018.05.007.

Holm, P., Buck, B. H. and Langan, R. 2017. Introduction: new approaches to sustainable offshore food production and the development of offshore platforms. In: Buck, B. H. and Langan, R. (Eds). *Aquaculture Perspective of Multi-use Sites in the Open Ocean: the Untapped Potential for Marine Resources in the Anthropocene*. New York: Springer International Publishing.

IIRC 2021. The international integrated reporting framework. Available at: https://integratedreporting.org/wp-content/uploads/2021/01/InternationalIntegratedReportingFramework.pdf.

IOD 2016. King IV Report on Corporate Governance in South Africa. Johannesburg, South Africa: Lexis Nexus South Africa.

Kaiser, M. J., Yu, Y. and Snyder, B. 2010. Economic feasibility of using offshore oil and gas structures in the Gulf of Mexico for platform-based aquaculture. *Marine Policy* 34(3), 699-707. DOI: 10.1016/j.marpol.2010.01.002.

Kibria, G. and Haroon, A. K. Y. 2017. Wetland science: perspectives from South Asia. In: Prusty, B., Chandra, R. and Azeez, P. (Eds). *Wetland Science*. New Delhi: Springer.

King, M. and Atkins, J. 2016. *The Chief Value Officer. Accountants Can Save the Planet*. Abingdon, Oxon: Greenleaf Publishing Limited.

King, M., Atkins, J. and Maroun, W. 2022. Establishing a framework for extinction accounting, accountability and finance. In: Atkins, J. and Macpherson, M. (Eds). *Extinction Governance, Finance and Accounting* (1st edn.). London: Routledge.

Marín Del Valle, T., Wu, J., Xu, C., Chen, Q., Wu, Y. and Yang, W. 2022. Spatiotemporal dynamics and resource use efficiency in mariculture production: A case study in Southeastern China. *Journal of Cleaner Production* 340, 130743-130763. DOI: 10.1016/j.jclepro.2022.130743.

Maroun, W., Usher, K. and Mansoor, H. 2018. Biodiversity reporting and organised hypocrisy: the case of the South African food and retail industry. *Qualitative Research in Accounting and Management* 15(4), 437-464. DOI: 10.1108/QRAM-07-2017-0066.

NOAA 2021. Habitat benefits from shellfish and seaweed aquaculture [Online]. *Fisheries*. Available at: https://www.fisheries.noaa.gov/feature-story/habitat-benefits-shellfish-and-seaweed-aquaculture#:~:text=An%20adult%20oyster%20can%20filter,oxygen%20levels%20in%20the%20water. [Accessed January 2023].

Puszkarski, J. and Śniadach, O. 2022. Instruments to implement sustainable aquaculture in the European Union. *Marine Policy* 144, 105215-105225. DOI: 10.1016/j.marpol.2022.105215.

Rees, S. E., Sheehan, E. V., Stewart, B. D., Clark, R., Appleby, T., Attrill, M. J., Jones, P. J. S., Johnson, D., Bradshaw, N., Pittman, S., Oates, J. and Solandt, J.-L. 2020. Emerging themes to support ambitious UK marine biodiversity conservation. *Marine Policy* 117. DOI: 10.1016/j.marpol.2020.103864.

Rinaldi, L. 2020. Chapter 19. Integrated thinking for stakeholder engagement. A processing model for judgments and choice in situations of cognitive complexity. In: De Villiers, C., Hsiao, P.-C. K. and Maroun, W. (Eds). *The Routledge Handbook of Integrated Reporting*. Abingdon, Oxon: Routledge.

Rinaldi, L., Unerman, J. and de Villiers, C. 2018. Evaluating the integrated reporting journey: insights, gaps and agendas for future research. *Accounting, Auditing and Accountability Journal* 31(5), 1294–1318. DOI: 10.1108/AAAJ-04-2018-3446.

Siddiki, S. and Goel, S. 2015. A stakeholder analysis of U.S. marine aquaculture partnerships. *Marine Policy* 57, 93–102. DOI: 10.1016/j.marpol.2015.03.006.

Smith, T., Holmes, G. and Paavola, J. 2020. Social underpinnings of ecological knowledge: business perceptions of biodiversity as social learning. *Organization and Environment* 33(2), 175–194. DOI: 10.1177/1086026618803723.

Sumesh, K. and Roshan, D. 2020. Aquaculture market by environment (marine water, fresh water, and brackish water), and fish type (carps, mollusks, crustaceans, mackerels, sea bream and others): global opportunity analysis and industry forecast. In: A. M. (Ed.). *Research, 2021-2027*. Coherent Market Insights: India. 1–292.

The Nature Conservancy 2023. A healthy ocean depends on sustainably managed fisheries [Online]. Available at: https://www.nature.org/en-us/what-we-do/our-priorities/provide-food-and-water-sustainably/food-and-water-stories/global-fisheries/#:~:text=Fish%20and%20other%20seafood%20products,percent%20of%20the%20world's%20population. [Accessed January 2023].

Van Eck, N. J. and Waltman, L. 2017. Citation-based clustering of publications using CitNetExplorer and VOSviewer. *Scientometrics* 111(2), 1053–1070.

Wan, N. F., Su, H., Cavalieri, A., Brack, B., Wang, J.-Y., Weiner, J., Fan, N.-N., Ji, X.-Y. and Jiang, J.-X. 2020. Multispecies co-culture promotes ecological intensification of vegetable production. *Journal of Cleaner Production* 257. DOI: 10.1016/j.jclepro.2020.120851.

Zhang, Y. and Kendall, A. 2019. Consequential analysis of algal biofuels: benefits to ocean resources. *Journal of Cleaner Production* 231, 35–42. DOI: 10.1016/j.jclepro.2019.05.057.

Chapter 6

Assessing the impact of pesticides on natural capital and biodiversity

Mira Lieberman, University of Sheffield, UK

1 Introduction

Wild animals are rapidly vanishing from the Earth with the negative impact of human activity on the environment highlighted in many reports (Rockström et al., 2009; World Wide Fund For Nature (WWF), 2018; IPBES, 2019). Intensive farming has been linked to biodiversity loss, partly due to increasing pesticide use. Almost half of the land in Britain is enclosed for farming, and pesticides are applied over a significant portion of this area. In addition to the use of pesticides for farming, pesticides are used in public areas such as parks, golf courses, road verges, private gardens, schools and public institutions. The relationship between agriculture, green spaces and the use of pesticides is therefore crucial if we are to understand their impact on wildlife.

Pesticides are defined as chemicals used to kill or control living organisms seen as 'pests' because of their threat to human health or the economic damage they cause to crops (Mesnage et al., 2019). Pesticides include insecticides (to control insects), fungicides (to control plant diseases) herbicides (to control weeds) and rodenticides (to control mammalian pests such as rats and mice). Pesticides are often used prophylactically to prevent biotic threats such as insects and plant pathogens from becoming established and starting to cause

http://dx.doi.org/10.19103/AS.2023.0128.11

damage. Pesticide effects on wild animals can be direct or indirect. Direct effects occur through physical contact via inhalation, touch and ingestion. Indirect effects occur via secondary exposure through consuming contaminated prey or carcasses (Harrington and Macdonald, 2002). There is now significant evidence that extensive, prophylactic use of pesticides is linked to 'the reduction of biodiversity in terrestrial and aquatic ecosystems' (Liess et al., 2019, p. 2).

It was only from the middle of the twentieth century that agricultural practices began to use the first generation of synthetic pesticides. For example, dichlorodiphenyltrichloroethane (DDT) was first developed in the 1930s (initially to control mosquitos and prevent malaria) and then used as an agricultural insecticide from the 1940s, before growing concerns about its risks to human and environmental health (including wildlife) led to its ban in the 1970s (Carson, 1962). Globally, about 3 million tons of pesticides are applied annually, with a market value of $40–65 billion (BBC Research, 2012; Gillam, 2017; Silva et al., 2019). The intensive, prophylactic, widespread and under-monitored use of pesticides is increasingly seen as having serious negative effects on the health of the environment, soils, insects, birds and mammals as well as on human health (Goulson, 2019; Pesticide Action Network UK (PAN), 2020). The impact of such intensive practices on wildlife is particularly damaging: 'Despite relative benefits of pesticide use on crop yields, industrial intensive farming is wiping out wildlife' (Goulson, 2019, p. 228).

The UK has an area of approximately 174 million hectares of agricultural land. Pesticide use in UK agriculture has risen in the past three decades by more than 63%, with the UK's Pesticide Action Network (PAN) reporting increases in the area of land and volume of crops treated (Pesticide Action Network UK (PAN), 2018, 2019b). The area on which pesticides are applied has increased by almost two-thirds between 1990 and 2016 and, while the overall volume of pesticides has decreased as pesticides have become more concentrated, their overall toxicity has increased (ibid). The frequency of application has also increased. For example, in 1990, only 21% of oilseed rape and 30% of cereals were treated with pesticides more than four times in the growing season. By 2016, this had increased to 80% for oilseed rape and 55% for cereals. The ubiquitous use of pesticides has replaced alternative methods of pest control to the point where 'many farmers know no other way' (Gillam, 2017, p. 237).

Pesticides do not just affect the fields to which they are applied. Almost all methods of application of pesticides risk drift of pesticide sprays and/or run-off of pesticide residues into neighbouring soil or watercourses. It has been estimated that up to 75% of pesticides miss the target crop when applied using aerial spraying by aircraft while 10–35% of pesticides miss their target when applied using ground-based techniques (typically using boom sprayers) (Gillam, 2017). Pesticide manufacturers like Bayer have developed drift reduction technologies to minimise this problem (Bayer AG, 2015). However, solutions such as adjuvants applied to pesticide formulation to help droplets better adhere to and penetrate leaves may themselves (by improving pesticide efficacy) pose a risk to the environment, animals and humans (Mesnage and Antoniou, 2018).

As noted earlier, the problem of pesticide use is not confined to farmland but extends to green spaces in urban areas. Local authorities across the UK spray 'huge amounts of herbicides (mainly glyphosate) to keep pavements tidy' (Goulson, 2019, p. 90) as well as using fungicides, insecticides and rodenticides in playgrounds, road verges, cemeteries, pavements and around council houses, university campuses, car parks, hospitals, private housing developments, shopping centres and schools (Pesticide Action Network UK (PAN), 2017). The exact amount is nondisclosed and unrecorded, but an estimate for a global, yearly use of glyphosate for non-agricultural use is about 80 000 tonnes.

This chapter discusses limitations in current pesticide authorisation and monitoring systems. It then reviews research on pesticide effects on mammalian and other wildlife in the context of climate change, before looking at current pesticide regulation and management in the UK and options for improvement. The chapter also includes a case study discussing the role of pesticides in the decline of the hedgehog, a key indicator species for the general health of the natural environment. It is based on a literature search undertaken using Scopus, Web of Science, and Google Scholar. Search terms were [product] [pesticide], where product included neonicotinoid, imidacloprid, glyphosate, Roundup, adjuvants, surfactants, formulation, fipronil, clothianidin, thiacloprid, metaldehyde, anticoagulant rodenticides, second-generation anticoagulant rodenticides (SGARs); [taxon] was either *vertebrate, mammal*, hedgehog* and followed up by references cited in the publications.

2 Limitations in current pesticide authorisation and monitoring systems

Concern has been expressed at how effective current processes for authorising and monitoring pesticides are. In the field of ecotoxicology, there remains an over-reliance on laboratory direct toxicity (acute) testing which limits the ability to understand how pesticides impact species in real environmental conditions. The most commonly studied mammals are the rat (*Rattus norvegicus*) and the mouse (*Mus musculus*). Individuals are fed an estimated dose of pesticides associated with the death of half of the subjects which is then recorded as the lethal dose. However, sub-lethal effects are not included in the classification of toxicity. A related problem is that studies of pesticide effects on no-target species are not always comprehensive, e.g. 'studies on exposure of non-targets to anticoagulant rodenticides have largely focussed on predatory birds and mammals and insectivores have rarely been studied' (Dowding et al., 2010).

Given the limited range of species tested, artificial testing conditions and the limited assessment of sub-lethal toxicity, there is concern that testing does not capture all the potential effects pesticides may have on wildlife. Another failing of risk assessment as Gibbons et al. (2015) highlight is 'the underestimation of interspecies variation in insecticide susceptibility [...] too few species are typically tested' which impacts the thresholds set for 'hazard doses (HD5)'. Hayes et al. (2006) have also argued that testing a single pesticide mixture and a single 'active' ingredient is insufficient to assess sub-lethal, long-term effects of pesticides on wildlife such as amphibians. These issues have, e.g. led Defarge et al. (2018, p. 156) to argue that current safety and toxicity testing measures 'are insufficient to ensure safety'.

It has also been suggested that agrochemical corporations may exert too much influence as a lobby group in the development of government policy on pesticide use and regulation (Rothstein et al., 1999; Clappison and Solomon, 2016). An example is their role in discussions by the Canadian Province of Ontario about banning neonicotinoids to protect declining numbers of pollinators. This resulted in a partial ban in 2015, studied by Ellis (2019). The author concluded that 'the partial ban on neonicotinoids in Ontario does not go far enough to address the pollinator crisis, partly due to the role of agrochemical corporations such as Bayer CropScience and Syngenta, manufacturers of neonicotinoids, in the development of the policy'. PAN (2018) also viewed the partial ban as a 'win' for the agrochemical companies in allowing continued use of neonicotinoids despite evidence of their harm to bees and other pollinators. Another example is the case of the so-called Monsanto Papers when internal documents produced by Monsanto were declassified by the US courts in 2017. These appeared to show that, from as early as 1999, Monsanto was aware of the potentially carcinogenic effect of its leading herbicide product glyphosate but chose instead to finance research which suggested otherwise (Donati, 2020). More broadly, it has been argued that a fundamental issue in improving pesticide regulation and policy is the need to challenge an established agrochemical 'narrative' which emphasises a central role for pesticides in pest management (Confino, 2012; Ellis, 2019).

Two particular issues that are not covered in the current authorisation processes are adjuvants and the 'cocktail' effect. These are discussed in more detail in the following sections.

2.1 The adjuvant problem

Applications of modern pesticides involve a cocktail of chemistries, often including an 'active' ingredient and formulants considered 'inert'. Proprietary spray adjuvants, also called co-formulants, are designed to increase the efficacy of the 'active' ingredient in a pesticide either by increasing adhesion or absorption or by increasing the half-life of 'active' ingredients in the soil (Mesnage and Antoniou, 2018). Adjuvants were estimated to have a market value of $2.51 billion in 2015, increasing by 5.8% to $3.18 billion by 2019 (Mesnage and Antoniou, 2018) and $4.7 billion by 2026 (Markets and Markets, 2018). Adjuvants are defined by the Health and Safety Executive (HSE) under Article 2 Scope (3d) of 1107/2009 (HSE, 2011) as:

> Substances or preparations which consist of co-formulants or preparations containing one or more co-formulants, in the form in which they are supplied to the user and placed on the market to be mixed by the user with a plant protection product and which enhance its effectiveness or other pesticidal properties, referred to as 'adjuvants'.

Adjuvant products are formulated combinations of surfactants, penetrant enhancers, activators, spreaders, stickers, co-solvents, wetting agents, pH modifiers, defoaming agents, drift retardants or nutrients (Witt, 2012). Usually, adjuvants are less expensive than the active ingredient, and their use can help reduce costs by improving application efficiency and efficacy of the active ingredient (Mullin et al., 2016).

Agrochemical risk assessment does not account for spray adjuvants and only tests 'active' ingredients. Currently, evaluation of toxicity effects which may be detrimental to non-target species is not undertaken due to the 'lack of disclosure of adjuvant and formulation ingredients coupled with a lack of adequate analytical methods which constrains the assessment of total chemical load on beneficial organisms and the environment' (Mullin et al., 2016, p. 1). Most adjuvants and 'inert' ingredients are also currently exempt from human safety risk assessments and are generally recognised as safe, with no environmental monitoring required (Mesnage et al., 2019). According to the UK's HSE (2011), 'there is no direct control over the content of product labels, other literature or the containers in which the product is marketed'. Only their reactions in relation to possible risk for eyes and skin are needed in current assessments.

Final pesticide products are made up of various chemical mixtures that, for authorisation purposes, are referred to as either 'inert' or 'active' ingredients. Defarge et al. (2018, p. 161) argue that the distinction made between inert and active ingredients is a regulatory one which is not based on toxicology: 'The difference between 'active ingredient' and 'inert compound' is a regulatory assertion with no demonstrated toxicological basis. This is despite the fact that adjuvants can be toxic in their own right (Mesnage and Antoniou, 2018). It has been argued that regulation should require disclosure of the full chemical composition of pesticide products in determining maximum residue levels (MRLs) to ensure the safety of food and feed (Mesnage et al., 2019).

2.2 The 'cocktail' effect

Goulson et al. (2019) have estimated that there are over 400 different pesticide products authorised for use in the UK. Pesticides are registered individually. Each 'active' ingredient needs to be authorised for the product to be put on the market. However, practical applications of pesticides in fields may result in multiple applications simultaneously of insecticides, herbicides and fungicides. Wang et al. (2020) note that synergistic interactions between fungicides and insecticides may increase risks to pollinators, especially when farmers routinely use tank mixing of fungicides with other insecticides to reduce spraying costs. In addition, to combat the problem of growing resistance to individual pesticides by target organisms, different pesticides may be combined to ensure that at least one active ingredient is effective. The combined use of different pesticides may lead to new kinds and degrees of toxicity through the so-called 'cocktail effect' (PAN, 2019).

Pesticide Action Network UK (2019) have examined the cocktail effect in the UK and its potential impact on the environment and human health. Their report, e.g. found that 'one sample of raspberries contained one "known carcinogen," one "probable carcinogen," two possible carcinogens and two endocrine disruptors, one developmental toxin which can have adverse effects on sexual function and fertility and one neurotoxin that can negatively affect the nervous system' (ibid.: 3). According to PAN UK, pesticide cocktails were found in more than half of rice and a quarter of bread samples tested by the UK Government in 2017. Grapes have been found to contain more than 19 (!) different pesticides. There is no current official monitoring of the effects of cocktails of pesticides on the environment and non-target wildlife species. There are, e.g. no ecotoxicological end points required for pesticide mixtures by the European Food Standards Agency (European Commission Directorate-General for Health and Food Safety, 2017). Silva et al. (2019, p. 1542) argue that 'urgent attention is required to address the toxicity of the mixtures of residues present in soil'.

Research on birds and mammals has focused on single pesticides, and there are no UK studies that have investigated the impact of pesticide mixtures on mammals (PAN, 2019a). Academic studies on bumblebees found that 43% had detectable levels of more than one pesticide with traces of seven pesticides in one single individual (PAN, 2019). Daniele et al. (2018) found ten pesticides in a small sample of earthworms, with as many as five different pesticides in one earthworm. Silva et al. (2019) examined soils of 11 EU countries, finding the UK to have the second highest level of pesticide mixtures as residues in soil. The authors find that 67% of the UK samples had multiple residues, with 25% having more than 6. Silva et al. (ibid) found glyphosate and its metabolite aminomethylphosphonic acid as well as DDT residues and broad-spectrum fungicides. The authors found that, while some chemical compounds were below toxic level end points for in-soil organisms, 'the presence of mixtures of pesticide residues in soils are the rule rather than the exception, indicating that environmental risk assessments should be adapted accordingly to minimize related risks to soil life and beyond' (Silva et al., 2019, p. 1532).

Pollution from mixtures of pesticides is seen as a major threat to soil health (Keesstra et al., 2016; Payá Pérez and Rodríguez Eugenio, 2018). Soil contamination raises concerns for soil biodiversity, soil function and delivery of ecosystem services as well as food safety. Soil contamination is also a source of water contamination through leaching and run-off as well as wind-based movement of soil particles (Payá Pérez and Rodríguez Eugenio, 2018). Sixty-six per cent of samples taken from UK waterways were found to contain more than ten pesticides (Pesticide Action Network UK (PAN), 2019b, p. 3). Silva et al. (2019) have found pesticide residues in soil dust which can be inhaled by humans and nonhumans.

Lentola et al. (2017) found that plants sold as 'bee-friendly' were grown in greenhouses that rely heavily on pesticides. Pollen, nectar and leaves were extracted from plants bought in nurseries and large retailers such as Aldi, B&Q and Homebase. Most of the plants contained a cocktail of pesticides, commonly fungicides and insecticides. Only 2 of the 29 plants examined were found to be pesticide-free. Seventy-six percent contained at least 1 pesticide, and 38% contained 2 or more insecticides. Seventy per cent of the plants under investigation contained neonicotinoids. Other insecticides found were pyrethroids and chlorpyrifos on 10% and 7% of plants; both were more toxic via contact than consumption. Concentrations of neonicotinoids found harmful to bees are as low as one to ten parts per billion (ppb) which can still impair bee navigation and reproduction and suppress their immune system (Goulson, 2019). Lentola et al. (2017) found the following concentrations of neonicotinoids: imidacloprid at up to a maximum of 29 ppb, clothianidin at 13 ppb, and thiamethoxam at 119 ppb. Di Prisco et al. (2017) found that neonicotinoid insecticides such as clothianidin can also have a negative effect on the immune system of vertebrates.

3 Pesticide effects on mammalian and other wildlife in the context of climate change

The threat to wildlife from pesticides has long been a subject of debate. Harrington and Macdonald (2002) have summarised the range of potential threats to species. They suggest that habitat loss is the most important factor in species extinction, closely followed by pesticides and other intensive agricultural practices (Fig. 1). As the authors indicate, it is not always possible to separate out pesticide use from other practices associated with intensive agriculture and

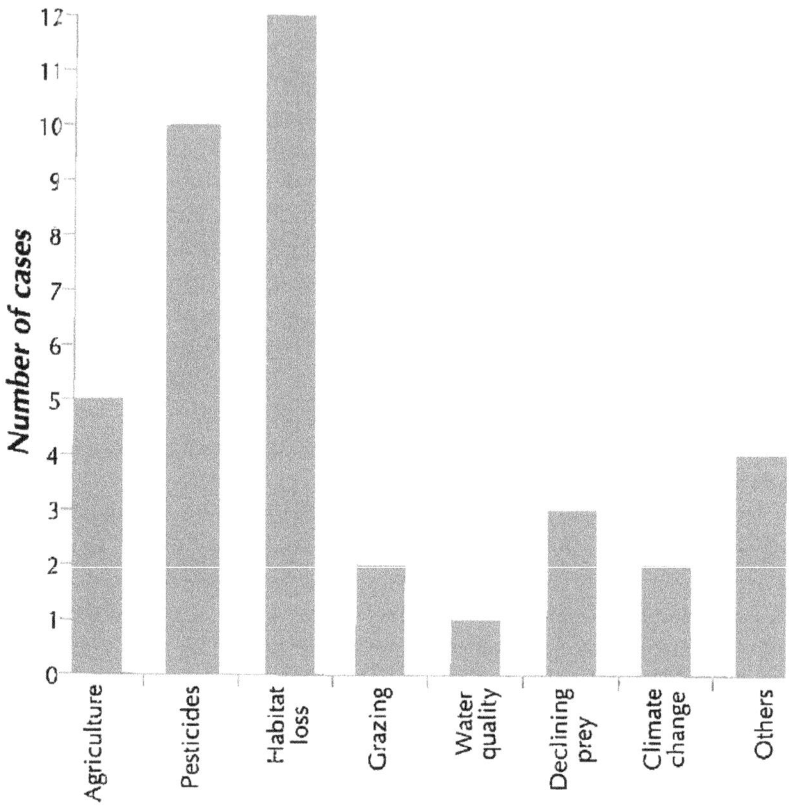

Figure 1 Possible threats to mammals that are declining. Adapted from: Harrington and Macdonald (2002, p. 2).

show a direct link with the decline of species such as hedgehogs. However, there is strong evidence supporting such a link.

A number of studies have shown that, by reducing insect pest numbers, insecticides reduce food sources for species such as hedgehogs (Peveling et al., 2003) and partridges (Berny, 2007) and, by disrupting the food chain, contribute to their decline. A particular focus of research is sub-lethal effects of insecticides such as neonicotinoids and fipronil. These include reproductive problems such as reduced sperm production and fertilisation, reduced rates of pregnancy, higher rates of embryo death, stillbirth, premature birth and reduced weight of offspring (Goulson, 2013; Gu et al., 2013). Other potential effects include neuro-behavioural disorders. Gibbons et al. (2015) reviewed 150 studies assessing direct and indirect effects of two neonicotinoids (imidacloprid and clothianidin) and a third insecticide (fipronil) and found that all three substances exerted sub-lethal effects, including impairing the immune system, reduced growth and reproductive success. They concluded: 'Although vertebrates are the intended target of only 2% of pesticides on the market, the unintentional impacts of pesticides on vertebrate populations have been marked and are well documented' (Gibbons et al., 2015, p. 103).

Imidacloprid (a neonicotinoid insecticide) and chlorpyrifos (an organophosphate insecticide) have been found to damage songbirds' ability to migrate due to exposure to treated seeds, sprays, sprayed soils and insect prey (Eng et al., 2017; Leahy, 2019). Examining the effects of these two insecticides on the white-crowned sparrow, Eng et al. (2017, p. 1) found that birds dosed with imidacloprid 'exhibited significant declines in fat stores and body mass and reduced ability to orient correctly' and that birds dosed with chlorpyrifos also failed to orient north. These findings have negative implications for wild bird population levels since unsuccessful migration means less breeding opportunities and higher mortality. Humann-Guilleminot et al. (2019, p. 2) carried out a large-scale survey of the house sparrow, concluding their research 'highlights how ubiquitous neonicotinoid insecticides have become in agricultural habitats, and reveals generalised exposure of house sparrows, and potentially other species inhabiting farmlands, to neonicotinoids'. The authors' results reveal that 'house sparrows are seldom exposed to a single neonicotinoid insecticide, but, instead, are exposed to cocktails of two or more compounds, at frequencies as high as 86% for thiacloprid in combination with any one other [neonicotinoid]' (Humann-Guilleminot et al., 2019, p. 14).

Hayes et al. (2006) examined the cocktail effect on frogs of nine pesticides used on cornfields in the USA. The authors found that pesticides acted as an immunosuppressant and endocrine disruptor. Köhler and Triebskorn (2013) also found immunosuppression in oysters, seals, amphibians and plankton, suggesting that pesticides affect all levels of the ecosystem, increasing susceptibility to parasites and disease. Even doses as low as 0.21 and 2.0 mg a day of imidacloprid have been shown to reduce immune function sperm production in rats (Gibbons et al., 2015, p. 110).

3.1 Pesticides and climate change

Since synthetic pesticide production is resource and energy intensive, pesticide manufacturing and distribution also contribute to greenhouse gas emissions responsible for climate change (Fig. 2). Increases in atmospheric CO_2 mean temperatures are expected to increase by more than 5°C in several regions of the world (Matzrafi, 2019). A warmer climate may well promote the persistence and spread of insect pests and diseases, increasing the need for pesticide applications (Delcour et al., 2015). Higher CO_2 levels may, e.g. increase the tolerance of invasive grass species to glyphosate (PAN, 2018). By affecting precipitation patterns and thus pesticide dispersal, Köhler and Triebskorn (2013) argue that 'global warming is decisively expected to affect the ecotoxicological

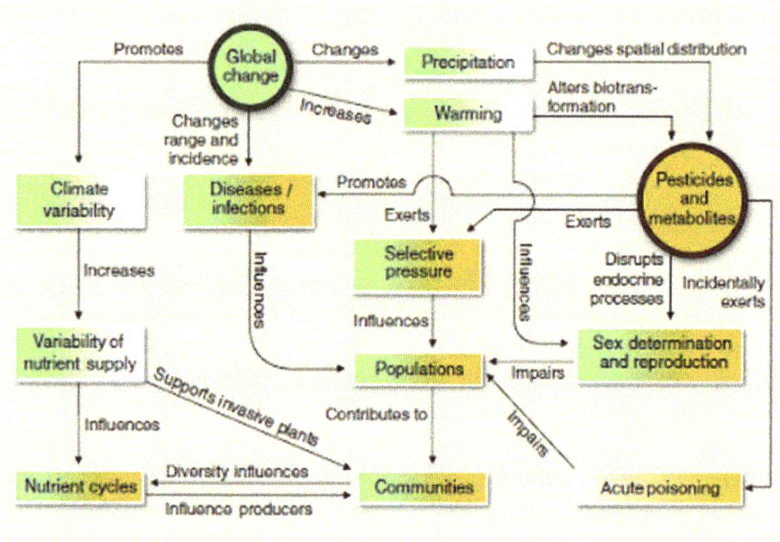

Figure 2 Anticipated interactions between climate change and pesticide effect on the physiology and ecology of wildlife. Adapted from: Köhler and Triebskorn (2013, p. 763).

potency of pesticides' (Fig. 2). As well as affecting pesticide use and efficacy, the negative effects of pesticides may also be exacerbated by climate change (Delcour et al., 2015). In reviewing over 150 studies, Holmstrup et al. (2010) found synergistic relationships between chemical pollutants and environmental stressors such as increased heat in affecting, e.g. human immune function that may continue in this trend in the long term 2030–2050.

4 Current pesticide regulation and management: options for improvement

This section discusses policy challenges relating to pesticide use and will attempt to provide alternative suggestions to current legislation and authorisation processes. The UK will be used as an example. In the EU, pesticides fall under Directive 91/414 (Authorisation Directive) which provides a community-wide framework for the regulation of pesticides that member states can follow. However, until 1986, the United Kingdom had no statutory regulations for the approval and registration of pesticides. Instead, a voluntary scheme known as the Pesticides Safety Precautions Scheme operated between government, manufacturers and suppliers (Rothstein et al., 1999, p. 244).

Pesticide production then came under the authority of the Food and Environmental Protection Act 1985 and the Control of Pesticides Regulation 1986 (COPR) in the UK. Part III of The Food and Environment Protection Act

1985 aims to ensure safe use; safeguard the environment; secure safe, effective and humane methods of controlling pests and make pesticide information available to the public. This is implemented by the COPR, as amended by the Control of Pesticides (Amendment) Regulation 1997, and by the various MRL Regulations. The COPR (as amended) states that all pesticides must gain approval before their sale, supply, storage and use. The Pesticides (Maximum Levels in Crops, Food and Feeding stuffs) (England and Wales) Regulations 1999 and The Pesticides (Maximum Levels in Crops, Food and Feeding stuffs) (Scotland) Regulations 2000 all define the MRLs of pesticides allowed in food.

The Control of Substances Hazardous to Health Regulations 1994 requires that exposure to substances hazardous to health is either prevented or, where this is not reasonably practical, adequately controlled. Anyone wishing to use a pesticide should determine:

1 whether its use is necessary;
2 which product poses least risk but is still effective; and
3 what measures are needed to minimise exposure to the pesticide.

Codes of good practice have been issued by Department for the Environment, Food and Rural Affairs (DEFRA) (some of which have statutory status) to assist farmers to meet their legal obligations. However, failure to comply is not an offence.

4.1 Toxicity measures and indicator organisms

Current pesticide risk assessment relies on the comparison of toxicity exposure rations. Trigger values are set according to ecotoxicological concentrations tested on indicator organisms such as earthworms, mites and nitrogen transformation microorganisms (Silva et al., 2019, p. 1533). This is usually achieved by a mortality threshold designated at LC50 - a concentration sufficiently toxic to kill 50% of the indicator organisms exposed to the substance. Doses are given at 24 and 48h. Chronic exposure is measured over 90 days.

PAN (2019b) and others argue that the UK regulatory system is ill-equipped to protect the natural environment from pesticide cocktails and does not fully monitor the longer term effects of residues to which humans and wildlife are exposed to. Pesticide mixtures are also not monitored in soil nor are their effects on soil health (Silva et al., 2019; Payá Pérez and Rodríguez Eugenio, 2018). Straub et al. (2020, p. 1) argue that current ecotoxicological risk assessments (ERAs) are misleading because they primarily measure death/survival of target organisms rather than long-term harm through sub-lethal effects that are 'far more common and likely having severe consequences' on wildlife. The sensitivity of indicator organisms may also differ from the sensitivity of other species from the same taxonomic group, potentially underestimating pesticide toxicity (Silva et al., 2019). Liess et al. (2019, p. 2) have argued that ERA fails to 'prevent adverse effect of pesticides on non-target organisms'. Another aspect of pesticide toxicity not taken into account by registration regimes relates to community shifts. Community shifts relate to the way in which pesticides may affect a community of organisms, including both beneficials and pathogens. These communities can be easily disturbed by pesticide residues, promoting harmful organisms that may reduce soil health and promote pathogens affecting crops and/or wildlife (Silva et al., 2019).

4.2 Doses and testing

PAN (2019b) have proposed new agricultural legislation in the UK that not only significantly reduces overall use of pesticide in agriculture and urban settings but also assesses combined toxic loads rather than testing one element and provides adequate soil protection policies. Mesnage et al. (2019, p. 143) argue for labelling all the ingredients in a pesticide formulation: 'the toxicity of co-formulants should be fully acknowledged and misleading terms such as "inerts" should be avoided'. In addition, the authors recommend that farm animals and human populations should be monitored for their daily intake of adjuvants in food in order to monitor potential health issues from exposure to these chemicals.

Some solutions to the policy issues presented previously include switching to a different metric to measure pesticide use, as weight no longer indicates toxicity. PAN UK is calling for the UK to adopt two new metrics:

- Treatment frequency index; and
- Number of doses.

Using these new metrics would help indicate to what extent UK citizens and the natural environment are exposed to the harmful impacts of pesticides. It would also assist in developing a reliable indicator for the UK's post-Brexit agricultural subsidy payment scheme which is focussed on delivering ecosystem services and promoting biodiversity. Adopting meaningful metrics for measuring pesticide use will be vital to underpin such a system (Pesticide Action Network UK (PAN), 2018).

4.3 Improving surveillance

Very few countries have a wildlife surveillance programme tracking wildlife exposure to pesticides. Among direct exposure cases are accidental poisoning resulting from intended (licensed) use and illegal poisoning (Berny, 2007).

Reviewing wildlife poisoning investigation systems and their monitoring of wildlife exposure to pesticides in Britain and France, Berny (2007) found that poisoning can occur even if the product has been used according to the product label. Avian mortality, e.g. has been documented by Mineau (2002, in Berny, 2007) after 'regular' use of pesticides in agricultural fields.

There are no post-registration schemes to monitor exposure of wild mammals to pesticides (Vijver et al., 2017). One partial exception in the UK is The Wildlife Incident Investigation Scheme (WIIS) managed by the Scottish Agricultural Science Agency in Scotland on behalf of the Environment and Rural Affairs Department of the Scottish Executive. This conducts post-marketing surveillance of pesticide poisoning in wildlife though it is not part of the official pesticide registration process (Berny, 2007; Taylor and Giela, 2006).

In the UK as a whole, pesticide poisoning surveillance falls under the competence of the DEFRA. DEFRA delegates this duty to the HSE who delegates it in turn to its Chemical Regulation Division (CRD) which is responsible for approval and safe use of pesticide products. The CRD liaises with the organisation Natural England which advises DEFRA on ways of protecting and managing the natural environment. Natural England makes use of schemes such as WIIS to highlight potential damage to the natural environment from pesticide use. Chemical analyses of pesticide residues are conducted on behalf of the Wildlife Incident Unit (WIU) by Fera Science (formerly a government body: the Food and Environmental Research Agency). In 2015, 75% of Fera Science was privatised and sold to Capita plc, with DEFRA retaining 25%. This has raised concerns that commercial considerations may skew Fera's testing priorities.

WIU assessments determine which residues are present in the animal's tissues and whether the cause of death can be established as a result of exposure to these chemicals. This causal link is only made, however, when there is information on baits and local use of pesticides, usually organophosphates since they are fast-acting compounds and cause death quickly. However, it is much more difficult to establish the cause of death and link it to pesticide poisoning in the case of chronic, sub-lethal exposure where death may not immediately result from a single contamination incident (Berny, 2007; Köhler and Triebskorn, 2013). There is also a clear bias in the WIIS data towards more visible and common species or those of public interest (Berny, 2007). These surveillance schemes struggle to monitor chronic effects of cumulative exposure to pesticides which may constitute the principal threat to wildlife (Berny, 2007; Mason et al., 2013).

According to the WIIS, the most commonly reported pesticides involved in acute poisoning of wildlife are insecticides such as organophosphates and carbamates, the rodenticides strychnine and SGARs, as well as molluscicides such as metaldehyde and methiocarb. The WIIS 2018 (WIIS, 2019) excel spreadsheet analysis indicated the cause of death in 204 incidents. One hundred and thirty-eight incidents (30.07% of all 459 incidents accepted into the WIIS) tested positive for pesticides and other chemicals. Fifty-four of these incidents were attributed to pesticide poisoning. Starvation was noted in 36 cases as the cause of death which may have resulted indirectly from poisoning (Holmstrup et al., 2010).

Current WIIS analysis has several shortcomings. Firstly, a full report analysing WIIS data and drawing conclusions has not been available since 2006. Only the excel spreadsheet with the basic data and quarterly summaries are available, limiting the ability to influence policy. While animals are listed in the spreadsheet, the particular species is not named nor is the conservation status. Cross-checking entries of poisoning in WIIS with conservation databases by the author showed that 11 species are classified as amber, 12 as red and 16 as green. This emphasises the need for the WIIS to produce more detailed reports, e.g. to assess how vulnerable species are affected by pesticides and biocides.

Vijver et al. (2017) argue that the registration and re-registration processes should be an iterative process in which assessment relies on data collected from post-registration monitoring of wildlife exposed to pesticides 'and/or cross-ecosystem calculations and discuss how such monitoring can be incorporated as an essential safety lock in environmental risk assessment and management to curtail the ongoing impairment of ecosystems, including their biodiversity, their functioning, and the services they provide' (ibid.: 860).

5 Case study: pesticides and hedgehog decline

Hedgehogs are a valuable bioindicator species, through which the state and health of the natural environment can be seen: 'Their presence is indicative of sustainable populations of important invertebrates, especially earthworms and the many insects whose larvae and adults perform vital ecological functions. If hedgehogs are present and correct, then all is well. If they are declining or absent, something has gone wrong' (Morris, 2018, p. 378). Current research suggests declines in hedgehog numbers and the need for conservation measures such as improving cover from predation and connectivity between rural villages, e.g. by increasing hedgerow length and quality (Pettett et al., 2017; Hof and Bright, 2012).

The effect of pesticides on hedgehog decline is largely unexplored in hedgehog ecological research. In an earlier study, Dickmann (1988) found 78% of his sample of 109 hedgehog corpses were roadkills, while 2.8% died from eating slug pellets. Reeve and Huijser (1999) suggested that although most hedgehog deaths resulted from parasitic infestation, other causes could have a significant effect on hedgehog populations. Reeve and Huijser (1999, p. 10) emphasise that although the data from wildlife rescues are biased by the nature of the source, they are 'the only practical source of substantial data'. The authors note that while poisoning from pesticides and pollution is listed as a minor cause of mortality in their study, many deaths go unrecorded which means deaths linked to pesticides may be under-represented. They conclude that hedgehogs are 'under considerable pressure from anthropogenic mortality factors' (Reeve and Huijser, 1999, p. 18).

Hedgehogs are omnivore, generalist feeders which makes them potentially vulnerable to a wide range of poisons such as pesticides designed to eliminate invertebrate pests (Morris, 2018). Hedgehogs may be particularly vulnerable to molluscicides such as metaldehyde and methiocarb designed to control slugs. As one author concludes, 'the use of agricultural pesticides cannot have been benign, if only because their purpose was to reduce or eliminate the very prey that hedgehogs rely on for most of their food' (Morris, 2018, p. 254). He further argues that 'Pesticides that appear safe and irrelevant for hedgehogs are potentially implicated in declining hedgehog populations even if there are no direct effects on the individual animals as a result of eating contaminated prey' (Morris, 2018, p. 256). A possible factor identified by Mesnage et al. (2018) is that neonicotinoids accumulate particularly in fats which, given reliance on fat reserves in the winter, may be a factor responsible for the high mortality of hedgehogs during hibernation (Rautio et al., 2016). The closest study to date to demonstrate a population-level impact of systemic insecticides has been on the lesser hedgehog tenrec, native to Madagascar (Peveling et al., 2003). The study monitored the effects of the insecticides fipronil and triflumuron and found that fipronil caused sufficiently high mortality of locusts (a target pest) to have a substantial impact on lesser hedgehog tenrec populations where fipronil was sprayed. While lack of prey was the principal factor in the decline of hedgehogs, direct toxicity may also have been a factor.

Rodenticides are widely used on farms in rural Britain as well as urban and suburban landscapes. However, 'studies on exposure of non-targets to anticoagulant rodenticides have largely focussed on predatory birds and mammals and insectivores have rarely been studied' (Dowding et al., 2010; see also Shore et al., 1999). High proportions of dead hedgehogs are found to contain rodenticide residues (Morris, 2018). Rodenticide residues have also been found in up to 91% of barn owls (Dowding et al., 2010).

Dowding et al. (2010) investigated the exposure of 120 European hedgehogs in the UK to first-generation anticoagulant rodenticides (FGARs) and SGARs. The authors investigated some FGARs (warfarin and coumatetralyl) and all of the SGARs (difenacoum, bromadiolone, brodifacoum, and flocoumafen) that are licensed for use in Britain. They demonstrated that hedgehogs have been negatively affected by rodenticides as a result of eating carrion. The authors conclude that 'predation of contaminated invertebrates

[...] is likely to be a major pathway by which hedgehogs are exposed to anticoagulant rodenticides' (Dowding et al., 2010, p. 165). They emphasise that 'given that many ecological communities typically contain larger numbers of insectivorous vertebrates relative to predators, the contamination of invertebrates potentially poses the greater risk of non-target poisoning in terms of species and individuals' (Dowding et al., 2010, p. 162). The authors conclude that 'contamination of hedgehogs with anticoagulant rodenticides is commonplace' and is similar to the risk at which non-target predatory birds and mammals were found, such as the tawny owl, barn owl and polecat (Dowding et al., 2010, p. 165). This is of particular importance considering the residues found in most animals in the Natural England monitoring of pesticide poisoning of wild and domestic animals (Taylor and Giela, 2006). It has been stated that 'All first and second-generation anti-coagulant rodenticides fail the environmental risk assessment for use outdoors' (RSGOG, 2020), prompting the EU to change legislation to restrict the application of SGARs to farmers and professional pest controllers. Morris (2018, p. 264) concludes that 'rodenticides seem to be a likely contributor to recent declines in hedgehog numbers'.

6 Conclusion

The chapter has discussed the negative impact of pesticides on biodiversity and the natural environment in the face of increasing use and toxicity of pesticides. Prophylactic use of pesticides has been linked to soil degradation, disrupted food webs due to less food availability as well as reproductive and endocrine problems. It has been argued that farmers are 'stuck' on a pesticide treadmill (Mason et al., 2013; Gillam, 2017) and that 'the chemical companies are "like the drug cartel warlords that get their people addicted to their drugs"' (Gillam, 2017, p. 236). Resistance of plants and target species incurs larger and larger doses of applications.

The chapter has highlighted weaknesses in current pesticide authorisation and monitoring processes, including two areas that are not currently taken into account during the authorisation process: adjuvants and pesticide 'cocktails'. It also highlights the problems in distinguishing between 'active' and 'inert', which then determines the set of tests used to assess the toxicity of pesticide ingredients (Mesnage et al., 2019). However, research suggests that sub-lethal effects of pesticide formulations weaken the immune system of many species, as has been demonstrated in bees, fish, amphibians, birds and mammals (Mason et al., 2013; Hayes et al., 2006). Regulators need to balance the negative impact of pesticides with the potential economic benefit and health risk to society. Mitigation requires a shift from relying on pesticides to an ecological paradigm focused on non-chemical methods of control.

7 Where to look for further information

Good introductions to the subject for non-specialists can be found in the book by Gillam, C. 2017. *Whitewash: The Story of a Weed Killer, Cancer, and the Corruption of Science*. Washington: Island Press.

Pesticide Action Network are a world-wide NGO dedicated to tackling problems resulting from pesticides.

The following article provides a good overview of the subject: Mesnage, R. and Antoniou, M. N. 2018. Ignoring Adjuvant Toxicity Falsifies the Safety Profile of Commercial Pesticides. *Frontiers in Public Health*, 5(January), pp.1-8.

8 References

Bayer, AG. 2015. *New Way of Spraying Pesticides Further Minimizes Exposure of Pollinators Lower Application for*. [online] BEENOW. Available at: https://beecare.bayer.com/bilder/upload/dynamicContentFull/BEENOW/BEENOW_GB_dropleg_higher-protectioniwrsi02d.pdf (accessed 22 July 2019).

BBC Research. 2012. *Global Market for Pesticides to Reach $65.3 Billion in 2017*. [online] BBC Research. Available at: http://www.bccresearch.com/pressroom/chm/global-market-pesticides-reach-$65.3-billion-2017 (accessed 25 July 2019).

Berny, P. 2007. Pesticides and the intoxication of wild animals. *Journal of Veterinary Pharmacology and Therapeutics* 30(2), 93–100.

Carson, R. 1962. *Silent Spring*. London, New York: Penguin Books.

Clappison, M. and Solomon, A. 2016. Bees and pesticides: The Ontario Controversy. In: Atkins, J. and Atkins, B. (Eds), *The Business of Bees: an Integrated Approach to Bee Decline and Corporate Responsibility*. Sheffield: Greenleaf Publishing Limited, pp. 151-169.

Confino, J. 2012. *Campaign Challenges Destructive Power of Agro-Chemical Industry*. [online]. The Guardian. Available at: https://www.theguardian.com/sustainable-business/blog/anna-lappe-activist-food-campaign (accessed 26 April 2020).

Daniele, G., Lafay, F., Pelosi, C., Fritsch, C. and Vulliet, E. 2018. Development of a method for the simultaneous determination of multi-class pesticides in earthworms by liquid chromatography coupled to tandem electrospray mass spectrometry. *Analytical and Bioanalytical Chemistry* 410(20), 5009-5018.

Defarge, N., Spiroux de Vendômois, J. and Séralini, G. E. 2018. Toxicity of formulants and heavy metals in glyphosate-based herbicides and other pesticides. *Toxicology Reports* 5, 156–163. https://doi.org/10.1016/j.toxrep.2017.12.025.

DEFRA. 2018. *Further Restrictions on Neonicotinoids Agreed*. [online] DEFRA. Available at: https://www.gov.uk/government/news/further-restrictions-on-neonicotinoids-agreed (accessed 28 April 2020).

Delcour, I., Spanoghe, P. and Uyttendaele, M. 2015. Literature review: Impact of climate change on pesticide use. *Food Research International* 68, 7–15. http://dx.doi.org/10.1016/j.foodres.2014.09.030.

Di Prisco, G., Iannaccone, M., Ianniello, F., Ferrara, R., Caprio, E., Pennacchio, F. and Capparelli, R. (2017). The neonicotinoid insecticide Clothianidin adversely affects immune signaling in a human cell line. *Scientific Reports* 7(1), 13446. http://dx.doi.org/10.1038/s41598-017-13171-z.

Dickmann, C. R. 1988. Age-related dietary change in the European hedgehog, Erinaceus europaeus. *Journal of Zoology* 215(1), 1–14.

Donati, A. 2020. The glyphosate saga, A further but not a final step: the CJEU confirms the validity of the regulation on plant protection products in light of the precautionary principle. *European Journal of Risk Regulation* 11(1), 148–154.

Dowding, C. V., Shore, R. F., Worgan, A., Baker, P. J. and Harris, S. 2010. Accumulation of anticoagulant rodenticides in a non-target insectivore, the European hedgehog (Erinaceus europaeus). *Environmental Pollution* 158(1), 161–166. http://dx.doi.org/10.1016/j.envpol.2009.07.017.

Ellis, R. 2019. Save the bees? Agrochemical corporations and the debate over neonicotinoids in Ontario. *Capitalism Nature Socialism* 30(4), 104–122.

Eng, M. L., Stutchbury, B. J. M. and Morrissey, C. A. 2017. Imidacloprid and chlorpyrifos insecticides impair migratory ability in a seed-eating songbird. *Scientific Reports* 7(1), 15176. http://dx.doi.org/10.1038/s41598-017-15446-x.

Gibbons, D., Morrissey, C. and Mineau, P. 2015. A review of the direct and indirect effects of neonicotinoids and fipronil on vertebrate wildlife. *Environmental Science and Pollution Research International* 22(1), 103–118.

Gillam, C. 2017. *Whitewash: the Story of a Weed Killer, Cancer, and the Corruption of Science* Washington, DC: Island Press.

Goulson, D. 2013. REVIEW: An overview of the environmental risks posed by neonicotinoid insecticides. *Journal of Applied Ecology* 50(4), 977-987.

Goulson, D. 2019. *The Garden Jungle or Gardening to Save the Planet*. London: Penguin Random House UK.

Gu, Y. H., Li, Y., Huang, X. F., Zheng, J. F., Yang, J., Diao, H., Yuan, Y., Xu, Y., Liu, M., Shi, H. J. and Xu, W. P. 2013. Reproductive effects of two neonicotinoid insecticides on mouse sperm function and early embryonic development in vitro. *PLoS ONE*, 8(7), e70112.

Harrington, L. A. and Macdonald, D. W. 2002. *A Review of the Effects of Pesticides on Wild Terrestrial Mammals in Britain*. WildCRU Oxford.

Hayes, T. B., Case, P., Chui, S., Chung, D., Haeffele, C., Haston, K., Lee, M., Mai, V. P., Marjuoa, Y., Parker, J. and Tsui, M. 2006. Pesticide mixtures, endocrine disruption, and amphibian declines: are we underestimating the impact? *Environmental Health Perspectives* 114 (Suppl. 1), 40–50.

Health and Safety Executive (HSE). 2011. *Legislative Controls over Adjuvants Controls on Storage of Adjuvants Controls on Use of Adjuvants*. Available at: http://www.hse.gov.uk/pesticides/topics/pesticide-approvals/pesticides-registration/applicant-guide/the-applicant-guide-adjuvan.htm (accessed 5 July 2019).

Hof, A. R. and Bright, P. W. (2012). Factors affecting hedgehog presence on farmland as assessed by a questionnaire survey. *Acta Theriologica* 57(1), 79–88.

Holmstrup, M., Bindesbøl, A. M., Oostingh, G. J., Duschl, A., Scheil, V., Köhler, H. R., Loureiro, S., Soares, A. M. V. M., Ferreira, A. L. G., Kienle, C., Gerhardt, A., Laskowski, R., Kramarz, P. E., Bayley, M., Svendsen, C. and Spurgeon, D. J. 2010. Interactions between effects of environmental chemicals and natural stressors: A review. *Science of the Total Environment* 408(18), 3746–3762. http://dx.doi.org/10.1016/j.scitotenv.2009.10.067.

Humann-Guilleminot, S., Clément, S., Desprat, J., Binkowski, Ł. J., Glauser, G. and Helfenstein, F. 2019. A large-scale survey of house sparrows feathers reveals ubiquitous presence of neonicotinoids in farmlands. *Science of the Total Environment* 660, 1091–1097.

Intergovernmental Science-Policy Platform on Biodiversity and Ecosystem Services – IPBES. 2019. *Intergovernmental Science-Policy Platform on Biodiversity and Ecosystem Services*. [online] IPBES. Available at: http://bit.ly/IPBESReport.

Keesstra, S. D., Bouma, J., Wallinga, J., Tittonell, P., Smith, P., Cerdà, A., Montanarella, L., Quinton, J. N., Pachepsky, Y., Van Der Putten, W. H., Bardgett, R. D., Moolenaar, S., Mol, G., Jansen, B. and Fresco, L. O. 2016. The significance of soils and soil science towards realization of the United Nations sustainable development goals. *SOIL* 2(2), 111–128.

Köhler, H.-R. and Triebskorn, R. 2013. Wildlife ecotoxicology of pesticides: Can we track effects to the population level and beyond? *Science* 341(6147), 759–765.

Leahy, S. 2019. *Huge Decline in Songbirds Linked to Neonicotinoids*. [online]. National Geographic. Available at: https://www.nationalgeographic.com/environment/2019/09/widely-used-pesticide-makes-birds-lose-weight/ (accessed 28 April 2020).

Lentola, A., David, A., Abdul-Sada, A., Tapparo, A., Goulson, D. and Hill, E. M. 2017. Ornamental plants on sale to the public are a significant source of pesticide residues with implications for the health of pollinating insects. *Environmental Pollution* 228, 297–304.

Liess, M., Ratte, T., Ebke, P. and Hollert, H. 2019. 20 years SETAC GLB: Increasing realism of pesticide risk assessment. *Environmental Sciences Europe* 31(1), 13–15. https://doi.org/10.1186/s12302-019-0197-x.

Markets and markets 2018. *Agricultural Adjuvants Market by Function (Activator and Utility), Application (Herbicides, Insecticides, and Fungicides), Formulation (Suspension Concentrates and Emulsifiable Concentrates), Adoption Stage, Crop Type, and Region - Global Forecast*. [online] Markets and Markets. Available at: https://www.marketsandmarkets.com/search.asp?search=adjuvant (accessed 22 April 2020).

Mason, R., Tennekes, H., Sánchez-Bayo, F. and Jespen, U. P. 2013. Immune suppression by neonicotinoid insecticides at the root of global wildlife declines. *Journal of Environmental Immunology and Toxicology* 1(1), 3–12.

Matzrafi, M. 2019. Climate change exacerbates pest damage through reduced pesticide efficacy. *Pest Management Science* 75(1), 9–13.

Mesnage, R. and Antoniou, M. N. 2018. Ignoring adjuvant toxicity falsifies the safety profile of commercial pesticides. *Frontiers in Public Health* 5(January), 1–8.

Mesnage, R., Benbrook, C. and Antoniou, M. N. 2019. Insight into the confusion over surfactant co-formulants in glyphosate-based herbicides. *Food and Chemical Toxicology* 128(March), 137–145. https://doi.org/10.1016/j.fct.2019.03.053.

Morris, P. 2018. *Hedgehog*. London: HarperCollins.

Mullin, C. A., Fine, J. D., Reynolds, R. D. and Frazier, M. T. 2016. Toxicological risks of agrochemical spray adjuvants: Organosilicone surfactants may not be safe. *Frontiers in Public Health* 4(May), 92.

PAN. 2018. *Glyphosate: A Victory for Corporate Lobbying, Not Science*. [online] PAN. Available at: https://www.pan-uk.org/glyphosate-victory-corporate-lobbying-not-science/ (accessed 26 April 2020).

Payá Pérez, A. and Rodríguez Eugenio, N. 2018. *Status of Local Soil Contamination in Europe*. [online]. Available at: https://ec.europa.eu/jrc/en/publication/status-local-soil-contamination-europe-revision-indicator-progress-management-contaminated-sites.

Pesticide Action Network UK. 2017. *Pesticide-Free Towns Campaign*. [online] PAN Pesticide Action Network. UK: PAN. Available at: https://www.pan-uk.org/pesticide-free/ (accessed 7 May 2020).

Pesticide Action Network UK. 2018. *The Hidden Rise of UK Pesticide Use: Fact-Checking an Industry Claim*. [online] PAN. Available at: http://www.pan-uk.org/pesticides-agriculture-uk/.

Pesticide Action Network UK. 2019a. *Impact of Pesticides*. [online] PAN Pesticide Action Network. UK: PAN. Available at: http://www.pan-uk.org/key-issues/ (accessed 1 August 2019).

Pesticide Action Network UK. 2019b. *The Cocktail Effect: How Pesticide Mixtures May Be Harming Human Health and the Environment*. [online] PAN. Available at: https://www.pan-uk.org/the-cocktail-effect/.

Pesticide Action Network UK. 2020. *Pesticide News: An International Perspective on the Health & Environmental Effects of Pesticides*. [online] PAN. Available at: https://issuu.com/pan-uk/docs/pesticide_20news_20123_20-_20october_202020?fr=sYTBhOTExOTMxNQ.

Pettett, C. E., Moorhouse, T. P., Johnson, P. J. and Macdonald, D. W. 2017. Factors affecting hedgehog (Erinaceus europaeus) attraction to rural villages in arable landscapes. *European Journal of Wildlife Research* 63(3), 54–65.

Peveling, R., McWilliam, A. N., Nagel, P., Rasolomanana, H., Raholijaona, R., Rakotomianina, L., Ravoninjatovo, A., Dewhurst, C. F., Gibson, G., Rafanomezana, S. and Tingle, C. C. D. 2003. Impact of locust control on harvester termites and endemic vertebrate predators in Madagascar. *Journal of Applied Ecology* 40(4), 729–741.

Rautio, A., Isomursu, M., Valtonen, A., Hirvelä-Koski, V. and Kunnasranta, M. 2016. Mortality, diseases and diet of European hedgehogs (Erinaceus europaeus) in an urban environment in Finland. *Mammal Research Institute* 61(2), 161–169.

Reeve, N. and Huijser, M. 1999. Mortality factors affecting wild hedgehogs: A study of records from wildlife rescue centres. *Lutra* 42, 7–24.

Rockström, J., Steffen, W., Noone, K., Persson, Å., Chapin, F. S. I., Lambin, E., Lenton, T. M., Scheffer, M., Folke, C., Schellnhuber, H. J., Nykvist, B., de Wit, C. A., Hughes, T., van der Leeuw, S., Rodhe, H., Sörlin, S., Snyder, P. K., Costanza, R., Svedin, U., Falkenmark, M., Karlberg, L., Corell, R. W., Fabry, V. J., Hansen, J., Walker, B., Liverman, D., Richardson, K., Crutzen, P. and Foley, J. 2009. Planetary boundaries: Exploring the safe operating space for humanity. *Ecology and Society* 14(2), 472–475.

Rothstein, H., Irwin, A., Yearley, S. and McCarthy, E. 1999. Regulatory science, Europeanization, and the control of agrochemicals. *Science, Technology, and Human Values* 24(2), 241–264.

RSGOG. 2020. *Report on the Rodenticides Stewardship Regime Assessment of Implementation*. [online]. Available at: https://www.hse.gov.uk/biocides/eu-bpr/Rodenticides-Stewardship-Regime-GOG-rev-Mar2020.pdf.

Shore, R., Birks, J. D. S. and Freestone, P. (1999). Exposure of non-target vertebrates to second-generation rodenticide in Britain, with particular reference to the polecat Mustela putorius. *New Zealand Journal of Ecology* 23, 199–206.

Silva, V., Mol, H. G. J., Zomer, P., Tienstra, M., Ritsema, C. J. and Geissen, V. 2019. Pesticide residues in European agricultural soils - A hidden reality unfolded. *Science of the Total Environment* 653, 1532–1545. https://doi.org/10.1016/j.scitotenv.2018.10.441.

Taylor, M. J. and Giela, A. 2006. *Pesticide Poisoning of Animals in 2006*. [online]. Edinburgh: Scottish Agricultural Science Agency. Available at: https://www.sasa.gov.uk/sites/default/files/Animal Poisoning Report 2006.pdf.

Vijver, M. G., Hunting, E. R., Nederstigt, T. A. P., Tamis, W. L. M., van den Brink, P. J. and van Bodegom, P. M. 2017. Postregistration monitoring of pesticides is urgently required to protect ecosystems. *Environmental Toxicology and Chemistry* 36(4), 860–865.

Wang, Y., Zhu, Y. C. and Li, W. 2020. Comparative examination on synergistic toxicities of chlorpyrifos, acephate, or tetraconazole mixed with pyrethroid insecticides to honey bees (Apis mellifera L.). *Environmental Science and Pollution Research International* 27(7), 6971–6980.

WIIS. 2019. *Summary of Quarterly Reports Summary of 2019*. [online]. Available at: https://www.hse.gov.uk/pesticides/reducing-environmental-impact/wildlife/summary-of-quarterly-reports.pdf.

Witt, J. M. (2012). Agricultural spray adjuvants. Oregon State University. Available at: http://psep.cce.cornell.edu/facts-slides-self/facts/gen-peapp-adjuvants.aspx [Accessed 22 April 2020].

World Wide Fund for Nature (WWF). 2018. *Living Planet Report*. [online] Environmental Conservation. Available at: http://awsassets.panda.org/downloads/lpr_2012_summary_booklet_final.pdf.

Part 2

Protecting biodiversity and natural capital in agri-food supply chains: the role of accounting and finance

Chapter 7

Advances in corporate natural capital accounting

Joël Houdet, The Biodiversity Footprint Company and University of Pretoria, South Africa

1 Introduction

Nature, a Eurocentric concept based on the opposition between nature and culture (one of four existing worldviews according to Descola, 2005), is perceived and valued in starkly different and often conflicting ways by people (Pascual et al., 2017). For business, nature also has various meanings (Houdet et al., 2012). The mainstream twentieth-century perspective focused on the negative impacts of companies on the natural world (environmental impact perspective), e.g. through pollution or land use change. More recently, the natural capital concept has attempted to change the narrative towards a more holistic approach in which nature is also seen as a stock of resources on which business depends, directly or indirectly, and from which it can capture benefits to produce goods and services (MA, 2005; Natural Capital Coalition, 2016).

The agriculture, food and beverages (hereafter referred to 'agri-food') industries are among the largest ones dependent on nature (MA, 2005; World Economic Forum, 2020). This dependency typically revolves around making three main types of decisions about natural capital:

- Promoting or sustaining certain components of nature (e.g. active selection of crop varieties, capturing and storing water);
- Importing ecosystem services, embedded in purchased goods, captured by other industries (e.g. fertilisers); and

http://dx.doi.org/10.19103/AS.2023.0128.01

- Suppressing biodiversity elements incompatible with adopted production models (e.g. wetland drainage, extensive use of herbicides and pesticides, exclusion fencing against herbivory).

In other words, these trade-offs are generating agro-ecosystems managed for the delivery of specific ecosystem services, often against what is perceived as non-useful or detrimental biodiversity (i.e. ecosystem services versus disservices dichotomy; Saunders et al., 2016; Saunders and Luck, 2016; Wu et al., 2021). Agri-food industries have thus become the main driver of natural capital loss (Benton et al., 2021). Given the projected growth in food and beverages demand in the coming decade (McKinsey & Company, 2020), the need to better understand and manage their impacts and dependencies on nature should not be underestimated. To that end, natural capital accounting can play a significant role.

This chapter aims to introduce recent advances in corporate[1] natural capital accounting through the lens of the agri-food industries. It includes the following:

- A short overview of approaches available to date, highlighting:
    - Their diversity in framing, methods and applications and
    - Their benefits and limitations for companies and their stakeholders;
- Presenting recent attempts to standardise corporate natural capital accounting (CNCA), namely:
    - Why a new definition is warranted?
    - What should its scope and core components be? and
    - What would be the opportunities and challenges for its mainstreaming within the agri-food sector?

2 Accounting for natural capital from a business perspective: a brief overview of approaches

Several attempts have been made to classify business accounting approaches with respect to environmental issues (e.g. Burritt et al., 2002; Richard, 2009; Schaltegger and Burritt, 2017; UNEP-WCMC et al., 2022). This has always been a challenge given the broad diversity of comparison criteria, such as follows:

- The organisational focus: company, product/service, portfolio of assets, etc.;
- Value chain boundaries: suppliers, direct operations or clients;

[1] Natural capital accounting for nations is excluded from the scope of this chapter.

- Target audiences, from management, suppliers, clients, funders, auditors to external stakeholders, among others;
- Application (end-use) type: from risk assessment, permitting, compliance, auditing, planning, budgeting, comparing options, scenario modelling, product eco-design, certification, ex-post performance assessment, internal reporting to voluntary and regulatory disclosures, among others;
- Temporal focus: past or future, short- vs. long-term focus;
- Value framing or materiality perspective (i.e. for the interests of whom is it undertaken?): impacts on or consequences for the company, its internal stakeholders (e.g. board members, management, staff, etc.), its external stakeholders (shareholders, rating agencies, etc.) and/or its silent non-human stakeholders (i.e. nature or some of its components), with many typically adopting a single materiality perspective (e.g. a greenhouse gas footprint assessment aims to assess the scale of greenhouse gas emissions for a particular organisational scope over a given period, it does not assess the impact of climate change on the business);
- Source of input data: primary, secondary, artificially generated or modelled;
- Whether information is gathered *ad hoc* or routinely;
- The type of metrics used: measurements (e.g. Tons equivalent CO_2, hectares equivalents, m^3 of H_2O), values (qualitative - e.g. low, medium, high; quantitative - e.g. percentage of untransformed areas remaining; monetary) or both; and
- The chosen measurement and/or valuation method(s).

Yet, two key groupings of environmental accounting approaches for business should be highlighted:

- Biophysical vs. monetary accounting; and
- Environmental management vs. environmental financial accounting.

The first grouping differentiates approaches according to the type of metrics used, with measurements on the one side (e.g. Water Footprint Standard and the greenhouse gas (GHG) Protocol that help organisations account for green, blue and grey water footprints and greenhouse gas emissions, respectively) and monetary values on the other, which may involve actual transactions or externalities (i.e. the uncompensated costs of one's actions on others).

The second differentiates environmental accounting approaches considering the traditional management vs. financial accounting divide, whereby environmental management accounting includes all approaches that serve internal management decisions (i.e. flexible approach) whereas environmental financial accounting is a sub-field of traditional financial accounting (i.e. rules based for comparability across companies, accountability and tax-related purposes).[2] On the one hand, environmental management accounting may involve diverse applications, such as environmental cost accounting, environmental long-term financial planning or environmental life-cycle budgeting/target pricing. On the other hand, environmental financial accounting would involve segregating and making visible financial events due to their environmental materiality (e.g. contingent liability due to pending environmental litigation). This would typically lead to producing annotations within financial disclosures or tables of environmental expenses (in the statement of financial performance) and liabilities (in the statement of financial position) broken down per environmental category (e.g. waste, water, emissions, biodiversity).

From the perspective of the agri-food industries, one may highlight a few popular environmental accounting approaches, namely:

- Life-cycle assessments to assess the environmental impacts of food commodities (e.g. Cucurachi et al., 2019; Ruiz-Salmón et al., 2021);
- Environmental certification schemes (e.g. Roundtable on Sustainable Palm Oil supply chain certification systems or standards);
- Environmental management accounting at the farm level (e.g. Halber et al., 2005; Viere et al., 2011);
- Natural capital risk mapping and assessments (e.g. Ascui and Cojoianu, 2019); and
- Ecosystem services valuation (e.g. Bostian and Lundgren, 2022; Houdet et al., 2012; Yuan et al., 2022) at the farm/agricultural landscape level.

While this diversity of approaches shows that a great diversity of needs may be satisfied through various accounting approaches, it also shows that any approach attempting to assess, measure or value business environmental impacts and dependencies can be labelled as environmental accounting. This is in line with a broad accounting theory perspective whereby any record (e.g. a photograph, measurements, maps, etc.) of environmental or natural capital phenomena can be considered as a form of accounting (see, e.g. Atkins et al., forthcoming).

Yet, this presents several challenges for both businesses and their stakeholders for the business organisational focus area (the company as a legal entity) and its value chain boundaries, namely the lack of the following:

[2] One should note that environmental management accounting requires biophysical data for its implementation towards producing information such costs or revenues per unit of natural resource used. On the other hand, environmental financial accounting typically relies on the purpose of the contract or transaction for financial events to be labelled as such.

- Standardised conceptual models explaining the interactions between business and nature and the rationale for assessing them;
- Standardised data and metrics for comparative analysis and benchmarking; and
- Standardised presentation formats to track changes and/or demonstrate performance.

In the end, this prevents companies and their stakeholders from understanding what the real issues are, what is the scale of these problems and whether meaningful progress is being made to address them.

3 Standardising corporate natural capital accounting

In recognition of the limitations and challenges that flexible environmental accounting approaches present to our transition towards nature-positive economies and companies, several papers have called for greater standardisation of natural capital measurement and valuation approaches (Houdet et al., 2022; Natural Capital Coalition, 2016), notably in the biodiversity space (objectives of the ALIGN project; UNEP-WCMC, 2020).

Building on work ecosystem accounting undertaken for nations (UNEP, 2021) and various corporation-focused projects (primarily the Biological Diversity Protocol - BD Protocol and ALIGN project), a paper by the Capitals Coalition (2022) recently proposed to put the viability of nature at the core of the corporate accounting process. It argues that CNCA should be defined as 'the systematic process of identifying, measuring, recording, summarising and reporting the periodic and accumulated net changes to (a) the biophysical state of natural capital assets and (b) the associated values of natural capital to business and wider society'. The definition further requires the following:

- 'An asset inventory recognising the biophysical properties and dynamics of each asset category,
- Measurement techniques that use spatially explicit data and apply the principle of ecological equivalency (like-for-like),
- Recording rules based on double-entry bookkeeping[3] from financial accounting,

[3] Double-Entry Bookkeeping (DEBK) comes from financial accounting and was first popularised in the late thirteenth century by Luca Pacioli, who formalised the long-established accounting methods practiced by Venetian traders to keep track of their intricate web of transactions. With DEBK, every financial event involves recording each transaction in an account with an equal and opposite effect in at least one other account. These transactions are summarised in the preparation of financial statements, including the Statement of Financial Position (or Balance Sheet) and the Statement of Financial Performance (or Profit & Loss Statement), which are based on two inter-dependent equations. DEBK thus enables organisations to record both periodic and cumulative changes in transactions of a financial nature and to aggregate individual financial events at the organisational level. Because accounting journal entries must balance out, DEBK reduces the likelihood of errors and fraud and helps improve transparency and financial management. Trotman, K., Gibbins, M. (2003). *Financial Accounting: An Integrated Approach*. 2nd edition, Thomson Nelson Australia.

- Asset-specific biophysical statements of performance and position,
- A defined scope according to organisational and value chain boundaries'.

Focusing the 'going concern'[4] of nature (beyond the mere double materiality concept; e.g. Adams et al., 2021) becomes the underlying rationale for the accounting process is reflected via several complementary foundational elements:

- It recognises the incommensurability of natural capital phenomena (Farrell, 2007) by entrenching the ecological equivalency principle[5] and making explicit the need to associate biophysical data about the state of nature with geospatial information (Bezombes et al., 2017; Brondizio et al., 2019; Carton et al., 2021; Maseyk et al., 2016);
- It focuses the attention on natural capital stocks (i.e. 'natural capital assets') (Dickie et al., 2020; Finisdore et al., 2020; Fu et al., 2011; Hancock, 2013; Hoffmann and Chamie, 1999; Overhage and Suico, 2001; Wu, 1999), instead of narrowly measuring environmental flows (i.e. beyond mere input[6]/output[7] accounting), which, at best, only explain part of the story (e.g. annual water use statistics do not explain the state of water resources in a watershed); and
- The clear statement that any valuation exercise (i.e. expressing whether/how natural capital is important to business and its stakeholders, which is intrinsically contingent to the beholder or the one who holds decision-making power; Pascual et al., 2017) is secondary to the process of accounting for the state of nature. It may be undertaken, as an option, following any valuation approach (which are often in opposition to one another: e.g. intrinsic vs. monetary valuation), after completing biophysical accounts.

Beyond these foundational elements putting stocks of nature at the forefront, this new definition of CNCA heavily draws from financial accounting rules. This is critical for two main correlated reasons:

- It allows to account for both the periodic (e.g. annual or monthly) and accumulated (i.e. historical, since a reference pre-development or pristine state) changes in the state of natural capital, whereas other environmental accounting approaches only account for periodic changes (Houdet et al.,

[4] Traditionally, going concern is an accounting term for a company that is financially stable enough to meet its obligations and continue its business for the foreseeable future.
[5] Of course, the scale of implementation of the like-for-like principle is key. Ecological equivalency at the biome level should be understood as a parody of the principle, as it would mean, e.g. that almost any terrestrial temperate ecosystem in Western Europe is equivalent.
[6] Such as accounting for water or materials used by business.
[7] Such as accounting for wastewater or greenhouse gases emitted by the organisation.

2020), notably for greenhouse gas emissions (Houdet et al., 2022), allowing for environmental amnesia (e.g. Kahn and Weiss, 2017) and shifting baselines (e.g. Pauly, 1995; Soga and Gaston, 2018). This leads to distinguishing the statement of natural capital performance from the statement of natural capital position. In other words, using and adapting financial accounting conventions, notably double-entry bookkeeping, emulates what financial accounting theory did for the recording of financial transactions between economic agents. It provides an overall framework for organising natural capital measurements to record a comprehensive picture of the exchanges between business and nature at a point in time.

- It also enables corporations to conceptualise their interactions with nature from an asset perspective, which involves formalising natural capital debt-like[8] (i.e. accumulated negative impacts) and equity-like (i.e. accumulated positive impacts) accounts. In financial accounts, the statement of position explains how assets are financed by debt and equity. In CNCA, the statement of position explains how the state of natural capital assets is the result of both accumulated positive and negative impacts (Houdet et al., 2020).

Overall, among the purported benefits of this vision for CNCA, one may include (Capitals Coalition, 2022) the following:

- The improvement of the measurement foundations of any target, dependency and impact-related valuation method, when coupled with appropriate natural capital classification systems (e.g. Global Ecosystem Typology, 2021, Linnean taxonomy or PhyloCode);
- The establishment of meaningful quantitative, biophysical targets for natural capital state, notably for biodiversity conservation and recovery, by distinguishing reference states (which are embedded in condition/integrity measurements) from baselines (which are a political decision) through (a) appropriate methods to measure the state of natural capital (i.e. condition/integrity-based metrics) and (b) performance (changes from periodic baseline) from position (accumulated changes from reference state); and
- The improvement of the interoperability of data that support comparability of results across sites, business units, sectors, geographies, scales and times via the use of new conventions to double-entry bookkeeping. Organised in the right formats (e.g. chart of accounts, journal entries, statement of position and performance), measurement data should be more readily understandable to any internal or external stakeholder

[8] Without implying any financial liability.

for comparative analysis, benchmarking, monitoring, decision-making, planning or reporting.

However, Capitals Coalition (2022) argues that undertaking CNCA based on this definition requires further stakeholder consultation, research and developments through 'seven sequential standardisation steps':

1. 'Defining the natural capital stock as an asset', which implies that not all natural capital stocks have been conceptualised as assets (alongside the associated concepts of debt and equity). To date, such concepts have only been developed for ecosystems, species and greenhouse gases (Endangered Wildlife Trust, 2021; Houdet et al., 2020, 2022).
2. 'Developing natural capital asset registers of stocks using classification systems', explaining that existing classification systems have been developed for portions of natural capital stocks (e.g. ecosystems, species, soils, waterways, etc.) and implying that an overarching classification applicable to business (not nations, as partially done in the UN SEEA) has yet to be formalised.
3. 'Measuring changes in the state of natural capital using appropriate methods specific to each asset category', implying that such methods have yet to be adopted for all types of natural capital stocks (e.g. only recently the concept of condition or integrity has been put at the core of biodiversity impact assessment and the associated metrics - e.g. surface area equivalents; see case study in Box 1, many actors still being satisfied with land use change metrics as illustrated by the ALIGN project).
4. 'Employing recording rules for accounting events and linked journal entries', which highlights the need for new conventions to double-entry bookkeeping for all natural capital asset categories, including the underlying equations for the statements of position and performance. To date, this has been implemented for ecosystems (Houdet et al., 2020), species (Endangered Wildlife Trust, 2021) and greenhouse gas emissions and removals (Houdet et al., 2022).
5. 'Summarising the biophysical state of natural capital with a statement of natural capital position', which highlights the need to find consensus on how to present the accumulated net state of each natural capital category at a given point in time, including outputs from the current statement of performance and the accumulated net state of natural capital from previous periods for each asset category.
6. 'Summarising net periodic biophysical changes in natural capital with a statement of natural capital performance', which highlights the need to find consensus on how to present periodic gains and losses

for equivalent asset categories (i.e. a single metric covering all natural capital assets is not feasible).

7. 'Using valuation or other analytical methods to link the stock-based accounts and integrated datasets with complementary value framing perspectives', acknowledging that widely divergent valuation perspectives exist (Pascual et al., 2017) and suggesting that new ways of articulating the worth of priceless phenomena are warranted (see Farrell, 2007), e.g. through in-depth consultation and consensus.

Box 1 Corporate natural capital accounting in practice - Sibanye-Stillwater's biodiversity footprint

This case study illustrates the ongoing biodiversity impact accounting process by Sibanye-Stillwater, a leading producer of platinum, palladium, rhodium and other strategic metals in South Africa and the United States of America. This includes a baseline (Houdet and Teren, 2022) and a follow-up assessment (Houdet et al., 2023). The key CNCA elements are as follows:

- **Organisational and value chain boundaries**: The biodiversity footprint assessment was undertaken for the direct impacts[9] of the direct operations[10] of Sibanye-Stillwater. This included the following:
    - South African (SA) operations: Beatrix, Blueridge, Burnstone, Driefontein, Ezulwini, Kloof, Kroondal, Marikana, Rand Uranium and Rand Platinum Mines (RPM);
    - United States of America (USA) operations: East Boulder Mine and Stillwater Mine (SWM).
- **'An asset inventory recognising the biophysical properties and dynamics of each asset category'**: Following the BD Protocol (2021), two types of biodiversity impact accounts were identified: those that record impacts on ecosystems and those that account for impacts on taxa (species and sub-species).

[9] As per the BD Protocol (2021) *'For biodiversity impact accounting and reporting within the context of the BD Protocol, direct impacts constitute changes in the state of biodiversity which are caused directly by your business activities. In other words, direct impacts involve business impact drivers which can be traced to specific, verifiable biodiversity features, that is direct causal link between your company's actions (e.g. land clearing or ecosystem restoration measures) and a change in the state of ecosystems or taxa (e.g. decrease/increase in ecosystem condition, habitat loss/gain for several species)'.*

[10] The value chain boundaries of the BD Protocol differ from the three scopes of the GHG Protocol. In line with the Natural Capital Protocol (Natural Capital Coalition 2016, the BD Protocol first recognises three major parts of the value chain:

- Direct operations (gate-to-gate), which cover activities over which your business holds ownership or control;
- Upstream (cradle-to-gate), which covers the activities of suppliers; and
- Downstream (gate-to-grave), which covers activities linked to the purchase, use, re-use, recovery, recycling and final disposal of your business' products and services.

Thirty-eight ecosystem types and three material species (e.g. Canada lynx, grizzly bears and whitebark pine) were identified across the operations.

- **'Measurement techniques that use spatially explicit data and apply the principle of ecological equivalency (like-for-like)'**. Each ecosystem asset was mapped [geographical information system (GIS), extent in surface area metric] and assessed for condition (surface area adjusted for condition, i.e. Ha. eq.) across the operations. The habitat size for each material species was assessed (extent in surface area metric) and mapped, separately. Gains and losses from the chosen baselines, in Ha eq. for ecosystems and Ha for species, were assessed separately for each ecosystem and species asset.
- **'Recording rules based on double-entry bookkeeping from financial accounting'**: The study used the accounting foundations of the BD Protocol, which is based on two simple equations, adapted from double-entry bookkeeping, which ensures that the total biodiversity impacts of a company are equal to the sum of its accumulated positive and negative impacts. These equations are as follows:
    - Statement of Biodiversity Position: (A accounts) total impacts on biodiversity features = (B accounts) accumulated positive impacts on biodiversity + (C accounts) accumulated negative impacts on biodiversity (for all periods to date) and
    - Statement of Biodiversity Performance: (X accounts) net biodiversity impacts on biodiversity features over the accounting period = (Y accounts) periodic positive biodiversity impacts or gains – (Z accounts) periodic biodiversity negative impacts or losses.

As a result, accounting journal entries recorded (a) the reference state of assets, (b) their baseline state (e.g. based on present condition or integrity scores), (c) any changes in state against that baseline (e.g. to account for any management action since the baseline) and (d) the closure of the statement of performance.

- **'Asset-specific biophysical statements of performance and position':** Through this double-entry bookkeeping process, Sibanye-Stillwater produced the following:
    - For ecosystems:
        - Biophysical statements of performance for each operation, which listed all changes (gains and losses) in asset condition from (a) the reference state and (b) the chosen baseline;
        - Biophysical statements of position for each asset (A, B and C accounts), grouped per operation; and
        - Consolidated statements of performance and position (Table 1) at the group level, for USA and for SA operations, to show how data (surface area equivalents)

Table 1 The consolidated total biodiversity footprint (TBF), positive biodiversity footprint (PBF) and negative biodiversity footprint (NBF) of Sibanye-Stillwater (a) at acquisition, (b) for the 2021 baseline assessment and (c) the 2022 update assessment

	At acquisition (different dates)	2021 baseline	2022 update	Changes from 2021 assessment
Total biodiversity footprint (TBF, in ha)	49913.71	49928.30	52186.10	2257.80
Percentage of TBF/ TBF (%)	100.00	100.00	100.00	4.52
Positive biodiversity footprint (PBF, in ha eq.)	6753.45	6424.12	8048.15	1624.03
Percentage of PBF/ TBF (%)	13.53	12.87	15.42	25.28
Negative biodiversity footprint (NBF, in ha eq.)	43160.26	43504.18	44137.95	633.77
Percentage of NBF/ TBF (%)	86.47	87.13	84.58	1.46

can be consolidated for decision-making and disclosure, without losing the data specific to each asset (GIS, maps, extent and condition).

- For material species (USA operations only to date):
    - Biophysical statements of performance for each operation, which listed all changes (gains and losses) in habitat extent from (a) the reference state and (b) the chosen baseline;
    - Biophysical statements of position for each asset (A, B and C accounts), grouped per operation; and
    - Separate statements of performance and position at the group level to show how data can be consolidated (per species, no consolidation across species) for decision-making and disclosure.

---

4 Corporate natural capital accounting in the agri-food industries: opportunities and challenges

As existing CNCA approaches (SEEA EA, BD Protocol) are updated and improved, various opportunities and challenges arise for agri-food industries. Among the opportunities one could mention the following:

- Answering the call to go beyond commodity impact screening, mapping and assessment (risk avoidance approach), towards accounting for the state of biodiversity throughout supply chains (stewardship-based approach), making the point that all end-user companies of food-related products and services depend on agroecosystems. CNCA would enable companies to account for their impacts and dependencies on natural capital stocks.

- Segmenting and mapping natural capital asset registers per value chain boundary, making visible how value chains operate and rely/impact on real, verifiable natural capital assets. This would make natural capital stocks as tangible as more conventional financial assets (e.g. equipment, property, etc.).
- The need to collaborate and pool resources to assess (identify, map and measure the state) natural capital stocks globally, moving away from datasets which focus on flows and rely on case studies to extrapolate impacts to other situations. Assessing the actual changes in the state of natural capital is the only way to know whether a company manages its resources sustainably.

Yet, this is not without its challenges, including the following:

- Developing and adopting natural capital classification systems that apply to agri-food industries;
- Developing the double-entry bookkeeping foundations for all relevant natural capital stocks for which it has yet to be done;
- Agreeing and adopting the best methods for measuring different natural capital categories and components, apart from those for which there is broad agreement (e.g. GHG emissions and removals, ecosystems);
- Agreeing on apportionment rules[11] for the responsibility of changes to natural capital assets across value chain boundaries;
- Financing the upskilling of relevant staff members with respect to these emerging accounting techniques, notably at the farm or cooperative level; and
- Developing cost-effective audit systems throughout value chain boundaries, as unverified accounts would not be sufficient to build trust in the industry.

[11] Selecting an apportionment method for the biodiversity impacts of suppliers or clients is not anecdotal. It may lead to incorrect estimations of the overall biodiversity impacts of a company. The BD Protocol recommends selecting the method which makes most sense given both the business and biodiversity context, in accordance with the relevant accounting and reporting principle. For instance, a food retailer buying fruits directly from a farmer might have purchased 70% of the fruits produced during the period, but only contributed to 40% of the farmer's annual sales/revenues. The two apportionment methods would have significantly different implications for biodiversity accounting: Applying apportionment method 1 would lead the retailer to account for 70% of the farmers' biodiversity impacts over the period; and applying apportionment method 2 would involve accounting for only 40% of the farmers' biodiversity impacts. Yet, because annual production can be linked directly to the ecosystem assets controlled by the fruit farmer (assuming fruit production occurs uniformly across the property), the first method would constitute the best apportionment option for the food retailer (i.e. satisfying the relevance principle). In the case of financial institutions (e.g. loans to any company), the more relevant apportionment method would be the second one.

5 Conclusion

CNCA, despite its infancy, provides tangible opportunities for putting the 'going concern' of natural capital stocks at the core of the accounting process for companies, irrespective of their industries or location. While there will be significant barriers to broad-scale adoption in the short term, ranging from ongoing research needs, significant capacity constraints/upskilling requirements, material transition costs (information systems, data collection) and obvious political and socio-economic opposition, the need to give the same 'quality of attention' (i.e. double-entry bookkeeping and comprehensive audit trails) to natural capital as is done for financial capital has never been clearer and more warranted.

6 References

Adams, C. A., Alhamood, A., He, X., Tian, J., Wang, L. and Wang, Y. (2021). The double materiality concept: Application and issues, published by the global reporting initiative. Available at: https://www.globalreporting.org/media/jrbntbyv/griwhite paper-publications.pdf.

ALIGN Project. Available at: https://ec.europa.eu/environment/biodiversity/business/align/index_en.htm.

Ascui, F. and Cojoianu, T. (2019). *Natural Capital Credit Risk Assessment in Agricultural Lending: an Approach Based on the Natural Capital Protocol*. Oxford: Natural Capital Finance Alliance.

Atkins, J. F., Doni, F., McBride, K. and Napier, C. (forthcoming). Exploring the historical roots of environmental and ecological accounting from the dawn of human consciousness. *Acc. Aud. Acc. J.*

Benton, T. G., Bieg, C., Harwatt, H., Pudasaini, R. and Wellesley, L. (2021). Food system impacts on biodiversity loss: Three levers for food system transformation in support of nature. *Energy, Environment and Resources Programme*. Chatham House. Available at: https://www.chathamhouse.org/sites/default/files/2021-02/2021-02-03-food-system-biodiversity-loss-benton-et-al_0.pdf.

Bezombes, L., Gaucherand, S., Kerbiriou, C., Reinert, M. E. and Spiegelberger, T. (2017). Ecological equivalence assessment methods: What trade-offs between operationality, scientific basis and comprehensiveness? *Environ. Manag.* 60(2), 216–230. https://doi.org/10.1007/s00267-017-0877-5.

Bostian, M. and Lundgren, T. (2022). Valuing ecosystem services for agricultural TFP: A review of best practices, challenges, and recommendations. *Sustainability* 14(5), 3035. https://doi.org/ 10.3390/su14053035.

Brondizio, E., Diaz, S., Jose, S. and Ngo, H. T. (2019). Global assessment report on biodiversity and ecosystem services of the intergovernmental science-policy platform on biodiversity and ecosystem services. https://doi.org/10.5281/zenodo.3831673.

Burritt, R. L., Hahn, T. and Schaltegger, S. (2002). Towards a comprehensive framework for environmental management accounting – Links between business actors and environmental management accounting tools. *Austr. Acc. Rev.* 12(28), 39–50.

Capitals Coalition. (2020). Improving nature's visibility in financial accounting. Full report. URL: https://naturalcapitalcoalition.org/wp-content/uploads/2020/04/NatCap_VisFinAccount_final_20200428.pdf.

Capitals Coalition (2022). Time to take stock version 2.0.

Carton, W., Lund, J. F. and Dooley, K. (2021). Undoing equivalence: Rethinking carbon accounting for just carbon removal. *Front. Clim.* 1, 664130. https://doi.org/10.3389/fclim.2021.664130.

Cucurachi, S., Scherer, L., Guinée, J. and Tukker, A. (2019). Life cycle assessment of food systems. *One Earth* 1(3), 292–297. https://doi.org/10.1016/j.oneear.2019.10.014.

Descola, P. (2005). *Par-delà nature et culture*. Éditions Gallimard, Collection Folio Essais.

Endangered Wildlife Trust (2021). Biological diversity protocol (BD protocol). National Biodiversity and Business Network of South Africa. Available at: https://nbbnbdp.org/biodiversity-protocol/.

Farrell, K. N. (2007). Living with living systems: The co-evolution of values and valuation. *Int. J. Sustain. Dev. World Ecol.* 14(1), 14–26. https://doi.org/10.1080/13504500709469704.

Finisdore, J., Rhodes, C., Haines-Young, R., Maynard, S., Wielgus, J., Dvarskas, A., Houdet, J., Quetier, F., Lamothe, K. A., Ding, H., Soulard, F., Van Houtven, G. and Rowcroft, P. (2020). The 18 benefits of using ecosystem services classification systems. *Ecosyst. Serv.* 45. https://doi.org/10.1016/j.ecoser.2020.101160.

Fu, B. J., Su, C. H., Wei, Y. P., Willett, I. R., Lu, Y. H. and Liu, G. H. (2011). Double counting in ecosystem services valuation: Causes and countermeasures. *Ecol. Res.* 26(1), 1–14 https://doi.org/10.1007/s11284-010-0766-3.

GHG Protocol. Available at: https://ghgprotocol.org/companies-and-organizations.

Halber, N., Verschuur, G. and Goodlass, G. (2005). Farm level environmental indicators; are they useful? An overview of green accounting systems for European farms. *Agric. Ecosyst. Environ.* 105(1–2), 195–212.

Hancock, A. (2013). Best practice guidelines for developing international statistical classifications. *Expert Gr. Meet. Int. Stat. Classif.* New York, 13–15 May 2013, 16.

Hoffmann, E. and Chamie, M. (1999). Standard statistical classifications: Basic principles. In Statistical Commission 10 February 1999 Thirtieth Session New York, 1–5 March 1999 Items 8 of the Provisional Agenda, 35.

Houdet, J., Trommetter, M. and Weber, J. (2012). Understanding changes in business strategies regarding biodiversity and ecosystem services. *Ecol. Econ.* 73, 37–46.

Houdet, J., Ding, H., Quetier, F., Addison, P. and Deshmukh, P. (2020). Adapting double-entry bookkeeping to renewable natural capital: An application to corporate net biodiversity impact accounting and disclosure. *Ecosyst. Serv.* 45, 101104. https://doi.org/10.1016/j.ecoser.2020.101104.

Houdet, J., Quetier, F., Deshmukh, P. and Atkins, J. F. (2022). *Greenhouse Gas Accounting and Net-Zero Target Setting: The Case for Organisational GHG Double-Entry Book-keeping*. Pretoria: National Biodiversity & Business Network – Endangered Wildlife Trust /Sibanye-Stillwater.

Houdet, J. and Teren, G. (2022). Sibanye-Stillwater's consolidated biodiversity footprint. Pilot Assessment as per the Biological Diversity Protocol – Group Level Consolidated Report. National Biodiversity & Business Network – Endangered Wildlife Trust/Sibanye-Stillwater. Available at: https://407264.p3cdn1.secureserver

.net/wp-content/uploads/2022/09/Sibanye-Stillwater-Biodiversity-Footprint-May-2022.pdf.

Houdet, J., Teren, G. and Nelson, B. (2023). Sibanye-Stillwater's consolidated biodiversity footprint. Update Assessment as per the Biological Diversity Protocol - Group Level Consolidated Report. National Biodiversity & Business Network - Endangered Wildlife Trust/Sibanye-Stillwater. Available at: https://407264.p3cdn1.secureserver.net/wp-content/uploads/2023/06/ssw-ewt-biodiversity-footprint-update-2022.pdf.

Kahn, Jr., P. H. and Weiss, T. (2017). The importance of children interacting with big nature. Children, youth and environments. *Nat. Spaces Dev.* 27(2), 7–24 (18 pages) Published By: University of Cincinnati.

Maseyk, F., Barea, L., Stephens, R., Possingham, H., Dutson, G. and Maron, M. (2016). A disaggregated biodiversity offset accounting model to improve estimation of ecological equivalency and no net loss. *Biol. Conserv.* 204, 322–332. https://doi.org/10.1016/j.biocon.2016.10.016.

McKinsey & Company (2020). Agriculture and climate change Reducing emissions through improved farming practices. Available at: https://www.mckinsey.com/~/media/mckinsey/industries/agriculture/our%20insights/reducing%20agriculture%20emissions%20through%20improved%20farming%20practices/agriculture-and-climate-change.pdf.

Millennium Ecosystem Assessment (2005). *Ecosystems and Human Well-Being: Synthesis*. Washington, DC: Island Press.

Natural Capital Coalition (2016). Natural capital protocol (Online). Available at: www.naturalcapitalcoalition.org/protocol.

Overhage, J. M. and Suico, J. G. (2001). Sorting things out: Classification and its consequences. *Ann. Intern. Med.* 135(10), 934. https://doi.org/10.7326/0003-4819-135-10-200111200-00030.

Pascual, U., Balvanera, P., Diaz, S., Pataki, G., Roth, E., Stenseke, M., Watson, R. T., Başak Dessane, E., Islar, M., Kelemen, E., Maris, V., Quaas, M., Subramanian, S. M., Wittmer, H., Adlan, A., Ahn, S. E., Al-Hafedh, Y. S., Amankwah, E., Asah, S. T., Berry, P., Bilgin, A., Breslow, S. J., Bullock, C., Caceres, D., Daly-Hassen, H., Figueroa, E., Golden, C. D., Gomez-Baggethun, E., Gonzalez-Jimenez, D., Houdet, J., Keune, H., Kumar, R., Ma, K., May, P. H., Mead, A., O'Farrell, P., Pandit, R., Pengue, W., Pichis-Madruga, R., Popa, F., Preston, S., Pacheco-Balanza, D., Saarikoski, H., Strassburg, B. B., van den Belt, M., Verma, M., Wickson, F. and Yagi, N. (2017). Valuing nature's contributions to people: The IPBES approach. *Curr. Opin. Environ. Sustain.* 26-27, 7–16. https://doi.org/10.1016/j.cosust.2016.12.006.

Pauly, D. (1995). Anecdotes and the shifting baseline syndrome of fisheries. *Trends Ecol. Evol.* 10(10), 430.

Richard, J. (2009). Classification des comptabilites environnementales. In Dans: Colasse, B. (Ed.). *Encyclopedie de Comptabilite*, controle de gestion et audit. Economica.

RSPO. Available at: https://rspo.org/standards.

Ruiz-Salmón, I., Laso, J., Margallo, M., Villanueva-Rey, P., Rodríguez, E., Quinteiro, P., Dias, A. C., Almeida, C., Nunes, M. L., Marques, A., Cortés, A., Moreira, M. T., Feijoo, G., Loubet, P., Sonnemann, G., Morse, A. P., Cooney, R., Clifford, E., Regueiro, L., Méndez, D., Anglada, C., Noirot, C., Rowan, N., Vázquez-Rowe, I. and Aldaco, R. (2021). Life cycle assessment of fish and seafood processed products – a review of

methodologies and new challenges. *Sci. Total Environ.* 761, 144094. https://doi.org/10.1016/j.scitotenv.2020.144094.
Saunders, M. E., Peisley, R. K., Rader, R. and Luck, G. W. (2016). Pollinators, pests, and predators: Recognizing ecological trade-offs in agroecosystems. *Ambio* 45(1), 4–14. https://doi.org/10.1007/s13280-015-0696-y.
Saunders, M. E. and Luck, G. W. 2016. Limitations of the ecosystem services versus disservices dichotomy. *Conserv. Biol.* 30(6), 1363–1365. https://doi.org/10.1111/cobi.12740.
Schaltegger, S. and Burritt, R. (2017). *Contemporary Environmental Accounting: Issues, Concepts and Practice*. Routledge: London. https://www.taylorfrancis.com/books/mono/10.4324/9781351282529/contemporary-environmental-accounting-stefan-schaltegger-roger-burritt.
Soga, M. and Gaston, K. J. (2018). Shifting baseline syndrome: Causes, consequences, and implications. *Front. Ecol. Environ.* 16(4), 222–230, https://doi.org/10.1002/fee.1794.
UN (2021). *System of Environmental-Economic Accounting– Ecosystem Accounting (SEEA EA).* New York, White cover publication. Available at: https://seea.un.org/ecosystem-accounting.
UN Environment Programme World Conservation Monitoring Centre (2020). Biodiversity measures for business: Corporate biodiversity measurement and disclosure within the current and future global policy context. Cambridge, 60 pp. Available at: https://www.unep-wcmc.org/featured-projects/aligning-biodiversity-measures-for-business.
UNEP-WCMC, Capitals Coalition, Arcadis, ICF, WCMC Europe (2022). Recommendations for a standard on corporate biodiversity measurement and valuation, aligning accounting approaches for nature. ALIGN Project.
Viere, T., von Enden, J. and Schaltegger, S. (2011). Life cycle and supply chain information in environmental management accounting: A coffee case study. In Burritt, R., Schaltegger, S., Bennett, M., Pohjola, T. and Csutora, M. (Eds). *Environmental Management Accounting and Supply Chain Management. Eco-efficiency in Industry and Science* (vol. 27). Dordrecht: Springer, 23–40. https://doi.org/10.1007/978-94-007-1390-1_2.
Yuan, Y., Xu, G., Shen, N., Nie, Z., Li, H., Zhang, L., Gong, Y., He, Y., Ma, X., Zhang, H., Zhu, J., Duan, J. and Xu, P. (2022). Valuation of ecosystem services for the sustainable development of Hani terraces: A rice–fish–duck integrated farming model. *Int. J. Environ. Res. Public Health* 19(14), 8549. https://doi.org/10.3390/ijerph19148549.
Water Footprint Standard. Available at: https://waterfootprint.org/en/water-footprint/global-water-footprint-standard/.
WEF & PWC (2020). Nature risk rising: why the crisis engulfing nature matters for business and the economy. Available at: https://www3.weforum.org/docs/WEF_New_Nature_Economy_Report_2020.pdf.
Wu, J. (1999). Hierarchy and scaling: extrapolating information along a scaling ladder. *Can. J. Remote Sens.* 25(4), 367–380 https://doi.org/10.1080/07038992.1999.10874736.
Wu, S., Li, B. V. and Li, S. (2021). Classifying ecosystem disservices and valuating their effects – A case study of Beijing, China. *Ecol. Indic.* 129, 107977. https://doi.org/10.1016/j.ecolind.2021.107977.

Chapter 8

Natural capital and biodiversity accounting in the Swedish agri-food sector

Kristina Jonäll, University of Gothenburg, Sweden

1 Introduction

Research shows that biodiversity loss is a critical issue. The international organization Intergovernmental Science-Policy Platform on Biodiversity and Ecosystem Services (IPBES) has released several reports on biodiversity. In one of their 2019 reports 550 experts stated that the loss of animals and nature is as significant a threat to humans as climate change. The IPBES report estimated that species are being eradicated at a rate that is 100 to 1000 times faster than before industrialization. Most of the total extinction risk to species is estimated to have arisen in the past forty years. Furthermore, IPBES found that one half to one million of the world's approximately eight million species are at risk of extinction unless strong measures are put in place to halt current trends (IPBES, 2019).

http://dx.doi.org/10.19103/AS.2023.0128.15

As early as 1999, the Swedish Parliament decided to implement an Environmental Objectives System which established national environmental goals for Swedish government policy. The environmental objectives system consists of an overall generational goal, sixteen environmental quality objectives and a number of milestone targets to achieve these goals (Naturvårdsverket, 2022). The Swedish Parliament stated that 'The value of the agricultural landscape and the agricultural land for biological production and food production must be protected, at the same time as biodiversity and cultural environmental values are preserved and strengthened'. Sweden, like many other countries in the world, has signed the UN Convention on Biological Diversity and thus committed the country to preserving biological diversity.

Efforts to protect biodiversity can be linked to almost all of Sweden's environmental goals. Either the goal strengthens biodiversity through protecting different landscape types, e.g. ensuring a diverse agricultural landscape or by fulfilling goal promoting biodiversity, e.g. preventing eutrophication. Eutrophication is a process by which excess nutrients in the soil due to the use of fertilizers and other agricultural practices leach into nearby watercourses. These nutrients enable algae to grow rapidly, depleting oxygen and killing other aquatic life.

For national environmental goals to be achieved and biological diversity to be ensured, efforts are required from many actors. Business and industry are important actors in protecting biodiversity. The conservation of biodiversity has now developed into an issue which, in addition to economic growth, is of increasing importance for companies (Johansson et al., 2022). The business community has a significant impact on biological diversity, either directly, e.g. through land use, air and water pollution, and generation of waste, or indirectly through the actions of others caused by business activities, e.g. supply chains. Companies are therefore important players in terms of the opportunity to help reverse biodiversity loss (Jonäll, 2021).

More broadly, both agriculture and food production depend on a well-functioning ecosystem. Agri-food companies play an important role in preserving biodiversity and creating long-term sustainable ecosystems for plants, animals, and humans. Without biodiversity, the ecosystem services on which agriculture depends will disappear, making it more difficult or even impossible to continue farming (Wallander et al., 2019). The International Finance Corporation (IFC) highlights that biodiversity is 'a fundamental component of long-term business survival' (IFC, 2022). It is therefore important for companies to understand the ways that biodiversity loss affects their activities but also how they should account for the impact they have on biodiversity (KPMG, 2020). Companies need to develop, implement and demonstrate the effect of measures to achieve environmental and sustainability goals. Companies also need to inform stakeholders about their work to protect biodiversity.

Studies such as Roberts et al. (2021), and the wider public debate that has taken place in recent years, has contributed to an increased understanding of what drives companies to report on biodiversity. We still need to know more about how companies account for biodiversity and what motivates them to report on the actions they take, what measures companies take to reduce their impact on biodiversity and how effective these are. More knowledge is important for society to be able to develop and improve laws and policies that help companies to be transparent and report their impact on biodiversity (KPMG, 2020).

This chapter reviews how companies in the agricultural sector in Sweden deal with biodiversity. The findings of the study contribute to increased knowledge of how companies that have a substantial impact on biodiversity develop policies to monitor and improve their impacts on biodiversity and how they report on their performance.

2 Sources and methods for investigating corporate biodiversity policies and accounting

Data collection for this study was carried out in two stages. First a literature review was conducted based on scientific articles about impacts and actions for biodiversity in agriculture in Sweden. The literature study summarizes the latest research on how agricultural affects biodiversity. Second a desk study was carried out to study how biodiversity is dealt with in financial reports from some of the largest food manufacturing companies involved in the agriculture sector in Sweden.

The selection initially consisted of the fifty largest food manufacturing companies involved in the agriculture sector in Sweden (Largestcompanies.se, 2022). Among these companies were some that were part of a group of companies included separately in the list of the fifty largest companies as well as some that were not involved in production of agricultural produce or food products. These companies were therefore omitted and are shown as crossed out in Table 1. The remaining 36 companies are included in the desk study. Of these 25 have publicly available sustainability reports.

Documents for the desk study consisted of Annual Reports, Sustainability Reports, other publications relating to biodiversity, and company websites. The desk study aimed to show how companies report on their impact on natural capital, ecosystem services and biodiversity, ways of measuring the impact of companies on ecosystem services and biodiversity, and on implications for improved reporting.

Table 1 The 50 largest food manufacturing companies from the agriculture sector in Sweden

	Company	Turnover (*1000 SEK)	Type of report
1	AAK AB (publ)	27 934 000	Sustainability report
2	Arla Foods AB	17 198 449	Sustainability report
3	Scandi Standard AB (publ)	9 940 000	Combined report
4	HKScan Sweden AB	7 319 412	Combined report
5	KLS Ugglarps AB	6 203 833	No report
6	Orkla Foods Sverige AB	6 062 485	Combined report
7	Cloetta AB	5 695 000	Combined report
8	Skånemejerier AB	4 272 534	Sustainability report
9	Svenska Foder AB	4 086 103	No report
10	Pågen AB	3 280 577	Sustainability report
11	Arvid Nordquist Handels AB	3 277 692	Sustainability report
12	Atria Sverige AB	3 049 816	Sustainability report
13	Santa Maria AB	2 961 226	Combined report
14	Gunnar Dafgård AB	2 887 000	Sustainability report
15	Nordic Sugar AB	2 283 656	Combined report
16	Oatly AB	2 276 068	Sustainability report
17	Lantmännen Cerealia AB	2 192 220	Combined report
18	Blentagruppen AB	2 087 377	No report
19	Lantmännen Unibake Sweden AB	1 775 451	No report
20	Findus Sverige AB	1 746 185	Sustainability report
21	Guldfågeln AB	1 719 572	No report
22	Anders Löfberg, AB	1 600 273	Sustainability report
23	Skövde Slakteri AB	1 566 888	No report
24	Semper AB	1 472 448	Sustainability report
25	Fazer Bageri AB	1 368 442	Combined report
26	Nordic Food Group AB	1 232 516	No report
27	Estrella AB	1 200 067	Sustainability report
28	Espresso House Sweden AB	1 179 028	Sustainability report
29	GELITA Sweden AB	1 176 809	Sustainability report
30	Zoo Support Scandinavia AB	1 148 108	No report
31	Charkuterifabriken Sverige AB	1 005 888	No report
32	Marenor AB	987 213	Sustainability report
33	Grönsakshallen Sorunda AB	985 682	Sustainability report
34	Mårtenssons Partiaffär AB	985 539	No report
35	O Kavli AB	985 058	Sustainability report
36	Abdon Group AB	874 736	No report

The desk study involved two stages of analysis. The first stage involved a content analysis which extracted themes from biodiversity-related disclosures by companies. Secondly, a search for biodiversity-related initiatives and why companies engage in biodiversity issues was performed. Key themes from this search are discussed in the chapter. A range of different qualitative content analysis techniques were applied such as meaning-oriented and interpretive-content analysis. This implies that a search for underlying themes in the texts was made on the basis of interpretative content analysis assuming that words derive their meanings when they are used in specific situations. By looking at words and phrases in their context (paragraph or whole text) themes that exist independently from the interpreter can be discovered (Smith and Taffler, 2000).

3 Biodiversity in the Swedish agricultural landscape

The agricultural landscape contains almost half of Sweden's total plants, mammals, reptiles and amphibians, a quarter of nesting birds and half of all insects. A particular feature of the Swedish agricultural landscape is natural pasture: land that has been grazed for a long time and which has not been ploughed, fertilized or sown for a very long time (Larsson et al., 2020). Many of the natural, extensively grazed pastures in the Swedish landscape are among the most species-rich environments in the world (Lindborg et al., 2021), as rich in species as the rainforest (Olsson, 2008).

The greatest diversity of species is found in cultivated landscapes that are heterogeneous in character. Biodiversity thrives most in a landscape where meadows and natural pastures are mixed, where there are natural corridors between forest and cultivated land (e.g. field margins and lines of trees which make it easier for insects and animals to move around the landscape) and where wetlands, arable islands and stone walls e.g. are part of the terrain (Olsson, 2008). A prerequisite for the evolution of Swedish pastures, the country's open landscape and its rich biodiversity is the impact of grazing animals. The more animals trample, graze and fertilize pasture, the higher the richness of plant and animal species in pastures and meadows.

Modern agriculture has brought about changes that have contributed to reducing landscape diversity (Lindborg et al., 2021). Species richness decreases with the rationalization characteristic of modern agriculture. Gradually, as habitat types that promote biodiversity have become rarer, there is a risk that species will disappear (Wallander et al., 2019). Small fields and area of pasture that are important habitats for plants, amphibians and invertebrates are disappearing or being replanted with spruce as old pastures are abandoned (Larsson et al., 2020).

Numbers of some species have declined so much that they are now red-listed, which means that they are on the lists of species that are at risk of extinction in Sweden. As many as 80% of all the country's red-listed vascular plants, 66% of red-listed amphibians and 46% of all red-listed insects are found in agricultural landscapes (Artdatabanken, 2022). At the species database (Artdatabanken) at the Swedish University of Agricultural Sciences, work is being done to map the state of biodiversity in Sweden. Artdatabanken is a database of traits for species and other taxon levels, including species characteristics, distribution, habitat and substrate preferences, red-list data, relevant legislation etc. The database also ranks the factors threatening red-listed species in Sweden.

The most significant factors impacting species decline in Sweden are deforestation/felling and overgrowth (Ahrné and Bjelke, 2022). Deforestation is a threat that is widely seen internationally while overgrowth is mainly a major problem in Sweden. Of the more than 4700 red-listed species in Sweden, of both flora (41%) and fauna (59%), about 1400 are strongly negatively affected by deforestation or felling, while about the same number of species are affected by overgrowth (Ahrné and Bjelke, 2022).

A major problem in large parts of the country is that the landscape is overgrown. Overgrowth is when a small number of tree, shrub, and coarse grass species outcompete other species that normally live in an open landscape. Overgrowth can occur in different environments: hayfields, meadows and pastures as well as beaches, forests and bogs. The causes of overgrowth are mainly discontinued grazing and mowing, lack of natural disturbances such as fire and flood (which disrupt potentially dominant species from taking over a landscape), practices such as fertilization and field drainage (which may favour some species over others), tree planting on discontinued pastures, invasive plant species and a warmer climate (Ahrné and Bjelke, 2022).

4 Practices that promote biodiversity

Natural pastures play a particularly important role in the open landscape. Pastures where the soil is not cultivated have the highest biodiversity of plants, both in conventional and organic farming. Grazing takes place on cultivated agricultural land, permanent grazing or on leys as part of a crop rotation. Grazing management needs to be conducted in a way that species have suitable habitats and can spread (Wallander et al., 2019).

Biodiversity is enhanced by animal grazing, trampling and fertilization, by the right stocking densities and by rotating grazing (e.g. by moving herds between fields) to allow pasture to recover (Lindborg et al., 2021). Management of natural pastures therefore requires specific grazing practices using suitable animals to ensure the right type of disturbance (Larsson et al., 2020). Natural pastures are currently not grazed to a sufficient extent. Analysis shows that there is no shortage of grazing animals in Sweden (Larsson et al., 2020). The problem is that grazing animals are not placed on natural pastures since they are not considered sufficiently profitable for farmers compared to other forms of grazing or farming (e.g. feeding indoors). Targeted compensation and other incentives are therefore being discussed to get farmers to use natural pastures to a greater extent (Wallander et al., 2019).

In addition to felling and overgrowth, agricultural methods affect biodiversity. Modern methods, rationalization and specialized agricultural systems produce food, animal feed and other raw materials with great efficiency (Elmqvist, 2021). However, modern cultivation systems are based on the use of a few crops, which often leads to a depletion of biodiversity and eventually loss of natural ecosystems. In consequence, wild pollinators have decreased in the Swedish agricultural landscape and today one third of Sweden's wild bees and bumblebees and one fifth of the butterflies are red-listed (Artdatabanken, 2022). The explanation is mainly a lack of food for these species, e.g. due to planting fewer species which may not provide sufficient flowers at different points in the year (ensuring a sufficient and regular food source for pollinators and other beneficial insects), and reduced wild flora due to more intensive cultivation. Actions to promote biodiversity include more diverse rotations (to provide a range of flowering species throughout the year) and planting wild flower margins designed for pollinators at the edge of fields. If such zones are properly placed, they can also reduce nutrient loss from crop fields, soil erosion and prevent unwanted spread of plant protection products (such as pesticides) while, at the same time, benefiting pollinating insects, animals and birds (Elmqvist, 2021). These practices can therefore simultaneously promote and protect the environment, biodiversity and agricultural production.

Another factor that has reduced biodiversity is the increased use of fungicides, insecticides and herbicides which, while they may control pests, can also have negative effects on non-target species and, in the long term, undermine the ecosystems on which agriculture depends. As an example, long-term use of herbicides to control weeds can have a negative impact on both farm productivity and biodiversity. New research shows that if farmers allow the right kind and a variety of weeds around the crop, they can stop problematic weeds from spreading (MacLaren et al., 2020). Tolerating a variety of weeds reduces the risk of one species taking over and thus competing with the crops that farmers cultivate. Another advantage of weeds is that they can contribute to biodiversity. Weeds help increase the supply of flowers and thus provide food for pollinators and other beneficial species such as predators of pests

(MacLaren et al., 2020). Weeds can also be used as trap crops which lure pests away from a crop, to help store carbon in soil, prevent soil erosion and nutrient loss from fields.

Some of these techniques form part of broader farming systems such as organic farming, agroecology and regenerative agriculture. The latter has seen a recent resurgence of interest (Giller et al., 2021). One of the core principles of regenerative agriculture is use of cultivation methods designed to improve the physical, chemical and biological health of soils. These techniques include replacing tillage with direct drilling (to minimize physical disturbance of soil), permanent cover to protect soil (e.g. by leaving crop residues after harvest), more diverse rotations (including cover crops) and use of organic amendments. By increasing levels of organic matter in soil, it is possible to sequester more carbon, helping to reduce and offset greenhouse gas (GHG) emissions whilst, at the same time, increasing the resilience of soil against the effects of climate change such as increased droughts or floods (Giller et al., 2021). Regenerative farming methods increase biodiversity both above and below ground. Many regenerative methods are also part of what many farmers would regard as good agricultural practice and are already used by many farmers in Sweden (Lantbrukarnas_Riksförbund, 2022).

5 The need to monitor biodiversity

A growing number of companies are interested in developing biodiversity targets. As a consequence, companies are looking for methodologies, tools and/or metrics to measure their environmental footprint and effects on biodiversity in a credible and comparable way (Lammerant et al., 2021). However, measuring and valuing natural capital risks and impacts on biodiversity is a huge challenge. One reason is that there is no clear definition or common view of what biodiversity means for companies (Johansson et al., 2022). Biodiversity is also very site-specific, so it can be difficult to establish benchmarks and monitor change in a uniform, quantitative, and comparative way (Palme, 2021). Existing reporting structures for sustainability reporting and measurement of environmental impacts are also not always suited to measuring biodiversity (Jonäll, 2021).

In 2020 a survey was conducted among companies in the agricultural sector on how to measure and assess biodiversity. The overall perception among the companies was that measuring impact on biodiversity is difficult, with no clear guidelines for how to measure, where to start, when to combine measurement methods and how to compile results (Lammerant et al., 2021).

In Sweden different projects have been carried out with practitioners to define biodiversity and develop ways to account for natural capital and biodiversity in the agri-food sector (Cederberg et al., 2016; Hedlund et al., 2017). However, there is still a long way to go. Guidelines are needed for how biodiversity should be measured and assessed. They also need to be adapted to the specific context in which individual companies belong.

In Sweden there are several organizations that seek to link food producers with sustainable development and biodiversity. These organizations work together with researchers to test and evaluate new methods. One example is Axfoundation, a non-governmental organization that works with over 300 businesses to improve sustainability, including the food industry. Another example is the Agri Food Economics Centre which is a platform for cooperation between the Swedish University of Agricultural Sciences (SLU) and Lund University. The centre aims to bridge science and policy. It carries out advanced economic analyses of agriculture, food, and fishing in both Swedish and international contexts. Another example is The Federation of Swedish Famers (LRF) which reports on sustainable agriculture to help farmers test new methods to promote biodiversity.

6 Biodiversity reporting among Swedish companies involved in the agri-food sector

Compared to other countries, Swedish companies have a generally good record in sustainability reporting. An international sustainability reporting study by KPMG in 2020 showed that the national rate of sustainability reporting among the top one hundred companies by revenue rose from 88% in 2017 to 98% in 2020 (Threlfal et al., 2020). The proportion of integrated reports, where companies combine economic, social and environmental performance data in company in a single annual report, has also increased.

One problem however is that most companies in the agri-food sector in Sweden are not listed on a stock exchange and are therefore not obliged to produce publicly available reports for shareholders and regulatory authorities, unlike the one hundred companies included in KPMG's study. Only three of the companies in the present study are listed on a stock exchange. However, 25 of the 36 remaining companies in the study have published an annual sustainability or combined report available for a public audience. Three companies have both an annual report and a sustainability report, fourteen companies have only a sustainability report, and eight companies have published a combined report.

Categories for assessing reports are:

- Reporting that does not contain any information on biodiversity or sustainable agriculture.
- Reporting that mentions biodiversity or sustainable agriculture.
- Reporting showing specific commitments and/or targets for promoting biodiversity or sustainable agriculture.

Reports contain varying amounts of information on biodiversity, from a complete absence of information to a maximum of 24 mentioned in a report. Sustainable agriculture is not mentioned to the same extent, with the maximum references being eight in one report. Only four companies failed to mention either biodiversity or sustainable agriculture in their reports. The reports were reviewed to examine whether biodiversity or sustainable agriculture were only mentioned in general terms or whether there were specific references (e.g. specific targets or commitments).

To be assigned to the category of 'commitment to working on biodiversity', the company needed to have specific goals and timeframes for dealing with biodiversity. Examples include:

- stating that by 2025 all palm oil and soy used in production of animal feed must be traceable and come from plantations that are deforestation- and conversion-free.
- requiring that farmers supplying the company implement a regenerative agricultural program in which they must have at least seven measurable actions per year which benefit biodiversity.

Although these commitments vary in scope, timeframe and potential impact, they can be seen as a step in the right direction from only talking about sustainability issues in general terms to taking some action even if modest.

Twenty-one companies mention biodiversity in their report, usually only in general terms. Specific, quantifiable data (e.g. on current impacts or targets for improving performance) is rare or non-existent. Only four of the 21 companies that mention biodiversity show a demonstrable commitment with targets/ actions that are specific, measurable and time bound. When sustainable agriculture is mentioned, ten out of sixteen companies disclose in a general way, e.g. in connection to a discussion of climate change or SDGs in general. Six of the companies have a more committed approach to sustainable agriculture, demonstrated by referring to a specific sustainability project the company is involved in. One company demonstrates commitment to both biodiversity and sustainable agriculture in its report. Eight companies mention both biodiversity and sustainable agriculture in a general way.

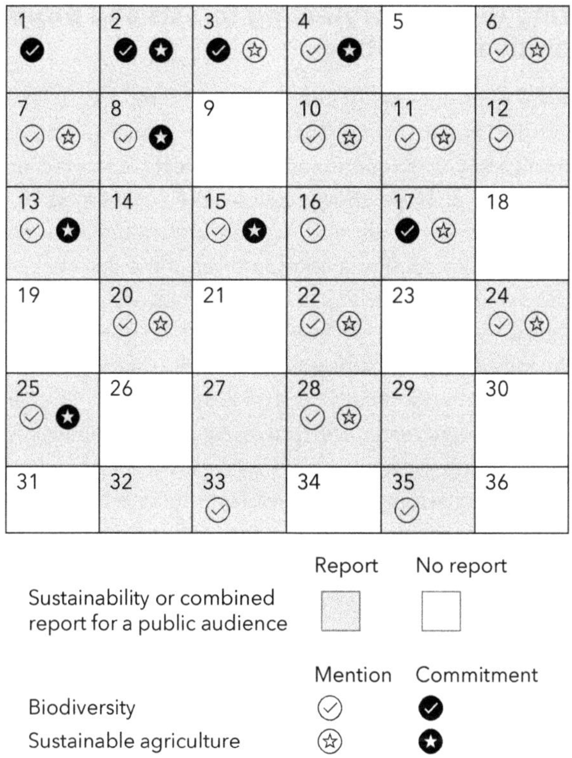

Figure 1 Summary of document study.

Figure 1 summarizes the results of the first report review:

- The numbers in the boxes represent the companies studied (see Table 1).
- Companies that have a publicly available sustainability or combined report are indicated by grey boxes.
- Companies without a report are indicated by white boxes.
- Companies that mention biodiversity in reports without commitment are shown by the symbol.
- Companies showing some type of commitment to biodiversity are shown by the symbol.
- Companies only mentioning sustainable agriculture are shown by.
- Companies showing some type of commitment to sustainable agriculture are shown by the symbol.

7 Reporting themes: protecting forests and promoting regenerative agriculture

After the first level of analysis of the companies' reports, a deeper level of analysis was carried out, involving detailed reading and re-reading of the text. Every time biodiversity was mentioned the context was noted and sorted by themes. The content analysis shows several themes related to biodiversity. In total, seven primary themes and sixteen secondary themes concerning biodiversity were identified. (For a complete list of the themes see Appendix). In the following sections the various primary themes are discussed.

Since a large proportion of Swedish food or animal feed production relies on palm oil or soybeans as raw materials, safeguarding forest was a key biodiversity theme identified in company reports. In many countries where palm trees or soybeans are grown as a crop, cultivation contributes to major environmental problems such as destruction of rainforests. Where biodiversity and 'safeguarding forests' were mentioned, companies referred to commitments to buying only palm and soy with relevant sustainability certification (i.e. grown without contributing to further deforestation and with a low carbon footprint). as well as initiatives to replace soy with locally sourced alternatives that were also grown sustainably. Eleven of the 25 companies in the study discussed safeguarding forest issues. AAK, the company that focused most on forest protection mentioned deforestation and action against deforestation sixty-seven times in its sustainability report. AAK mentions that they conduct risk assessments for raw materials such as palm oil, takes measures to stop deforestation in sensitive areas, uses satellite monitoring of plantations (to help monitor compliance), support reforestation schemes, and assess the financial impact of biodiversity protection (AAK, 2021). After AAK, Orkla Foods is the company that mentions deforestation the most, thirty-seven times. In their report they mention that they always ensure that the raw materials they buy come from deforestation-free suppliers:

> With regard to raw materials from tropical areas, we collaborate with selected suppliers who carry out their own programmes to ensure deforestation-free raw material production. In addition, we increasingly purchase raw materials certified in accordance with reliable third-party standards that include requirements of deforestation-free production. (Orkla, 2021)

Paulig, the group that includes Santa Maria AB, mentions deforestation 26 times. They state that they 'have identified biodiversity and deforestation as areas where we do not have sufficient understanding of our impacts and the related risks deeply enough' and are therefore planning to map the deforestation risks within their 'main forest risk commodities value chains'. During 2022 they have initiated a deforestation policy indicating that they 'will commit to working collaboratively to eliminate commodity-related deforestation from our supply chains as part of our sustainability approach' (Paulig, 2021).

When it comes to policies on farming, companies refer to topics such as how chemicals (such as pesticides) are used, regenerative agriculture, ways of preserving biodiversity (e.g. use of flowering corridors) and use of sustainability certification for grain crops.

Nine companies mention the secondary theme: regenerative agriculture/farming practices. Such accounts are for example, about research on regenerative methods in milk production. Regenerative practices include improving water and energy footprints (e.g. through more efficient use), reducing reliance on external inputs (such as animal feed) in favour of grazing, improving pasture quality (e.g. through more diverse grass mixes), promoting soil health, improving soil carbon sequestration, encouraging plant diversity and improving animal health (which contributes both to welfare and more productive animals producing fewer greenhouse gas (GHG) emissions such as methane). One company describes a newly started project where they will study regenerative diary.

> Along with 24 pilot farmers, we also set out to explore regenerative dairy farming practices and document their impact on climate, animal care, biodiversity and soil health (Arla, 2021).

In their sustainability report another company states they will support farmers to increase carbon sequestration through regenerative farming to transform farmland from a carbon source to a carbon sink (Fazer, 2021). Another company undertakes not just to encourage regenerative techniques but to document their impacts on nature and climate (Orkla, 2021). However, the company does not describe specific activities or how impacts will be measured. Not reporting ways to measure performance or results achieved in terms of biodiversity applies consistently to all companies in the study.

The secondary theme 'chemical use in agriculture and on crops' is mentioned by seven companies in their reports. One company mentions that they replacing chemical fertilizer with chicken litter as an organic fertilizer, thereby improving soil health (Scandi_Standard, 2021). Another company mentions that they work to produce non-chemical treated seed (Lantmännen, 2021). Others companies refer generally to environmental impacts due to chemical usage but do not go into detail. Most often chemical usage is mentioned in a very general way, e.g. 'The chemicals used in processing can cause water pollution' (Orkla, 2021).

The secondary theme 'new methods used in agriculture' are mentioned by seven companies in their reports. Most often it is mentioned in the context of how new methods can contribute to sustainability, whether measures to improve soil health or the use of digital technologies to target supply of feed to better match animal needs (with lower feed requirements, healthier animals and less animal waste) or to monitor animal health better (to improve animal welfare, productivity and reduce emissions) e.g. 'At Viken, Lantmännen's diary and trial farm, the aim is to make diary production even more sustainable through digitalization and new methods for measuring feed consumption and animal health' (Lantmännen, 2021) .

How to manage weeds and to grow flowers for pollinators in corridors between fields is a secondary theme that seems to be a popular way for companies to contribute to biodiversity. Six companies describe in their reports how they encourage flower strips or zones. Some examples are:

> *We made room and planted seeds to give pollinators access to important nectar and pollen (Arvid_Nordquist, 2021)*

> *Many of our farmers have started with flowering field edges (Pågen, 2021)*

> *The project is about getting more farmers to use the outer edges of their pastures and fields for growing flowering plants. The strips of flowers across the landscape are not only nice to look at, but also attract many of the important pollinating insects (Skånemejerier, 2021)*

Sometimes companies state a concern about promoting pollinators and biodiversity more generally but do not specify what activities they undertake.

'Verified grain' is the fifth secondary theme under the primary theme 'actions in farming'. Some companies mention that they make sure that the seeds or grain are locally grown. Four companies mention verified grain in their reports (i.e. grain that has been subject to some form of certification relating to quality or sustainability e.g. organic certification). One company writes in their report that they are:

> *a major player in the organic value chain, from agri-supply, feed and grain to food. The organic grain market has generally been positive with a good price scenario and Lantmännen continues to work actively to create businesses that make a positive difference in this area (Lantmännen, 2021)*

Another company writes that by choosing sustainable and verified grain and setting requirements for how raw materials are grown and processed, they can contribute to biodiversity. Most often verified grain is only something companies say they use but not what verification involves and how they monitor compliance.

8 Reporting themes: actions to improve and monitor biodiversity

A primary theme is 'Actions to improve biodiversity'. Secondary themes in this category include how companies develop methods and frameworks for monitoring and improving biodiversity, as well as specific projects for biodiversity restoration and protection of natural habitats.

Three companies write about developing methods or a framework for improving biodiversity. One is engaged in developing methods, definitions and a framework for sourcing milk produced using regenerative farming practices which contribute to biodiversity (Arla, 2021). A second has a project to draw up a common framework for regenerative agricultural practices and biodiversity (Orkla, 2021). The third company writes that its new framework will enable it to set more ambitious goals for improving biodiversity (Löfbergs, 2021). None of the companies specify what the framework consists of, how it will be implemented and performance measured.

Five companies write about the second secondary theme 'Restoration of biodiversity and ecosystem'. One mentions measures on farms to improve biodiversity and soil health but does not specify what the improvements are (Arla, 2021). Another company states that they wish to restore natural ecosystems and then mentions some general ways to get there e.g. promoting regenerative cultivation practices and measures to increase biodiversity (Orkla, 2021). Another company writes that 'protecting and restoring forests is an essential part of the effort to tackle the loss of biodiversity' (Paulig, 2021).

The third secondary theme concerns how biodiversity and natural habitat can be protected. Eight companies write about protecting biodiversity, all in a general way. One company writes that better sourcing will protect biodiversity and ecosystems and reduce environmental impact (AAK, 2021). Another company states that their 'work to reduce greenhouse gas emissions and water consumption throughout our value chain will also help to preserve nature and biodiversity' (Orkla, 2021).

The fourth primary theme is how companies takes action to measure biodiversity. The first secondary theme in this category is how companies use data. Five companies write about this theme. One example is a company that states how they use digital tools and data to explore how they can 'reduce climate impact of primary production, support biodiversity and animal welfare, and prevent water eutrophication' (H. K. Scan, 2021). This company mentions measures such as more diverse crop rotations as well as the use of precision agriculture techniques to optimize the use of fertilizers and plant protection products.

The second of the secondary themes relates to how companies measure their environmental impact, including their impact on biodiversity. Four companies discuss this theme. H. K. Scan (2021) writes that they were 'first in the world to calculate the water and environmental footprints of chicken'. Orkla (2021) mention that they monitor the impact they have on nature and biodiversity in every part of their value chain. Monitoring is used as a 'basis for further efforts to identify areas for improvement, analyse causality and consider actions to be taken'. Paulig (2021) is part of the Farming of the Future Programme which 'includes specific measures to reduce climate impact from cultivation and boost biodiversity, carried out by contracted farmers on their own farms'.

The third secondary theme is how companies set targets for biodiversity. One of the four companies that discuss this state that they are part of an 'initiative to set targets and action plans to reduce' their 'carbon footprint throughout the value chain' (Cloetta, 2021). What the targets includes are not disclosed but the company state that they will cooperate with their customers and suppliers to find solutions to tackle the challenge. Another company writes that 'for building trust in our work that we set science-based targets for nature and biological diversity, and that we report on them' (Orkla, 2021). The targets themselves are not disclosed.

Although many of the companies discuss biodiversity, the majority of the narrative is generic in nature. Examples include:

> *The most threatening challenges of our time are climate change and the loss of biodiversity and urgent measures are needed to mitigate the related risks (Paulig, 2021)*

> *Climate change, biodiversity loss and widespread social inequality pose major risks to our planet, our society and our business (Findus, 2021)*

> *Safeguarding forests and preventing conversion of native ecosystems is crucial for biodiversity and tackling climate change (AAK, 2021)*

Although these disclosures provide a type of mission statement, and highlight the companies' commitment to enhancing and conserving biodiversity, they are non-specific and do not provide any detail that shows how the company is addressing biodiversity issues in practice.

Eighteen companies mention a general commitment to biodiversity, seventeen mention progress in general while six mention both commitment and progress in connection to biodiversity. This general commitment is often linked to the UN Global Compact to improve corporate commitment to social and environmental sustainability and the UN Agenda 2030 to eradicate poverty. Some companies describe commitments in more detail. One example is from AAK (2021) which describes several targets for biodiversity. Under the heading 'Verified deforestation- and conversion-free' they set the year 2025 to have '100 percent verified deforestation-free and conversion-free palm and soy supply chains'. In 2025 they undertake to have '150 000 trees planted'. Another example is a company which sets sustainable sourcing targets for the amount of certified soy, palm oil and cocoa the company will use (Arla, 2021).

All companies in the study report on risk, which is not surprising since government regulations require Swedish companies to report on significant risks to company operations. Requirements for risk analysis in corporate sustainability reports have also increased recently. What is surprising, however, is that only nine of the companies in the study mention risks in relation to biodiversity, despite reporting requirements and the fundamental importance of biodiversity to sustainable agricultural production (Jones and Solomon, 2013).

One example of risk disclosure connected to biodiversity is a company which, under the heading risk and uncertainties, writes that *'New or changed conditions for running the business, for example, regulations on climate mitigation and adaptation, biodiversity and other sustainability areas, can result in unforeseen costs and require extensive investments'* (Scandi_ Standard, 2021). Another highlights the risk of *'failing to act on climate crisis, deforestation, or biodiversity'* which may *'influence relationships with customers and attractiveness to investors'* (Cloetta, 2021). Action to mitigate risk include more responsible sourcing of raw materials or that the 'planet priority involves work to mitigate climate change and improve energy efficiency' (Atria, 2021).

Jones (2014) stresses the need to account for biodiversity, including taking risk into account. The fact that so few Swedish companies in the agri-food sector identify biodiversity-related risks and then set specific priorities to deal with these risks confirms what previous research has shown: that biodiversity reporting by companies is limited, unsystematic and partial (Atkins et al., 2014).

9 Conclusion

This study only analysed reports from 2021. It is important to compare companies over time to see if reporting changes. This review of company reports has identified seven primary themes and sixteen secondary themes concerning biodiversity. Of the 25 companies, 21 that published a publicly available report mentioned biodiversity or biodiversity-related issues. Four of them have a clearly stated biodiversity commitment.

The companies examined in the document study are those that all have a direct impact on biodiversity. Many of the companies active in the agricultural sector describe in their reports how they make special efforts to support biodiversity e.g. by encouraging planting of floral field margins and corridors.

Companies in the agricultural sector all have a direct impact on biodiversity but in different contexts, which is why it is difficult to establish common benchmarks or to compare company policies, targets and performance. There are no common regulations or guidelines for reporting of biodiversity which makes it difficult for companies to report in a comparable way. Every company chooses its own way to report, making it difficult to compare or assess the information that companies provide. Swedish companies also express uncertainty about how biological diversity can be measured in a satisfactory and comparable manner.

A lack of a framework and a common definition or measure also makes it difficult to set clear, comparable and measurable goals. Standardized methods for measurement and follow-up would lead to increased transparency and the ability to compare companies. At a global level, various initiatives are underway to develop methods for measuring and assess biodiversity. There are also initiatives to find benchmarks against which to track progress (Lammerant et al., 2021).

Despite an increase in sustainability reporting, the issue of biodiversity still seems to be an under-prioritized theme for many companies since only a minority of the companies that produce a sustainability report review their impact on biodiversity. However, there seems to be growing acknowledgement of the importance of biodiversity in the business community. Some companies also mention the important work of non-governmental organizations (NGOs) in highlighting biodiversity issues, suggesting solutions and working with companies to help implement them.

10 Appendix

Themes

A) Safeguarding forest
- Work against deforestation;
- Verified palm oil/soy; and
- Locally produced beans to replace palm oil/soy.

B) Actions in farming
- Regenerative agriculture /farming;
- Less chemical use in farming;
- New methods/practices in farming to preserve biodiversity;
- Flowering corridors; and
- Verified grain.

C) Actions to improve biodiversity
- Developing method and framework for biodiversity
- Restoration of biodiversity and echo system; and
- Protecting biodiversity and natural habitat.

D) Actions to measure biodiversity
- Data input to work with/promote biodiversity;
- Measure impact on biodiversity in value chain; and
- Targets for biodiversity.

E) Generally, about biodiversity
- General comments and information; and

- Increased awareness of biodiversity.

F) Commitment and progress

G) Company risk and priorities

11 References

AAK (2021). Sustainability Report, Available at: https://www.aak.com/sustainability/sustainability-reports/.

Ahrné, K. and Bjelke, U. (2022). *Sammanfattning Rödlista 2020*. Artdatabanken, S. and Artdatabanken, S. L. U., Available at: https://www.artdatabanken.se/var-verksamhet/rodlistning/Sammanfattning-rodlista-2020/.

Arla (2021). *Sustainability Report*, Available at: https://www.arla.com/company/responsibility/csr-reports/.

Artdatabanken (2022). Sveriges Lantbruksuniversitet. Retrieved 06. 05. 2022. Available at: https://artfakta.se/

Arvid_ Nordquist (2021). *Sustainability Report*, Available at: https://www.arvidnordquist.se/om-arvid-nordquist/hallbarhet/.

Atkins, J., Gräbsch, C. and Jones, M. (2014). Corporate biodiversity reporting: exploring its anthropocentric nature. In: Jones, M. J. (Ed.). *Accounting for Biodiversity*. Routledge, New York.

Atria (2021). *Sustainability Report*, Available at: https://www.atria.fi/en/group/corporate-responsibility/corporate-responsibility-reporting/.

Cederberg, C., Landquist, B., Tidåker, P. and Molander, S. (2016). *Tjänster och gentjänster - hur kan ekosystemtjänster inkluderas i jordbruksföretagens hållbarhetsarbete?* SIK, Institutet för Livsmedel och Bioteknik, Available at: https://login.lantbruksforskning.se/sbs/projectbank/downloadPb?appFormId=402880f6485f65ff01485f670ce60b89.

Cloetta (2021). Annual and Sustainability Report, Available at: https://www.cloetta.com/en/annual-and-sustainability-report-2021/.

Elmqvist, H. (2021). Multifunctional buffer zones - an innovative method for optimization of environmental goals and production goals, Available at: https://www.odlingibalans.com/projekt/samzon-37709980. Swedish University of Agricultural Sciences, 2016-4866.

Fazer (2021). *Annual Review*, Available at: https://www.fazergroup.com/financials/annual-report-2021/.

Findus (2021). *Sustainability Report*, Available at: https://www.findus.se/hallbarhet/hallbarhetsrapporter/hallbarhetsrapport-2021.

Giller, K. E., Hijbeek, R., Andersson, J. A. and Sumberg, J. (2021). Regenerative agriculture: an agronomic perspective. *Outlook on Agriculture* 50(1): 13.

H. K. Scan (2021). *Annual and Responsibility Report*, Available at: https://www.hkscan.com/globalassets/hkscan.com/annual-report-2021---vuosikertomus-2021/hkscan_annual_and_responsibility_report_2021.pdf.

Hedlund, K., Brady, M., Hansson, H., Hristov, J., Alkan Olsson, J., Smith, H. G. and Wilhelmsson, F. (2017). *Värdering av Ekosystemtjänster Inom Jordbruket - för Effektivt Beslutsfattande, Naturvårdsverket, Rapport; Nr. 6753*, Available at: https://lucris.lub.lu.se/ws/portalfiles/portal/56830747/VESA_Rapport_2017.pdf.

IFC (2022). The relationship of business to biodiversity. Retrieved 05. 06. 2022, Available at: https://www.ifc.org/wps/wcm/connect/topics_ext_content/ifc_external_corporate_site/sustainability-at-ifc/publications/biodiversityguide_understanding_business.

IPBES (2019). *Global Assessment Report on Biodiversity and Ecosystem Services of the Intergovernmental Science-Policy Platform on Biodiversity and Ecosystem Services*. E. S. Brondizio, J. Settele, S. Díaz and H. T. Ngo, Bonn, Germany.

International Integrated Reporting Council. *The International <IR> Framework*.

Johansson, H., Mattsson, E. and Perjo, L. (2022). Näringslivets arbete med biologisk mångfald. En kartläggning av företagens strategiska arbete och uppföljning, Available at: https://www.naturvardsverket.se/om-oss/publikationer/7000/978-91-620-7039-7/. Naturvårdsverket, 978-91-920-7039-7.

Jonäll, K. (2021). Företagen och biologisk mångfald. In: Tunón, H. and Sandell, K. (Eds). *Biologisk Mångfald, Naturnyttor, Ekosystemtjänster. Svenska Perspektiv På Livsviktiga Framtidsfrågor*. SLU Centrum för biologisk mångfaldm Uppsala, & Naturvårdsverket, Stockholm.

Jones, M. J. (Ed.) (2014). *Accounting for Biodiversity*. Routledge, New York.

Jones, M. J. and Solomon, J. F. (2013). Problematising accounting for biodiversity. *Accounting, Auditing and Accountability Journal* 26(5): 668.

KPMG (2020). The time has come. The KPMG survey of sustainability reporting 2020. KPMG Impact. R. Threlfal, A. Kung, J. Shilman and W. Bartels, Available at: https://home.kpmg/sustainabilityreporting.html.

Lammerant, J., Driesen, K., Vanderheyden, G., Starkey, M., De Horde, A. and Bor, A.-M. (2021). *Assessment of biodiversity measurement approaches for businesses and financial institutions, update report 3 on behalf of the EU Business @ Biodiversity Platform*. European Union Business @ Biodiversity Platform, Available at: https://ec.europa.eu/environment/biodiversity/business/assets/pdf/EU%20B@B%20Platform%20Update%20Report%203_FINAL_1March2021.pdf

Lantbrukarnas_Riksförbund (2022). *Regenerativt Jordbruk: Biologisk Mångfald och Jordhälsa I Fokus* Retrieved 06. 06. 2022, Available at: https://www.lrf.se/om-lrf/organisation/branschavdelningar/lrf-mjolk/aktuellt-fran-lrf-mjolk/regenerativt-jordbruk-biologisk-mangfald-och-jordhalsa-i-fokus/.

Lantmännen (2021). *Annual Report with Sustainability Report*, Available at: https://www.lantmannen.com/siteassets/documents/01-om-lantmannen/finansiell-information/annual-report-with-sustainability-report-2021.pdf.

Largestcompanies.se (2022). The largest companies by turnover in Sweden. Retrieved 05. 06. 2022, Available at: https://www.largestcompanies.se/topplistor/sverige/de-storsta-foretagen-efter-omsattning/bransch/livsmedelsframstallning?includeChildCompanies=True.

Larsson, C., Boke-Olén, N. and Brady, M. (2020). *Naturbetesmarkens framtid - En Fråga Om Lönsamhet*. AgriFood Economics Centre, Lund, Available at: https://agrifood.se/Files/AgriFood_Rapport20201.pdf.

Lindborg, R., Lennartsson, T. and Smith, H. G. (2021). Naturbetesmarker - en resurs för biologisk mångfald och ekosystemtjänster. In: Tunón, H. and Sandell, K. (Eds). *Biologisk Mångfald, Naturnyttor, Ekosystemtjänster. Svenska Perspektiv På Livsviktiga Framtidsfrågor*. Saint Louis University, Uppsala & Stockholm Centrum för biologisk mångfald & Naturvårdsverket.

MacLaren, C., Storkey, J., Menegat, A., Metcalfe, H. and Dehnen-Schmutz, K. (2020). An ecological future for weed science to sustain crop production and the environment. A review. *Agronomy for Sustainable Development* 40(4).

Naturvårdsverket (2022). Environmental objectives. Retrieved 24. 05. 2022, Available at: https://www.naturvardsverket.se/en/environmental-work/environmental-objectives/.

Olsson, R. (2008). Mångfaldsmarker: naturbetesmarker - en värdefull resurs. Centrum för Biologisk Mångfald, Uppsala.

Orkla (2021). *Annual Report*, Available at: https://investors.orkla.com/English/financial-reports/annual-reports/default.aspx.

Palme, U. (2021). Att bedöma produkters påverkan på biologisk mångfald, Uppsala & Naturvårdsverket, Stockholm. In: Tunón, H. and Sandell, K. (Eds). *Biologisk Mångfald, Naturnyttor och Ekosystemtjänster. Svenska Perspektiv På Livsviktiga Framtidsfrågor*. SLU Centrum för biologisk mångfald.

Paulig (2021). *Annual Report*, Available at: https://www.pauliggroup.com/sites/default/files/2022-04/Paulig%20Annual%20Report%202021.pdf.

Pågen (2021). Sustainability Report, Available at: https://pagen.se/vi-bakar-med-karlek/vart-hallbarhetsarbete/.

Roberts, L., Hassan, A., Elamer, A. and Nandy, M. (2021). Biodiversity and extinction accounting for sustainable development: A systematic literature review and future research directions. *Business Strategy and the Environment* 30(1): 705.

Scandi (2021). *Annual Report*. Standard, Available at: https://investors.scandistandard.com/afw/files/press/scandistandard/202203291018-1.pdf.

Skånemejerier (2021). *Sustainability Report*, Available at: https://foretag.skanemejerier.se/content/uploads/2022/05/Skanemejerier_HR_2021_ENG.pdf.pdf.

Smith, M. and Taffler, R. J. (2000). The chairman's statement: A content analysis of discretionary narrative disclosures. *Accounting, Auditing and Accountability Journal* 13(5): 624.

Threlfal, R., King, A., Shulman, J. and Bartels, W. (2020). The time has come. The KPMG survey of sustainability reporting 2020, Available at: https://home.kpmg/xx/en/home/insights/2020/11/the-time-has-come-survey-of-sustainability-reporting.html. KPMG.

Wallander, J., Karlsson, L., Berglund, H., Mebus, F., Nilsson, L., Bruun, M. and Johansson, L. (2019). Plan för odlingslandskapets biologiska mångfald. Ett samverkansprojekt inom Miljömålsrådet. *Jordbruksverket* 2019: 1, Available at: https://www2.jordbruksverket.se/download/18.36d57baa168c704154d46f04/1549611543321/ra19_1.pdf.

Chapter 9

Biodiversity reporting in the fast-food industry

Dusan Ecim and Warren Maroun, University of the Witwatersrand, South Africa

1 Introduction

The global food industry is a significant driver of biodiversity loss (Mansoor and Maroun, 2016; Benton et al., 2021). According to the International Union for Conservation of Nature (IUCN), unsustainable farming practices currently threaten an estimated 24 000 species with the risk of extinction (Benton et al., 2021).

Current systems of production and consumption depend on producing more food at a lower cost. This 'cheaper food' paradigm is at the heart of the fast-food industry (Benton et al., 2021).

Fast food refers to convenient food made and served at quick service restaurants and street vendors. Fast food is usually less expensive than alternatives but is also less nutritious. The global fast-food market, valued at US$ 647.7 billion in 2019, is expected to reach US$ 1044.61 billion by 2027, but this is before taking into consideration the sector's health and environmental impacts (Global Fast Food Market, 2021). These include, for example,

- intensified agricultural production which degrades soils and, paradoxically, lowers the productive capacity of arable land;
- loss of biodiversity, including deforestation, because of the conversion of wilderness areas into farms;
- the emergence of new diseases and the more rapid spread of existing diseases because of habitat destruction and concentration of livestock;

http://dx.doi.org/10.19103/AS.2023.0128.22

- the direct and indirect cost of the energy needed to produce processed foods;
- toxic waste in the form of wrappers, boxes and single-use plastics;
- greenhouse gas emissions from animal rearing;
- food wastage and the breakdown in nutrient recycling; and
- increased water contamination because of irresponsible use of fertilisers in the growing of crops, chemicals in processing facilities, and antibiotics for animal rearing.

(National Biodiversity and Business Network, 2021; Benton et al., 2021; Wolff et al., 2017; Helm, 2020; PwC, 2022)

While biodiversity-related impacts and reporting have received significant attention from both the academic (see, for example, Blanco-Zaitegi et al., 2022; Büchling and Maroun, 2021; Hassan et al., 2020; Maroun et al., 2018) and practitioner communities (see, for example, National Biodiversity and Business Network, 2021; Benton et al., 2021; ISSB, 2021), few studies deal directly with the impact of the fast-food industry on biodiversity and related reporting (see Chan and Satterfield, 2013; Sponsel, 2001; Igoe et al., 2010 for examples of fast-food industry research). This chapter outlines the biodiversity-related risks applicable to the sector by examining how McDonald's Corporation (McDonald's), Yum! Brands Inc. (Yum!), and Restaurant Brands International Inc. (RBI) report on their biodiversity impacts and proposes an 'integrated thinking' model for managing biodiversity responsibly.

2 Background

'Biological diversity' or 'biodiversity', according to the United Nations Convention on Biological Diversity (CBD and U, 1992, p.5), refers to the

> 'variability among living organisms from all sources including, inter alia, terrestrial, marine and other aquatic ecosystems and the ecological complexes of which they are part: this includes diversity within species, between species and of ecosystems'.

Given the concerns about the state of the environment and related social challenges, organisations are coming under increasing pressure to apply an integrated thinking philosophy which balances financial imperatives with extra-financial considerations, including biodiversity conservation (Atkins et al., 2020; Tirado-Valencia et al., 2021; Malafronte and Pereira, 2021). The integrated thinking logic focuses on the close interconnection between long-term financial performance and the importance of managing social and environmental capitals responsibly. Stated simply, an organisation cannot generate financial returns sustainably for its investors if it ignores social and environmental inputs on which its business model is directly and indirectly dependent. Biodiversity is a key example (Buchling and Atkins, 2020; Wolff et al., 2017; De Villiers et al., 2020).

The agri-food industry impacts biodiversity and is well-placed to address these impacts. For example, several researchers propose a more sustainable food production cycle which operates within the bounds of greenhouse gas, energy, and waste contaminant limits and promotes local production to lower biodiversity impact (Soussana, 2014; Crenna et al., 2019; Schmitt et al., 2017). Others call for a change in consumer eating preferences. Plant-based diets place fewer demands on food production, lead to lower greenhouse gas emissions, and are associated with improved health (Castañé and Antón, 2017; Karlsson Potter and Röös, 2021). Biodiversity-related concerns can also be addressed by integrating new technology with sustainable farming practices (Head et al., 2014; Karlsson Potter and Röös, 2021; de Lima Vieira et al., 2021; PwC, 2022).

A combined effort from policymakers and stakeholders will drive a more responsible approach to addressing biodiversity concerns throughout the supply chain (Boccia et al., 2019). Organisations should first aim to achieve a 'no-net-loss' position where the biodiversity gains from mitigation activities match the biodiversity losses resulting from business activity (National Biodiversity and Business Network, 2021). A no-net-loss position minimises biodiversity impacts, restores affected areas to their original condition, and offsets residual impacts (Baker et al., 2019). This policy should, at a minimum, be the standard practice.

Once a no-net-loss policy is achieved by an organisation, core activities can be enhanced to achieve a 'net-gain' position in terms of which biodiversity gains exceed biodiversity losses (National Biodiversity and Business Network, 2021). Although a net-gain position is difficult to maintain, organisations should aim to leave biodiversity 'in a better state than before' and develop 'an approach where developers work with local governments, wildlife groups, landowners and other stakeholders to support their priorities for nature conservation' (Baker et al., 2019, p.16). In this way, the organisation can ensure that it has a long-lasting sustainable impact on the environment in which it operates (refer to Baker et al., 2019 for good practice principles to achieve a net gain).

Given the globalisation and integration of food supply chains across multiple jurisdictions, the growing number of consumers embracing environmentally sensitive practices and increased stakeholder scrutiny, biodiversity needs to be a strategic priority. Biodiversity must be embedded in an organisation's policies, the manner in which its operations are executed and how its business processes are evaluated (Maroun et al., 2023; Rossi and Luque-Vílchez, 2021; King et al., 2022). This type of integrated approach to biodiversity needs to be mindful of the industry-specific issues relating to the organisation, how the organisation addresses these issues and guidance provided by codes of best practice. In the interest of transparency and to enable accountability, how the organisation addresses its biodiversity impact needs to be communicated clearly to the relevant stakeholders (PwC, 2022).

2.1 Biodiversity practices and issues in the context of the fast-food industry

Four factors form the basis for an integrated thinking approach to managing biodiversity. The first deals with emissions, healthy forests, and access to safe water and soil quality (Osei-Owusu et al., 2021; Notarnicola et al., 2012). For example, reforestation projects and moving to paper-based packaging can play an important role in protecting the environment (Boccia et al., 2019). The responsible use of fertiliser, coupled with other sustainable farming techniques, is important for maintaining productive soils and avoiding nutrient and chemical run-offs into rivers (Auerbach, 2018; Boccia et al., 2019). New technologies should be explored and implemented to improve agri-processing, streamline distribution networks, and contribute to the lowering of greenhouse gas emissions throughout the value chain (de Lima Vieira et al., 2021).

The second factor is concerned with responsible sourcing of food. The fast-food industry is well placed to drive sustainable farming practices which include animal health and welfare. Possible interventions include

- using soy and palm oil only when it has been certified as being sustainably farmed;
- insisting on minimum standards for animal rearing and egg laying which ensure the ethical treatment of animals while minimising adverse environmental impacts;
- sourcing important foodstuffs (such as coffee and cocoa) from producers committed to conserving biodiversity and paying fair wages to farm workers; and
- relying on seafood suppliers who practise sustainable fishing techniques, including the screening of species which are under threat.

(de Lima Vieira et al., 2021; Boccia et al., 2019; Notarnicola et al., 2012)

The third factor deals with people and communities. An established culture of sustainability and biodiversity must form a core part of the organisation's strategy and operations and be integrated into the conduct of individual employees. High-quality training, benchmarking with peers, and engagement with researchers are essential for understanding biodiversity-related risks and equipping individuals to take mitigating action. Governance structures must be put in place to monitor activities, assess changing risk profiles, and hold an organisation's managers accountable for biodiversity conservation (Maroun and Atkins, 2021; King et al., 2022).

For example, an environmental committee can be established to direct the assessment of biodiversity risks, establish risk tolerance levels, and conclude on the adequacy of management's responses (Miras-Rodriguez et al., 2019; Tricker, 1984). Partnerships can be formed with non-governmental organisations, think tanks, co-operatives, and farming associations to share best practices, conduct joint research, and implement more sustainable production processes throughout the value chain (Büchling and Maroun, 2021; Atkins and Maroun, 2018; Raji and Hassan, 2021). Suitable key performance indicators (KPIs) should be established which specifically incorporate biodiversity impact. Variances should be carefully monitored by managers, reviewed by assurance providers and form part of the organisation's overall governance of business risk (Dumay and Dai, 2017; Oliver et al., 2016; Van Zijl et al., 2017).

The customer is the fourth consideration. The cost of food is an important consideration but so is the nutritional value of what consumers are eating and the indirect impact on the environment. If food is more nutritious, it requires less processing and contributes to improved health (Castañé and Antón, 2017). Consumers should be made aware of how food has been sourced and produced, the natural resources which were consumed to supply that food and the fact that sustainability standards have been adhered to in each part of the value chain (Head et al., 2014).

Transparent reporting is the final part of an integrated approach to managing biodiversity. Communicating with stakeholders is essential for maintaining key stakeholders' confidence in the sector, educating consumers and providing a feedback mechanism which can be used to drive further changes to production and governance processes. Reporting is explained in more detail in the following section.

2.2 Reporting on biodiversity

To yield any benefits, biodiversity reporting must be genuine. It cannot be treated as a compliance-driven exercise or a means for managing impressions (Haji and Anifowose, 2016). While biodiversity must be factored into an organisation's strategy, mission statements and a broader approach to risk identification, it must also be incorporated into how the organisation is managed (Maroun et al., 2018; King et al., 2022). Management should report on its achievements, but challenges encountered, additional risks identified, and weaknesses in existing production processes should also be explicated. This should be complemented by details on planned remedial action, post-implementation reviews, and the results of formal assurance and other monitoring activities (Donkor et al., 2021).

Several frameworks provide more detailed guidance on reporting on biodiversity. Examples include the Biological Diversity Protocol, Global Biodiversity Framework, Global Reporting Initiative (GRI), Sustainable Development Goals (SDGs), and the Task Force on Nature-Related and Climate-Related Financial Disclosures. Each can be applied to the fast-food industry and complemented by an extinction accounting framework (refer to Atkins and Maroun, 2018), which deals specifically with the loss of species. The aim is not to provide generic detail in corporate reports but to explain how biodiversity impacts the business model. For example, the loss of species can have serious implications for the quality and variety of food which can be supplied, the reliability of those supplies, and their costs (Hassan et al., 2020; Atkins et al., 2019).

Organisations can rely on narrative details, graphs, and images to contextualise the relevance of biodiversity for their business models and the associated risks and opportunities (Mansoor and Maroun, 2016; Atkins and Maroun, 2020; Sullivan and Hannis, 2017). Although assigning a monetary value to natural capital results in technical and moral challenges, stakeholders may be unable to understand the environmental impact of biodiversity loss from only a non-financial perspective (Atkins and Maroun, 2018). In practice, a mix of qualitative, quantitative, and monetary disclosures will be required to provide a complete and reliable account of how biodiversity is being managed.

Assurance can be used to bolster the utility of biodiversity reporting. Internal auditors working in conjunction with external assurance experts can attest to the accuracy, completeness, and validity of the information being included in different types of sustainability reports (Prinsloo and Maroun, 2021). This can be complemented by details in the internal controls, monitoring activities, and review functions overseen by managers and governing bodies. The integration of these internal systems, internal assurance functions, and the work of external assurance providers is sometimes referred to as 'combined assurance'.

3 International case studies of biodiversity reporting in the fast-food industry

To assess the state of biodiversity reporting in the fast-food industry, three prominent global fast-food companies were selected: McDonald's, Yum!, and RBI. McDonald's and Yum! rank first and second, respectively, in terms of market capitalisation globally, while RBI ranks fourth. Chipotle Mexican Grill, with a market capitalisation of US$43.03 billion, is ranked third. As the company operates in limited jurisdictions (United States, United Kingdom, Canada, Germany, and France), none of which are developing economies, it has not been included in the analysis. Refer to Table 1.

Table 1 Breakdown of fast-food chains assessed for the study

Global chain	Fast-food brands	Market capitalisation, US$ billion (per Google Finance as on October 2022)	Country base
McDonald's Corporation	McDonald's	187.28	USA
Yum! Brands Inc.	KFC, Pizza Hut, Taco Bell, The Habit Burger Grill	31.59	USA
RBI Inc.	Tim Hortons, Burger King, Popeyes, Firehouse Subs	17.36	Canadian-American

3.1 Method

The research was conducted in two phases.

3.1.1 Phase 1

A search was performed on the Scopus database to establish all recent academic literature relating to biodiversity. Scopus is the largest database of peer-reviewed academic literature and is recognised for the quality of its filtering criteria and the fact that it includes journals of good standing with robust peer-review processes in place (Dumay et al., 2016; Rinaldi et al., 2018; Ecim and Maroun, 2023).

'Biodiversity' was searched in the title, abstract, and keywords of academic papers. A total of 885 sources were identified. These were subject to the following filters:

- Only academic papers published from 2013 to 2022 were considered.
- Papers had to deal with business management and accounting related topics.
- The papers must have food (including food production and processing) as a core topic.

The final sample of 30 papers was analysed and used to determine key biodiversity themes, using VosViewer software (Van Eck and Waltman, 2017). The top 10 keywords by occurrence and link strength[1] are indicated in Table 2.

The keywords in Table 2 were used to develop a coding checklist which is summarised in Table 3. The keywords were complemented by the details in the Biodiversity Disclosure Project (see National Biodiversity and Business Network,

[1] The higher link strength is indicative of the number of cited references two items have in common, showing a stronger link between this keyword and biodiversity-related research.

Table 2 Core themes associated with biodiversity research

Keyword	Occurrences	Total link strength
Biodiversity	443	6514
Sustainable development	139	2141
Sustainability	95	1087
Conservation	78	1138
Climate change	70	1111
Ecosystems	68	1269
Ecotourism	66	747
Land use	59	1161
Environmental impact	53	999
Life cycle	52	1134

Table 3 Outline of biodiversity checklist

Question	Biodiversity reporting indicator
Q1	Does the company disclose its biodiversity impacts and have an action plan/ policy?
Q2	Does the company measure its biodiversity dependencies and impacts?
Q3	Does the company have a biodiversity strategy, biodiversity targets and associated KPIs?
Q4	Does the company disclose its biodiversity risks and performance?
Q5	Does the company have a biodiversity monitoring system in place for continuous improvement?
Q6	Is there a separate environmental/social committee dealing with biodiversity impacts?
Q7	Does the organisation have well-developed management control systems and information systems in place to collect and process data on environmental and biodiversity impact and performance?
Q8	Is a combined assurance model which relies on a mix of internal and external sources of assurance to address biodiversity disclosures/environmental risks in place?
Q9	Has the organisation disclosed how it has adapted quickly to changes in the external environment and environmental challenges?
Q10	Does the company make use of well-established ESG frameworks/guidelines to support the biodiversity assessment, evaluation and reporting including: • GRI • Sustainable Development Goals • IIRC framework • Capital coalition • Accountability • GHG protocol • Task Force on Climate-Related Financial Disclosures • Other

Table 4 Rating scale applied

0	No relevant disclosure or evidence of response
1	Boilerplate/superficial disclosure of interactions with biodiversity without any substantiation of relevance and/or application to value creation
2	Basic disclosure of interactions with biodiversity provided with a focus on impact avoidance and/or minimisation
3	Good level of disclosure of interactions with biodiversity provided with a focus on no-net-loss impacts
4	Comprehensive disclosure of interactions with biodiversity provided with a focus on net-gain/positive impacts

2021). The final checklist was calibrated using a limited number of interviews with environmental, social, and governance (ESG) specialists to ensure that no essential issues were omitted.

3.1.2 Phase 2

Phase 2 involved a content analysis of each organisation's corporate reports and webpages from 2019 to 2021 guided by the questions in Table 3. The information (irrespective of its source) was scored for each of the 10 questions using the scale in Table 4.

Coding sources required professional judgement to ensure that researcher bias and unintended restrictions on the scope of the analysis were avoided (informed by Rinaldi et al., 2018; de Villiers et al., 2019). As additional information from organisations was identified, disclosures from other organisations were revisited and revised. The scores for each year and organisation were allocated by the lead researcher and confirmed by a second coder to ensure validity and reliability of results. Nevertheless, the coding is inherently subjective and results should be interpreted accordingly.

3.2 Findings and analysis

3.2.1 Phase 1: Core biodiversity themes

Figure 1 illustrates the prevalence of the biodiversity-related keywords presented in Table 2. There is a notable increase in the frequency of the themes being discussed in the reports over a 3-year period across all companies. Yum! provides the highest frequency of biodiversity-related disclosures, while RBI makes the least references to core themes.

Figure 2 provides a more granular assessment. Product life cycle, ecotourism, environmental impact, and sustainable development are never or rarely mentioned. Although sustainable/sustainability is frequently referenced,

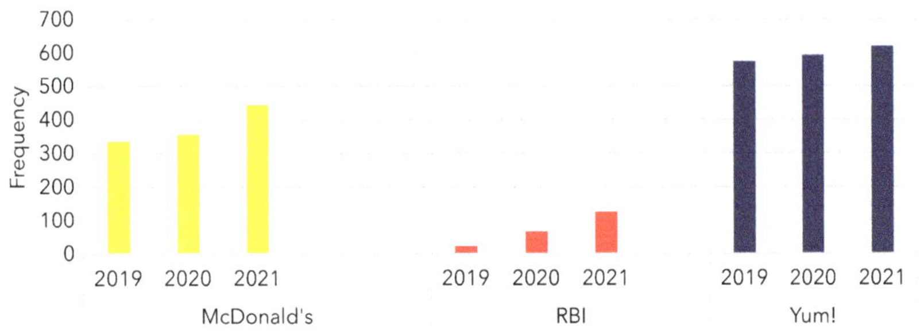

Figure 1 Prevalence of biodiversity-related keywords.

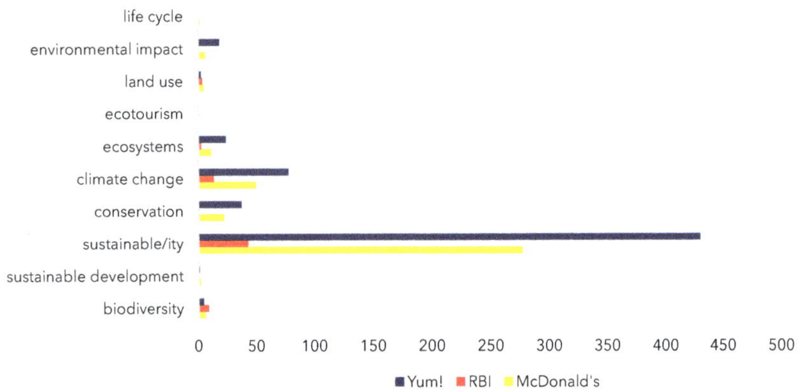

Figure 2 Average frequency of biodiversity-related keywords.

this is often linked to boilerplate disclosures rather than an active internalisation and communication of sustainable impacts.

The next section analyses the application of the biodiversity checklist to the respective disclosures of each company.

3.2.2 Phase 2: Biodiversity checklist scores

McDonald's, Yum!, and RBI address biodiversity-related concerns to various degrees. Yum! scored highest (68%) across the 10 questions in Table 3 followed by McDonald's (55%) and RBI (51%). The disclosure scores were consistent over a 3-year period (un-tabulated). These scores align with the frequency of disclosures presented in Fig. 1. To provide additional insights, the scores are analysed per question over a 3-year average. Results are presented in Fig. 3.

All three companies included a good level of disclosure on their biodiversity action plan/policy (Q1, score = 3). Each company included a broad range of plans on or in sustainability reports, biodiversity-specific policies and websites. Summaries of various initiatives and the associated outcomes were presented.

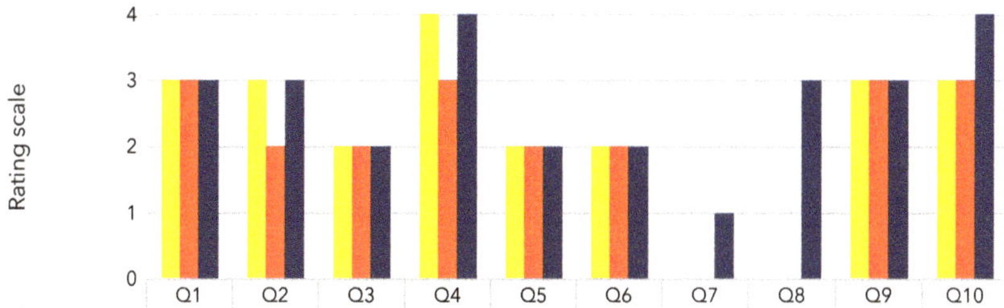

	Q1	Q2	Q3	Q4	Q5	Q6	Q7	Q8	Q9	Q10
McDonald's	3	3	2	4	2	2	0	0	3	3
RBI	3	2	2	3	2	2	0	0	3	3
Yum!	3	3	2	4	2	2	1	3	3	4

Figure 3 Average 3-year biodiversity indicator scores.

However, the focus was on no-net-loss impacts with minimal emphasis placed on net-gain/positive impacts. The companies are attempting to reduce or limit biodiversity impact but are not yet able to report net gains.

Both McDonald's and Yum! include a good level of disclosure on the measurement of biodiversity dependencies and impacts (Q2, score = 3). Each company has a clear statement explaining material direct and indirect biodiversity dependencies and impacts in different parts of the supply chain. RBI, however (Q2, score = 2), deals mainly with direct impacts and provides less information on indirect impacts.

Biodiversity targets (including KPIs) are addressed to only a limited extent by the three companies (Q3, score = 2). There is minimal disclosure on the link between performance evaluation and natural capital with only basic or high-level targets being set. KPIs are not frequently referred to and it is unclear if they are based on no-net-loss or net-gain targets. As a result, it is difficult to gauge the extent of the integration of KPIs with biodiversity priorities.

In contrast, the companies provide more detail on overall biodiversity performance. McDonald's and Yum! provide most disclosures (Q4, score = 4) on material direct and indirect biodiversity dependencies and impacts and make reference to net-gain strategies which were implemented. RBI has a slightly lower score (Q4, score = 3) but still provides a good level of disclosure on supply chain impacts.

All three companies included only basic disclosures (Q5, score = 2) dealing with monitoring biodiversity impacts and dependencies. The information focuses predominantly on direct impacts rather than indirect ones and is generic rather than context specific. Although there are separate committees dealing with biodiversity impacts, the role, functions, and outcomes are not sufficiently detailed to allow stakeholders to understand how the relative functions are executed. The use of biodiversity experts, ESG specialists, and a dedicated Chief Value Officer or Chief Sustainability Officer is not clearly defined (Q6, score = 2).

Although each company may have robust management control systems over biodiversity, this is not disclosed in detail (Q7, score = 0 or 1). Similarly, McDonald's and RBI provide no details on how combined assurance models are used to ensure the validity, accuracy and completeness of biodiversity-related disclosed (Q8, score = 0). Only Yum! refers explicitly to assurance (Q8, score = 3). It relies on KPMG to assure greenhouse gas emissions and water usage.

All organisations provided a good level of disclosure (Q9, score = 3) on how they responded to external changes and adapted to new industry-specific technology, new sustainable farming practices, and industry-specific biodiversity requirements. COVID-19 environmental responses, in particular, were prevalent throughout the 2020 reports. Buildings have moved to more sustainable designs. Other initiatives include

- equipment which uses less electricity;
- digital menu boards to reduce waste;
- high-efficiency rooftop units to improve heating and cooling;
- recycling and composting food waste;
- low-flow plumbing fixtures; and
- satellite technology to manage deforestation risks.

For each company, reporting is guided by the use of established biodiversity frameworks. McDonald's and RBI made use of the Carbon Disclosure Project (CDP) guidance, Global Reporting Initiative (GRI), and Sustainable Development Goals (SDGs) to support disclosures (Q10, score = 3). Yum! (Q10, score = 4) used the GRI, SDGs, CDP, World Resources Institute/World Business Council for Sustainable Development (WRI/WBCSD), and the Greenhouse Gas Protocol. Yum! is also a member of the New York Declaration on Forests, Tropical Forest Alliance, and UK Roundtable on Sustainable Soy. These bodies monitor legislation worldwide and participate in discussions about how governments and businesses can protect forests.

None of the companies explains succinctly how financial performance is interconnected with material social and environmental issues, including biodiversity impact. To appreciate how the three companies manage biodiversity-related matters, the following sources had to be considered:

- annual reports;
- corporate governance principles;
- Carbon Disclosure Project reports on forests, climate change, and water;
- livestock welfare report/sustainable animal policies;
- purpose and impact progress summaries;
- the Sustainability Accounting Standards Board (SASB) index (tracks energy management, water management, food & packaging waste

Table 5 Summary of best and worst practice disclosures in the fast-food industry

Factor	Disclosure examples: best practice	Disclosure examples: worst practice
Environment	The biodiversity action plan policy is well defined. An internal and external stakeholder biodiversity materiality matrix is developed. Adoption of recognised biodiversity frameworks. Measurement targets are disclosed, shortfalls are explained and performance is evaluated across a multi-timeframe analysis. Combined assurance of ESG-related information is disclosed. Management control systems are defined and explained. A balanced focus of no-net-loss *and* net-gain impacts are disclosed.	Boilerplate or generic ESG disclosures; Limited link to the KPIs of the company; Disclosures scattered across multiple reports that are difficult to aggregate and assess holistically; Only no-net-loss objectives are defined; and Biodiversity focuses only on direct impacts and neglects indirect impacts.
Responsible sourcing	A well-defined vendor code of conduct is available. All aspects of the supply chain and food production process are assessed with clear references to the sourcing of food. Clear animal health and welfare objectives are set out.	Sourcing of food products is vague with no food audits disclosed. The entire supply chain is not analysed. Sourcing unsustainable products is not clearly disclosed.
People and communities	The board structures, roles, responsibilities and performance have been disclosed. Initiatives, training, collaborations and social work are clearly disclosed with defined goals communicated.	No clear responsibility is assigned in the report. The company only focuses on financial stakeholders rather than a broad group of stakeholders across the supply chain.
Customer	Nutritional value is clearly disclosed. Food safety and quality are discussed. Marketing is informed and accurate and considers biodiversity impacts.	The organisation prioritises the 'cheaper food' paradigm. Profits are prioritised over values.

management, food safety, nutritional content labour practices, supply chain management, and food sourcing);

- climate risk and resilience summaries;
- supplier codes of conduct;
- charters of sustainability and corporate responsibility committees;
- codes of conduct; and
- website information.

This made it difficult to score and assess companies' biodiversity impact. On average, each company had approximately 12 documents (excluding various tabs/pages/links on the website), which needed to be analysed to understand biodiversity impact. An integrated report summarising all key information is recommended (see IIRC, 2021). Certain disclosure details are summarised in the following section.

3.3 Biodiversity best and worst practice disclosures

Using the various sources disclosing the biodiversity impact of McDonald's, RBI and Yum!, examples of best and worst practice disclosures have been summarised in Table 5.

4 Conclusion and future trends

This chapter provided a broad overview of the biodiversity impact of the fast-food industry. A review of academic literature was complemented by an analysis of three global fast-food companies' biodiversity reporting suite and responses, namely McDonald's, Yum!, and RBI.

The companies are addressing key issues grouped according to four factors impacting biodiversity which include environmental impacts, responsible sourcing, engaging with people and the community, and focusing on the customer. There has been progress on biodiversity reporting but details are lacking and the focus is on limiting the biodiversity impact through a no-net-loss strategy rather than net-gain strategies. Suggested areas for improvement include

- the integration of biodiversity-related metrics with the organisation's key performance indicators;
- the disclosure of robust management control systems to collect, analyse, monitor and disclose biodiversity data;
- robust governance structures with defined responsibilities to manage risks, implement strategies and evaluate performance;
- adopting a combined assurance model over biodiversity-related information;
- using biodiversity frameworks to support the measurement, evaluation and disclosure of biodiversity-related information; and
- providing a holistic, multi-capital summary of all biodiversity impacts to limit the number of reports that stakeholders need to access to gain information.

To assist with a more holistic approach to biodiversity reporting, a roadmap has been developed to enhance the measurement, evaluation, disclosure and

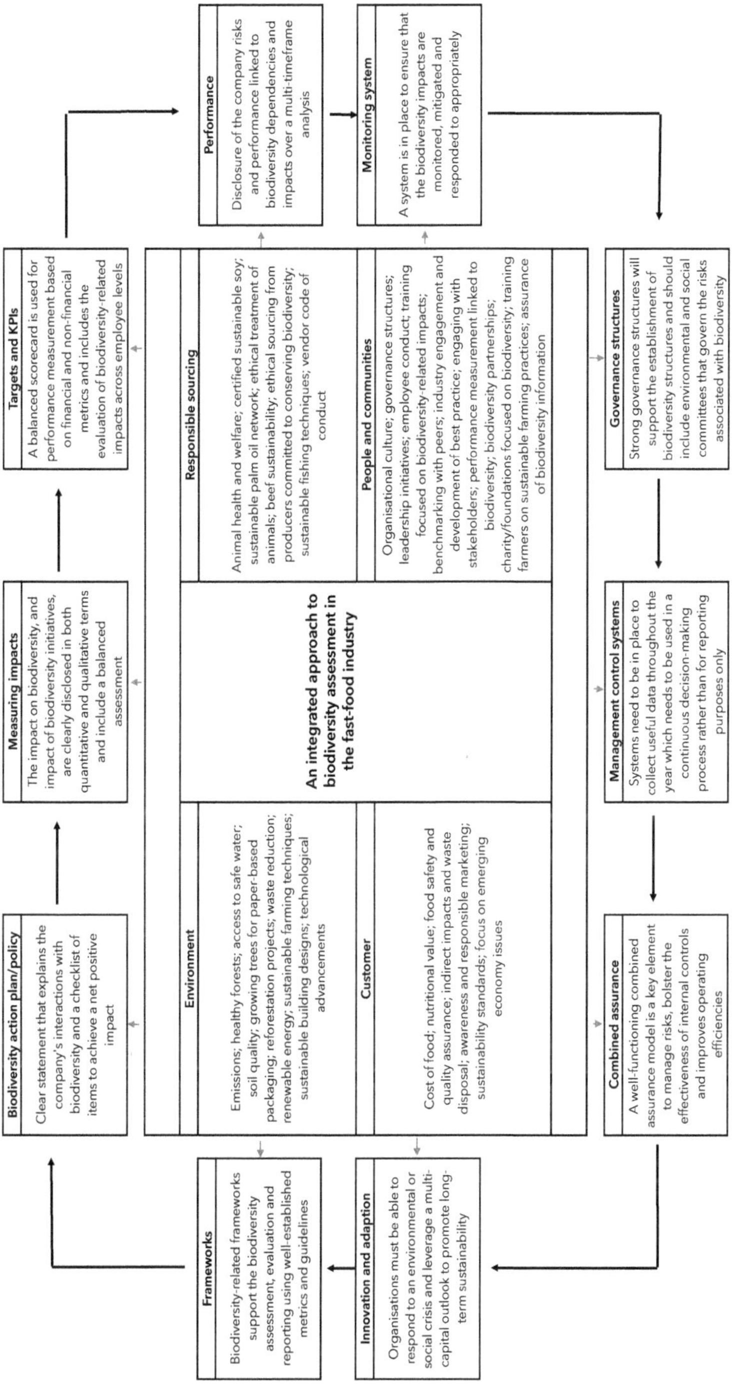

Figure 4 An integrated approach to biodiversity assessment in the fast-food industry.

monitoring of biodiversity impacts in the fast-food industry. This is developed using the core issues noted in the prior research (Section 2.1) in conjunction with the biodiversity checklist (Section 3.1). The approach is summarised in Fig. 4.

Ultimately, the goal is to ensure that organisations are operating in a manner to conserve and preserve biodiversity and to limit environmental footprints. The objective is to develop more sustainable and environmentally sensitive organisations capable of meeting financial needs while balancing a multi-capital outlook.

Future research will be needed to develop more robust biodiversity measurement or evaluation frameworks to measure the level of biodiversity reporting of an organisation. The impact of combined assurance in this process also requires further investigation. This is particularly relevant in light of the risk of impression management noted by earlier researchers (e.g. Leung et al., 2015; Solomon et al., 2013).

The impact of sustainability-related standards such as the International Sustainability Standards Board's (ISSB) sustainability requirements and disclosures (see ISSB, 2021) will need to be assessed in the context of biodiversity management. A more holistic and robust assessment of different biodiversity-related frameworks and guidelines can assist organisations with implementing, evaluating, and monitoring biodiversity impacts. Finally, further research will be required to understand the impact of governance structures and performance incentives on biodiversity issues.

5 Where to look for further information

Further information on biodiversity impacts, including specific issues encountered in the food industry, can be found using the following sources.

The following articles provide a good overview of biodiversity and sustainability reporting:

- Atkins, J. and Maroun, W. 2018. Integrated extinction accounting and accountability: building an ark. *Accounting, Auditing & Accountability Journal* 31(3), 750–786.
- Baboukardos, D., Mangena, M. and Ishola, A. 2021. Integrated thinking and sustainability reporting assurance: International evidence. *Business Strategy and the Environment*, 30, 1580–1597.

The following book chapter is useful to understand the link between accounting and biodiversity:

- Buchling, M. and Atkins, J. 2020. Chapter 29: reporting on more than just natural capital. In: De Villiers, C., Hsiao, P.-C. K. and Maroun, W. (Eds). *The Routledge Handbook of Integrated Reporting*. 2 Park Square, Milton Park, Abingdon, Oxon: Routledge.

The following articles deal with food-specific concerns related to biodiversity:

- Blanco-Zaitegi, G., Álvarez Etxeberria, I. and Moneva, J. M. 2022. Biodiversity accounting and reporting: a systematic literature review and bibliometric analysis. *Journal of Cleaner Production* 371.
- Notarnicola, B., Hayashi, K., Curran, M. A. and Huisingh, D. 2012. Progress in working towards a more sustainable agri-food industry. *Journal of Cleaner Production*, 28, 1–8.
- Wolff, A., Gondran, N. and Brodhag, C. 2017. Detecting unsustainable pressures exerted on biodiversity by a company: application to the food portfolio of a retailer. *Journal of Cleaner Production* 166, 784–797.

Finally, the following practitioner guidance provides an overview of sustainability and biodiversity frameworks/guidelines:

- IIRC. 2021. The international integrated reporting framework. Available: https://integratedreporting.org/wp-content/uploads/2021/01/InternationalIntegratedReportingFramework.pdf.
- ISSB. 2021. General requirements for disclosure of sustainability-related financial information prototype. In: CDSB, I., TCFD, VRF, WEF (Ed). CDSB, IFRS, TCFD, VRF, WEF.
- National Biodiversity and Business Network. 2021. *The 2020 Biodiversity Performance Rating of South African Companies*. Midrand South Africa: Endangered Wildlife Trust.
- Baker, J., Hoskin, R. and Butterworth, T. 2019. Biodiversity net gain. Good practice principles for development. CIRIA, Griffin Court, 15 Long Lane, London: Chartered Institute of Ecology and Environmental Management. Available: https://cieem.net/wp-content/uploads/2019/02/C776a-Biodiversity-net-gain.-Good-practice-principles-for-development.-A-practical-guide-web.pdf.

Key research in these areas can be found at the following organisations:

- Value Reporting Foundation. https://www.valuereportingfoundation.org/.
- Global Reporting Initiative. https://www.globalreporting.org/.
- International Union for Conservation of Nature. https://www.iucn.org/.
- Carbon Disclosure Project. https://www.cdp.net/en.

6 References

Atkins, J., Buchling, M., Cerbone, D., Lange, Y., Maroun, W., Kok, M., Ram, A., Sebastian, A., Wadee, Z. and Van Zijl, W. 2020. Accounting, governance and integrated thinking in the context of COVID-19. Available at: https://www.wits.ac.za/media/wits-university/faculties-and-schools/commerce-law-and-management/accountancy/documents

/Accounting%20governance%20and%20integrated%20thinking%20in%20the%20context%20of%20COVID-19%20(FInal).pdf [Accessed 21 May 2020].

Atkins, J. and Maroun, W. 2018. Integrated extinction accounting and accountability: building an ark. *Accounting, Auditing and Accountability Journal* 31(3), 750–786.

Atkins, J. and Maroun, W. 2020. The Naturalist's Journals of Gilbert White: exploring the roots of accounting for biodiversity and extinction accounting. *Accounting, Auditing and Accountability Journal* 33(8), 1835–1870.

Atkins, J., Maroun, W. and Atkins, B. 2019. How can accounting, integrated reporting and engagement prevent extinction? In: Atkins, K. and Atkins, B. (Eds). *Around the World in 80 Species. Exploring the Business of Extinction*. London, UK, Routledge.

Auerbach, R. 2018. Sustainable food systems for Africa. *Economia Agro-Alimentare* 20(3), 301–320.

Baker, J., Hoskin, R. and Butterworth, T. 2019. *Biodiversity Net Gain. Good Practice Principles for Development*. Griffin Court, 15 Long Lane, London, CIRIA. Chartered Institute of Ecology and Environmental Management.

Benton, T. and Bieg, C., Harwatt, H., Pudasaini, R. and Wellesley, L. 2021. Food system impacts on biodiversity loss. *Three levers for food system transformation in support of nature*. Chatham House, London. 1–74.

Blanco-Zaitegi, G., Álvarez Etxeberria, I. and Moneva, J. M. 2022. Biodiversity accounting and reporting: A systematic literature review and bibliometric analysis. *Journal of Cleaner Production* 371.

Boccia, F., Di Donato, P., Covino, D. and Poli, A. 2019. Food waste and bio-economy: A scenario for the Italian tomato market. *Journal of Cleaner Production* 227, 424–433.

Buchling, M. and Atkins, J. 2020. Chapter 29. Reporting on more than just natural capital. In: De Villiers, C., Hsiao, P.-C. K. and Maroun, W. (Eds). *The Routledge Handbook of Integrated Reporting*. 2 Park Square, Milton Park, Abingdon, Oxon OX14 4RN, Routledge.

Büchling, M. and Maroun, W. 2021. Accounting for biodiversity and extinction: the case of South African national parks. *Social and Environmental Accountability Journal* 41(1–2), 66–97.

Castañé, S. and Antón, A. 2017. Assessment of the nutritional quality and environmental impact of two food diets: A Mediterranean and a vegan diet. *Journal of Cleaner Production* 167, 929–937.

CBD and U. 1992. Abgerufen am [Online]. United Nations Convention on Biological Diversity (vol. 5).

Chan, K. M. A. and Satterfield, T. 2013. Justice, equity, and biodiversity. In: Levin, S. A. (Ed.). *Encyclopedia of Biodiversity*. Oxford, Elsevier.

Crenna, E., Sinkko, T. and Sala, S. 2019. Biodiversity impacts due to food consumption in Europe. *Journal of Cleaner Production* 227, 378–391.

De Lima Vieira, J., Dos Santos Nunes, L., Rodrigues De Menezes, F. G., Vieira De Mendonça, K. and Viana De Sousa, O. 2021. An integrated approach to analyzing the effect of biofloc and probiotic technologies on sustainability and food safety in shrimp farming systems. *Journal of Cleaner Production* 318.

De Villiers, C., Dumay, J. and Maroun, W. 2019. Qualitative accounting research: dispelling myths and developing a new research agenda. *Accounting and Finance* 59, 1449–1458.

De Villiers, C., Hsiao, P.-C. K. and Maroun, W. 2020. *The Routledge Handbook of Integrated Reporting*. 2 Park Square, Milton Park, Abingdon, Oxon OX14 4RN, Routledge.

Donkor, A., Djajadikerta, H. G. and Mat Roni, S. 2021. Impacts of combined assurance on integrated, sustainability and financial reporting qualities: evidence from listed companies in South Africa. *International Journal of Auditing* 25(2), 475–507.

Dumay, J. and Dai, T. 2017. Integrated thinking as a cultural control? *Meditari Accountancy Research* 25(4), 574–604.

Dumay, J., Bernardi, C., Guthrie, J. and Demartini, P. 2016. Integrated reporting: A structured literature review. *Accounting Forum* 40(3), 166–185.

Ecim, D. and Maroun, W. 2023. A review of integrated thinking research in developed and developing economies. *Journal of Accounting in Emerging Economies* 13(3), 589–612.

Global Fast Food Market 2021. Global Fast Food Market by Product (Pizza, Pasta, Burgers, Sandwich and Chicken. Asian/Latin American, seafood, and others), By End User (quick-service restaurants (QSRs), fast casual restaurants, and others), By Regional Outlook, Industry analysis report and forecast 2021-2027. In: Research. M. I. S. P. L. K. (Ed.).

Haji, A. A. and Anifowose, M. 2016. The trend of integrated reporting practice in South Africa: ceremonial or substantive? *Sustainability Accounting, Management and Policy Journal* 7(2), 190–224.

Hassan, A. M., Roberts, L. and Atkins, J. 2020. Exploring factors relating to extinction disclosures: what motivates companies to report on biodiversity and species protection? *Business Strategy and the Environment* 29(3), 1419–1436.

Head, M., Sevenster, M., Odegard, I., Krutwagen, B., Croezen, H. and Bergsma, G. 2014. Life cycle impacts of protein-rich foods: creating robust yet extensive life cycle models for use in a consumer app. *Journal of Cleaner Production* 73, 165–174.

Helm, D. 2020. The environmental impacts of coronavirus. *Environmental and Resource Economics* 76(1), 21-38.

Igoe, J., Neves, K. and Brockington, D. 2010. A spectacular eco-tour around the historic bloc: theorising the convergence of biodiversity conservation and capitalist expansion. *Antipode* 42(3), 486–512.

IIRC 2021. The international integrated reporting framework. Available at: https://integratedreporting.org/wp-content/uploads/2021/01/InternationalIntegratedReportingFramework.pdf.

ISSB 2021. General requirements for disclosure of sustainability-related financial information prototype. In: WEF (Ed.). *CDSB, IFRS, TCFD. CDSB, I., TCFD, VRF. VRF, WEF.*

Karlsson Potter, H. and Röös, E. 2021. Multi-criteria evaluation of plant-based foods -use of environmental footprint and LCA data for consumer guidance. *Journal of Cleaner Production* 280.

King, M., Atkins, J. and Maroun, W. 2022. Chapter 2: Extinction governance: establishing a framework for extinction accounting, accountability and finance. In: Atkins, J. and Macpherson, M. (Eds). *Extinction Governance, Finance and Accounting. Implementing a Species Protection Action Plan for the Financial Markets* (1st edn.). Milton Park, Abingdon, Oxon, UK, Routledge.

Leung, S., Parker, L. and Courtis, J. 2015. Impression management through minimal narrative disclosure in annual reports. *The British Accounting Review* 47(3), 275–289.

Malafronte, I. and Pereira, J. 2021. Integrated thinking: measuring the unobservable. *Meditari Accountancy Research* 29(4), 805–822. https://doi.org/10.1108/MEDAR-12-2019-0640.

Mansoor, H. and Maroun, W. 2016. An initial review of biodiversity reporting by South African corporates: the case of the food and mining sectors. *South African Journal of Economic and Management Sciences* 19(4), 592–614.

Maroun, W. and Atkins, J. 2021. A practical application of accounting for biodiversity: the case of soil health. *Social and Environmental Accountability Journal* 41(1–2), 37–65.

Maroun, W., Ecim, D. and Cerbone, D. 2023. Refining integrated thinking. *Sustainability Accounting, Management and Policy Journal*, Ahead of print 14(7), 1–25. DOI: http://dx.doi.org/10.1108/SAMPJ-07-2021-0268.

Maroun, W., Usher, K. and Mansoor, H. 2018. Biodiversity reporting and organised hypocrisy: the case of the South African food and retail industry. *Qualitative Research in Accounting and Management* 15(4), 437–464.

Miras-Rodriguez, M. D. M., Martnez-Martinez, D. and Escobar-Perez, B. 2019. Which corporate governance mechanisms drive CSR Disclosure practices in emerging countries. *Sustainability* 11(1), 61.

National Biodiversity and Business Network 2021. *The 2020 Biodiversity Performance Rating of South African Companies*. Midrand South Africa, Endangered Wildlife Trust.

Notarnicola, B., Hayashi, K., Curran, M. A. and Huisingh, D. 2012. Progress in working towards a more sustainable agri-food industry. *Journal of Cleaner Production* 28, 1–8.

Oliver, J., Vesty, G. and Brooks, A. 2016. Conceptualising integrated thinking in practice. *Managerial Auditing Journal* 31(2), 228–248.

Osei-Owusu, A. K., Wood, R., Bjelle, E. L., Caro, D. and Thomsen, M. 2021. Understanding the trends in Denmark's global food trade-related greenhouse gas and resource footprint. *Journal of Cleaner Production* 313.

Prinsloo, A. and Maroun, W. 2021. An exploratory study on the components and quality of combined assurance in an integrated or a sustainability reporting setting. *Sustainability Accounting, Management and Policy Journal* 12(1), 1–29.

PWC 2022. *The Novel Food Market*. https://www.pwc.com/it/it/publications/assets/docs/pwc-the-novel-food-market.pdf.

Raji, A. and Hassan, A. 2021. Sustainability and stakeholder awareness: A case study of a Scottish university. *Sustainability* 13(8).

Rinaldi, L., Unerman, J. and De Villiers, C. 2018. Evaluating the integrated reporting journey: insights, gaps and agendas for future research. *Accounting, Auditing and Accountability Journal* 31(5), 1294–1318.

Rossi, A. and Luque-Vílchez, M. 2021. The implementation of sustainability reporting in a small and medium enterprise and the emergence of integrated thinking. *Meditari Accountancy Research*, Ahead of print 29(4), 966–984.

Schmitt, E., Galli, F., Menozzi, D., Maye, D., Touzard, J. M., Marescotti, A., Six, J. and Brunori, G. 2017. Comparing the sustainability of local and global food products in Europe. *Journal of Cleaner Production* 165, 346–359.

Solomon, J. F., Solomon, A., Joseph, N. L. and Norton, S. D. 2013. Impression management, myth creation and fabrication in private social and environmental reporting: insights from Erving Goffman. *Accounting, Organizations and Society* 38(3), 195–213.

Soussana, J. F. 2014. Research priorities for sustainable agri-food systems and life cycle assessment. *Journal of Cleaner Production* 73, 19–23.

Sponsel, L. 2001. Human impact on biodiversity, overview. In: Levin, S. A. (Ed.). *Encyclopedia of Biodiversity*. Oxford, Elsevier.

Sullivan, S. and Hannis, M. 2017. "Mathematics maybe, but not money": on balance sheets, numbers and nature in ecological accounting. *Accounting, Auditing and Accountability Journal* 30(7), 1459–1480.

Tirado-Valencia, P., Cordobés-Madueño, M., Ruiz-Lozano, M. and De Vicente-Lama, M. 2021. Integrated thinking in the integrated reports of public sector companies. Evidence and contextual factors. *Sustainability Accounting, Management and Policy Journal* 12(2), 330–352.

Tricker, R. I. 1984. *Corporate Governance: Practices, Procedures, and Powers in British Companies and Their Boards of Directors*. Aldershot, UK, Gower Publishing Co.

Van Eck, N. J. and Waltman, L. 2017. Citation-based clustering of publications using CitNetExplorer and VOSviewer. *Scientometrics* 111(2), 1053–1070.

Van Zijl, W., Wöstmann, C. and Maroun, W. 2017. Strategy disclosures by listed financial services companies: signalling theory, legitimacy theory and South African integrated reporting practices. *South African Journal of Business Management* 48(3), 73–85.

Wolff, A., Gondran, N. and Brodhag, C. 2017. Detecting unsustainable pressures exerted on biodiversity by a company. Application to the food portfolio of a retailer. *Journal of Cleaner Production* 166, 784–797.

Chapter 10

Natural capital and biodiversity accounting in the dairy industry: the case of Valio Group

Hannu Schadewitz, Turku School of Economics at the University of Turku, Finland

1 Introduction

The interest in sustainability is gaining increasing attention globally. Sustainable development is, '... development that meets the needs of the present without compromising the ability of future generations to meet their own needs' (Brundtland's Report of the World Commission on Environment and Development: Our Common Future, 1987). Sustainability has a substantial impact on firms because they are both major producers and consumers of resources and goods with significant resulting impacts on the environment and biodiversity. Consequently, firms can also play an important role in contributing to more sustainable societies and preserving biodiversity. The EU

http://dx.doi.org/10.19103/AS.2023.0128.08

Sustainable Development Goals (SDGs) in its 2030 Agenda include 'inclusive and sustainable economic growth', 'full and productive employment' and 'sustainable production and consumption patterns' as goals. The EU also has a biodiversity strategy for 2030 to protect nature and reverse the degradation of ecosystems.[1] Achieving these important goals requires the active participation of the business sector.[2]

This chapter reviews corporate reporting of natural capital and biodiversity within the dairy industry. The chapter focuses on the dairy industry as it is heavily dependent on extensive use of natural and animal resources. Milk is the central raw material for a wide variety of dairy products given its wide range of nutritional and functional properties (Huppertz and Vasiljevic, 2022). A primary concern at the very beginning of the production chain is animal well-being (McLaren and Appleyard, 2020). Since milk and products derived from milk are vulnerable to spoilage and pathogenic contamination and may have a limited shelf-life, the sector also needs a robust, well-managed supply chain from farm to processor and on to wholesalers and retailers. Production of dairy products thus represents a significant source of greenhouse gases (GHGs) across the supply chain. The carbon footprint created by the dairy industry is therefore a major global issue. The Intergovernmental Panel on Climate Change (IPCC) is the United Nations body established to highlight and assess science related to climate change.[3] The IPCC has described the ways the production of animal-sourced food emits GHGs, its contribution to climate change and also offers guidelines on policy to address this problem (www.ipcc.ch/srccl). However, it needs to be acknowledged that the impact of food production is more far-reaching than GHGs. Biodiversity and ecosystem impacts are critically important dimensions of sustainability which are linked to but also distinct from climate change (Kok et al., 2020).

A broader perspective on, and a greater insight into, the areas where the food industry has an impact on sustainability can be gained from consulting the Sustainability Accounting Standards developed specifically for the meat, poultry and dairy industry (Sustainability Accounting Standards Board, 2018). This standard covers ten sustainability disclosure topics with related accounting metrics. These topics are

- greenhouse gas emissions;
- energy management;
- water management;
- land use and ecological impacts;
- food safety;

[1] See https://environment.ec.europa.eu/strategy/biodiversity-strategy-2030_en.
[2] See http://ec.europa.eu/environment/sustainable-development/SDGs/index_en.htm.
[3] See www.ipcc.ch.

- antibiotic use in animal production;
- workforce health and safety;
- animal care and welfare;
- environmental and social impacts of the animal supply chain; and
- animal and feed sourcing.

As this list demonstrates, the impact of sectors such as the dairy industry is multidimensional and, in turn, highlights several areas where management needs to adapt their operations accordingly. Dairy firms are important actors in the production/manufacturing chain where consumer demand strongly steers the output manufactured. Dairy firms also have a high demand for raw materials, primarily milk. All this creates an important role for dairy firms and their sustainability and biodiversity impact in society.

This chapter begins by looking at ways of studying business sustainability practices, both in terms of external pressures and, in particular, internal processes within individual businesses. It then focuses on the Valio Group as a case study. The chapter reviews general findings from in-depth interviews with senior Valio staff. Subsequent sections look at key interview themes related to how a sustainability agenda is pursued in Valio: disciplined corporate governance, market-driven product development, science-based sustainability targets, sustainable dairy farming, stakeholder/consumer-oriented communication and UN SDGs.

2 Ways of studying business sustainability practices

Prior research has placed great emphasis on sustainability reporting as a driver for organizational transformations towards a more sustainable approach. However, scholars note a failure to produce the desired organizational change (Thomson, 2014). Hence, a more comprehensive approach to examining sustainability within business organizations is called for. We need to understand better how sustainability is emerging within firms at both strategic and operational levels and how those levels interplay.

This section aims to address two research objectives to provide insight into how sustainability is approached in the Finnish dairy industry (see also Table 1):

1. How do the institutional (external) pressures imposed by environmental, social and governance (ESG) practices affect the strategies, organization and supporting systems of Valio?
2. How do the internal pressures imposed by ESG practices affect the strategies, organization and supporting systems of Valio?

Table 1 Research objectives, methods and data

Objective 1. Institutional (external) pressures influence on strategies, organization and supporting systems Methods: archival search, interviews. Data: regulations, SASB standard, recommendations, governance guidelines, Valio, interviews.
Objective 2. Internal pressures influence on strategies, organization and supporting systems Methods: interviews, analyses of firm's external documents. Data: Valio, interviews.

Sustainability practices have been mostly examined through the lens of legitimacy and stakeholder theories (Yang et al., 2015). Our first objective is understanding the external factors affecting Valio's sustainability. Objective 1 approaches the subject from an institutional (external) viewpoint. It involves characterization of institutional contexts, such as regulations and recommendations, corporate governance guidelines and potential reporting requirements regarding sustainability. This analysis is required to understand the context in analysing Valio's approach to sustainability (see also Senn and Giordano-Spring, 2020; Farooq and de Villiers, 2019).

As indicated in Objective 1, compliance can be related to the institutional setting, e.g. compliance with legislation or internationally/nationally recognized certification schemes. Relevant schemes Valio has committed to include ISO 14001 environmental certification, ISO 9001-certified quality system, FSSC 22000-certified food safety control system, self-monitoring plans compliant with HACCP guidelines (HACCP refers to Hazard Analysis Critical Control Point), accredited laboratories, Sedex (Supplier Ethical Data Exchange) membership and compliance of emissions to air calculations with the guidelines of the international GHG Protocol (www.ghgprotocol.org). In addition, milk and product purity is verified through stricter residue control than is required by law. (Sustainability Report 2020). The milk and product purity requirement at Valio is one example of how Valio has set the bar higher than required by the institutional setting (in this case legislation). Appendix 1 lists sustainability reporting regulations and recommendations in Finland relevant to Valio.

Objective 2 is more firm specific and looks at the firm's internal structure in relation to sustainability. Political, economic, legal, cultural, ideological, organizational and personal factors can affect how a firm undertakes sustainability activities. Prior research on determinants for corporate sustainability has almost exclusively focused on external reporting, leaving aside other important practices. However, external reporting is vulnerable to business concerns regarding reputation and image. Cooper and Owen (2007) have also suggested weaknesses in the links between those managing firms and external stakeholders. Those factors can bias external reporting. Internal practices, sustainability and accounting systems can be more meaningful for successful implementation of sustainability inside a firm. Analysing firm practices and related internal pressures can thus provide additional insights. These factors can be captured by interviews (see Hofman et al., 2017). The chapter analyses whether internal, sustainability-related pressures affect strategies and organization. This area is less well researched.

To study sustainability attitudes and systems within Valio itself, it is not be possible to collect quantitative data, as there are no publicly available data and no databases available. As a context, the author used relevant official data and articles, such as Hofman et al. (2017), Kautto (2001) and Marquis et al. (2011). However, interviews were used as a key tool to gain an understanding of what external and internal pressures influence managers' decisions regarding promoting sustainability at strategic, organizational and supporting system levels. Besides answering the 'what' question, this chapter answers also the 'how' question, i.e. ascertaining how those pressures are factored into strategies, organization and supporting systems within Valio.

Data were collected using semi-structured interviews. Interviews helped to map out general issues and identify key sustainability issues in response to research questions. The interview guide is shown in Appendix 2. It should be pointed out that the interview guide contains a few questions related to China. These questions are related to another project and are not used for this chapter. Interviews were targeted at managers responsible for sustainability issues. Interviewees were selected from three levels: strategic (e.g. executive board membership), organizational and supporting systems.

3 Case study: Valio Group

To gain a better understanding of the factors affecting dairy firms' approach to sustainability and sustainability (including biodiversity) accounting, this chapter focuses on one firm headquartered in Finland: Valio Group (henceforth Valio). Finland is a Nordic country, a member of the EU with a population of 5.5 million. As a country, Finland has been at the forefront of sustainability and biodiversity since concerns about sustainability and biodiversity arose (Sachs et al., 2021).

Valio was established in 1905 and is now the largest dairy firm in Finland with a workforce of 3500, processing roughly 1.8 billion litres of milk per year. It is a brand leader in Finland and also a significant player in the international dairy ingredients market. The company has subsidiaries in Sweden, all the Baltic countries, the USA and China. In 2020, Valio's net sales were 1.8 billion euros. Valio's exports from Finland account for about 25% of Finland's total food exports. Valio conducts intensive and successful product development: it holds over 300 patents in 50 countries.

Valio has a foundation as a cooperative: it is owned by dairy farmers. More specifically, it is owned by a group of regional dairy cooperatives (with a total of 13 shareholders) comprising some 4000 Finnish dairy farmers. Valio pays out all its profits to its member farmers through these regional dairy co-operatives. This means Valio can source locally produced milk with a clearly identified origin. The non-listed position of Valio also releases it from certain reporting requirements for listed firms and also from certain market pressures. Recent research by Allen, Lewis-Western and Valentine (2022) has shown that the burden of financial regulation can reduce research and development (R&D) spending and innovation. Allen et al. (2022) show that the consequences can also be a less innovative organizational culture. As a non-listed firm, Valio can bypass some of the regulatory reporting requirements for listed firms and seems to prefer more tailored, non-regulated forms of interaction with its stakeholders. Consistent with Allen et al.'s (2022) finding is the conclusion that flexibility in corporate reporting can sometimes support more R&D and innovation.

Valio is recognized, amongst other things, for its high-quality products, R&D innovation as well as its responsible production chain. Valio has achieved first place eight times in a row (years 2014–21) in Finland's most sustainable brand survey.[4] As this suggests, Valio has a high level of ambition in the area of sustainability, as shown in its 2020 Sustainability Report (p. 9): 'Our vision is to be the leader in innovative dairy and food solutions. Innovativeness is visible not only in product development but also in all the other activities we do'. The 2020 Board of Directors' Report specifies the vision further (p. 2): 'Valio's strategy was updated in 2020. Our vision is to be the world's most innovative dairy and food company'.

Valio has listed the following eight focus areas in its 2020 Sustainability Report:

- Regenerative agriculture;
- Circular operations;
- Emission reductions in farming and milk production;
- Biodiversity recovery;
- Animal welfare;
- Sustainable solutions for customers and society
- Environmentally smart packaging; and
- Innovations for well-being.

[4] See Sustainability Report 2020, link to sustainable brand survey https://www.sb-index.com/finland.

Valio applies Global Reporting Initiative (GRI) Standards in compiling its annual sustainability report. However, the report is not assured externally.

Three interviews were conducted in Valio with a total time spent of 4 h and 5 min. The first interviewee was the Senior Vice President - Group Sustainability (henceforth SVP - GS). The interview was conducted on 8 October 2020 with a duration of 69 min. The SVP - GS had a BA in business and administration and had also participated in several executive education courses covering such areas as management, digitalization and leadership. She had worked in Valio for over 30 years in various positions in different markets (countries), starting as an assistant in market research. The SVP - GS's duties focused on various marketing and brand-related responsibilities. With that background she had taken on her current sustainability role. This position was intentionally designed for a person with a marketing and brand background to help create a relative competitive edge. The purpose of the role is to integrate brands with actions promoting sustainability.

The second interviewee was the Executive Vice President - Sustainability and Stakeholder Relations (henceforth EVP - SSR). This interview was conducted on 26 August 2021 and lasted 86 min. The EVP - SSR is a lawyer with various responsibilities at Valio. She also had a MBA and had studied abroad. Her background is a business lawyer in non-listed firms. She had worked for around 20 years in various legal-oriented positions at Valio. Nowadays the EVP - SSR has six main areas of responsibility (each with a director reporting to her: sustainability, stakeholder relations, carbon-neutral milk chain, primary production, legal matters and group communication). EVP has been in her current position for almost 4 years. Previously she was director of legal matters, including responsibility for compliance matters. A more strategic orientation to sustainability and its management has been in place from 2018 when a carbon-neutral milk chain was developed.

The third interviewee was the Senior Vice President - Group Communications (henceforth SVP - GC). This interview was conducted on 12 October 2021 and lasted 90 min. The SVP - GC has an M.Sc. in business studies majoring in business communication. The SVP - GC's professional duties have focused on communication including sustainability issues. SVP has worked in Valio for about 4 years and in the current position for over 1 year.

4 Findings from interviews with senior Valio staff

In general, it can be said that all three interviewees were very uniform regarding their communication about sustainability matters and their importance in Valio. This is one indication that the overall roots of a sustainability culture in Valio are established and solid.

It is fair to start with a recognition of Artturi Virtanen, Professor of Biochemistry, and his fundamental and long-lasting role in Valio, as characterized by the SVP - GS:

> *At the time Virtanen was awarded Nobel Prize in Chemistry and he led our product development roughly 50 years and somehow his spirit lives on the walls of Valio and in the souls we feel passionate about product development. We continuously develop new products and new practices.*

The Nobel Prize in Chemistry in 1945 was awarded 'for his [Virtanen´s] research and inventions in agricultural and nutrition chemistry, especially for his fodder preservation method'. (https://www.nobelprize.org/prizes/chemistry/1945/virtanen/facts/). Specifically: 'Access to nutritious food year-round, especially in countries with cold winters, requires that the nutrients in food are preserved. Artturi Virtanen studied the chemistry of nutrients, including that in processes for dairy products. In 1932 he developed a method for better preserving the nutrients in the hay. Using hydrochloric and sulphuric acids, he impeded certain processes, preserving the grass´ proteins and vitamins. This helped cows produce more nutritious milk year-round without imported fodder'.

The first theme that emerged from SVP - GS was the importance of the competitive advantage of brands and how this advantage could be strengthened through sustainability. The SVP - GS emphasized that Valio is actively investing in sustainability even though the general public may not necessarily be aware of the sustainability issues. The name Valio can be translated into English as 'premier' or 'champion', indicating that, from the beginning, Valio has targeted high-quality products. Valio began by exporting Finnish butter to the UK which required high-quality products to ensure competitiveness. Finland's reputation for product safety and quality is well recognized and valued by customers, especially in China. A commitment to sustainability helps to reinforce this image.

Another theme that came up with the interview with the SVP - GS was disciplined corporate governance (henceforth CG). CG is the system by which a company is directed or controlled, including designation of appropriate responsibilities. Valio has a matrix organization with clear functional responsibilities, e.g. designated managers for functions such as Energy, Logistics and Packaging, There is, e.g. a designated senior vice president for the company's Carbon-Neutral Milk Chain Project. This is one indication of the commitment at Valio to sustainability and carbon reduction and, ultimately, elimination of carbon emissions.

The interview with the EVP - SSR reinforced the degree of cohesion and coherence on sustainability matters taking place at Valio. Examples include a traffic lights system, periodic sustainability forums and an initiative/innovation reward system. A traffic lights system provides a monitoring tool for all strategic projects. The green light indicates that the phase of the project is proceeding to plan, the yellow light indicates that the project requires additional attention and the red light indicates that the project is delayed and needs corrective actions.

The EVP - SSR also identified pressures from stakeholders in relation to sustainability, particularly consumers and farmers. Consumers are showing a greater interest in sustainability. Dairy farmers, in turn, are concerned about making their living and how sustainability requirements affect profitability. Communication issues with stakeholders will be discussed later in the chapter.

The SVP - GC emphasized the importance of management commitment to spread the sustainability theme within Valio. Practical tools to support this include mandatory digital courses for staff, forums, development discussions and specialized teams. The SVP - GC emphasized the importance of the whole production chain from field to end product, especially dairy farms. In terms of biodiversity, there are surveys and consultation visits to some farms by professionals related to biodiversity. Those visits assist dairy farmers in developing biodiversity best practices which are also communicated to other contract dairy farmers. Besides a focus on climate change and biodiversity, well-being and health are strongly emerging themes at Valio.

5 Interview themes: disciplined corporate governance and global-oriented market-driven product development

Sustainability is a multidimensional theme in an organization. As the EVP - SSR commented, 'doing right' is not always a strong enough argument supporting sustainability. There need to be stronger arguments for owners and top management to commit to sustainability. These include political pressure and consumer attitudes.

Well-structured and disciplined corporate governance is essential to achieve corporate objectives. In interviews, themes such as a matrix organization, use of professional teams and cooperation inside the organization and with outside stakeholders were emphasized. Besides logical governance structures and a commitment of management to sustainability, sustainability needs to be communicated through the organization and to farmers.

Disciplined CG can be seen in the company's annual activity clock which coordinates forums, monthly meetings and other events for relevant people to meet frequently. Sustainability forums are organized three times per year. Those forums serve as a platform for top management to present and review strategically important sustainability matters and their progress. Other examples include translating the company code of conduct into language employees can understand and related mandatory web-based courses. Valio has also a brand manual supporting and guiding brand-related work.

The SVP - GS works closely with the logistics director and packaging development director. She also works closely with the EVP - SSR. When e.g. renewable energy issues need to be discussed, the SVP - GS coordinates with the corporate energy director. Valio is organized as a matrix where thematic professional teams have an important role. Those teams are organized around key areas. Whilst the SVP - GR has overall responsibility for sustainability, she leads and helps to coordinate other teams that connect with sustainability. A company code of conduct helps to support teamwork. The company also has a well-designed intranet which provides an important communication channel. The SVP - GS and fellow experts are constantly seeking the right approaches where Valio has or could build competitive advantage, such as the Carbon-Neutral Milk Chain 2035 Programme.

5.1 Market-driven product development

A major external factor having an impact on product portfolio is market demand. Valio sells products to both private consumers and industrial customers (hotels, restaurants and catering - HoReCa). Market trends are monitored very intensively at Valio. The SVP - GS highlighted the importance of sensitivity to customer needs. The largest retailers and industrial customers in particular provide valuable information on consumer attitudes. There are also sustainability projects shared between Valio and its retailer customers. Valio also has a Consumer Insight Department that follows global market trends and carries out consumer research. Two examples of market-driven developments mentioned by the SVP - GS are lactose-free products and plant-based products. Valio's lactose-free technology is very advanced and highly valued abroad. The company's plant-based products category is a rapidly developing area in line with growing demand (regarding the potential for reducing climate impact and preserving biodiversity by producing oat-based milk (instead of cow's milk) (Röös et al., 2016)).

Finally, the SVP - GS emphasizes the importance of active dialogue between Valio and its customers. This dialogue is facilitated via the use of social media channels, consumer service operations, Valio's web pages, meetings with stakeholders and an active media presence. Enquiries directed to the consumer service team are, e.g. valuable in understanding consumer sustainability attitudes. The EVP - SRR also mentioned that Valio conducts systematic media watching activities. She stated that Valio wants to be active in discussions instead of being a bystander.

6 Interview themes: science-based sustainability targets and sustainable dairy farming

Science-driven decision-making is key to Valio regarding sustainability issues. The environmental impact of the dairy industry is multidimensional and measuring its impacts is challenging. A science-based approach is required to measure those impacts and to set appropriate science-based targets (SBTs). Valio sets climate targets aligning with the Paris Agreement (https://unfccc.int/process-and-meetings/the-paris-agreement/the-paris-agreement). This agreement 'is a legally binding international treaty on climate change. It was adopted by 196 Parties at COP 21 in Paris, on 12 December 2015 and entered into force on 4 November 2016. Its goal is to limit global warming to well below 2, preferably to 1,5 degrees Celsius, compared to pre-industrial levels'.

Valio's climate impact in 2020 was as follows (Sustainability Report 2020). GHG emissions from its operations in Finland were 2.3 million carbon dioxide equivalent tonnes (Mton Co_2eq).[5] The vast majority (79%) of GHG emissions originated from raw milk. Energy use at plants accounted for 7%, and logistics for 2%. Other emissions were related to other raw materials used in product manufacturing, packaging and waste processing (Sustainability Report 2020). The SVP - GC commented that the emission-reducing actions are targeted especially at functions causing most emissions. Some examples of these mitigation actions are described next.

The main example of emissions reduction is the Carbon-Neutral Milk Chain 2035 Programme which includes specific SBTs to track progress. Valio's climate programme involves collaboration between Valio, its dairy farms, agricultural and energy sector companies, and research institutes. Research institute partners include VTT Technical Research Centre of Finland, the Natural Resources Institute (Luke) and the University of Helsinki. Part of Valio's SBT setting process is an emission inventory using the International Greenhouse Gas Protocol (https://ghgprotocol.org/). The carbon footprint of raw milk has been calculated using Valio's lifecycle calculation model, certified in 2019 by the Carbon Trust (https://www.carbontrust.com/).

6.1 Sustainable dairy farming

In addition to innovative product development, the slogan 'Valio owned by farmers' has also helped to create a positive corporate image. According to the SVP - GS, this underlines the view that money is not the number one priority in Valio's operations. The carbon footprint of raw milk is also a highly important

[5] The computation followed the calculations guidelines of the international GHG Protocol (www.ghgprotocol.org). Renewable energy's Direct emissions are not calculated as emissions in the inventory, i.e. computationally, they are zero.

measure. Valio's carbon footprint is split as follows. Primary production on farm is responsible for 93% of the firm's carbon footprint:

- 49% from cows' rumination (methane);
- 26% from feed crop cultivation (nitrous oxide);
- 6% from energy use on farms (mainly carbon dioxide);
- 4.5% from fertilizer production (carbon dioxide);
- 4.5% from manure handling (methane and nitrous oxide); and
- 3% other inputs.

Beyond the farm:

- 5% is from logistics and factories (3.5% energy use at factories; 1.5% logistics); and
- 2% is from packaging.

Valio is making an ongoing effort to reduce its carbon footprint in all these areas with the ultimate target of achieving a carbon-neutral milk chain by 2035. It should be mentioned that all dairy farms are included in Valio's sustainability bonus programme (which rewards farmers for implementing sustainability practices) from the beginning of 2021. According to the SVP - GS, all dairy farms approved to supply milk to Valio must belong to the bonus programme.

One way to reducing the carbon footprint is feed additives that reduce methane emissions from cows. Valio and the Dutch company DSM have collaborated to find opportunities to utilize a new additive that can reduce methane emissions from cattle by about 30%. (Sustainability Report 2020). It should also be mentioned that imported soy as feed is forbidden in the Valio Group (Sustainability Report 2020).

Besides sustainability issues related to GHG emissions and climate change, farming also has an impact on biodiversity (Kok et al., 2020). The SVP - GC noted that biodiversity experts from the Rural Women's Advisory Organization https://www.maajakotitalousnaiset.fi/english have conducted dozens of visits to dairy farms. Together with farmers, these experts have looked for areas of existing high natural value, potential areas for biodiversity enhancement and targets to monitor progress. The organization offers, among other services, environmental landscape management services (Sustainability Report 2020; for biodiversity indicators and tools for evaluating biodiversity - see Sizemore, 2015).

In addition, during the summer of 2020, indicator species in agricultural environments were monitored as a joint project with the Finnish Biodiversity Information Facility (FinBIF) at the University of Helsinki (Sustainability Report 2020, link to FinBIF https://laji.fi/en). As its website states: 'FinBIF is an open access data repository for researchers, government, and the public. FinBIF consolidates many collections and datasets of living Finland in a single source. Our online portal, laji.fi, allows you to browse, search and download information about all forms of biological life, and to record and share your observations. FinBIF is committed to the sharing and promotion of open access data'.

Farmers have received hands-on advice on how to support biodiversity on their farms. This advice has then been distributed in various ways (such as magazine articles, videos and electronic learning platforms) to other dairy farmers as well. According to Valio Sustainability Report 2020, the best ways to secure biodiversity are related to carbon farming practices, organic milk production and grassland grazing of dairy cattle. Sustainability Report 2020 states also that several of Valio's climate programme measures support biodiversity (on harnessing biodiversity for healthy dairy farms, see Zhang et al., 2022).

According to the Sustainability Report 2020, Valio is leading the CARBO® project that focuses on regenerative agriculture. One result of the project is Valio's CARBO® environmental calculator. This gives dairy farmers precise information about environmental impacts of a farm's raw milk production. As noted, the Carbon Trust has certified the calculation model. The model is based on IPCC's recommendations, the European Commission's Product Environmental Footprint Category Rules (PEFCR) and Valio's long-term scientific work. Training is available for farmers to use the environmental calculator. Information regarding a farm-specific baseline and emission sources helps to target measures for reducing emissions more accurately. (Sustainability Report 2020). The CARBO® project has been a factor in Valio seeking independent Gold Standard certification (https://www.goldstandard.org/, Sustainability Report 2020).

According to Valio, a key challenge in cutting milk's carbon footprint is the draining of wetlands or peatland forests over decades (Sustainability Report 2020). Draining of peatlands and peat extraction releases carbon dioxide into the atmosphere. In 2020, these releases were measured at four Valio dairy farms in a project called climate change mitigation in organic soils on cattle farms (OMAIHKA). More project information and its stakeholders (such as Valio and Ministry of Agriculture and Forestry of Finland) are available from Natural Resources Institute Finland Luke at https://www.luke.fi/en. As an example, there has been research into using grass as a perennial crop on the surface of peatland to provide a more effective way to reduce emissions, compared to annual grain production. CARBO® seed mixes are another way to improve carbon sequestration in grasslands. The mix includes clovers, which help pollinators, as well as grass species that are more resistant to abiotic stress and improve soil structure. Five different seed mixes are available (Sustainability

Report 2020). These seed mixes are an important and practical way to support natural capital and biodiversity.

7 Interview themes: stakeholder/consumer-oriented communication and UN Sustainable Development Goals

The SVP - GS recognized that Valio's sustainability actions and reporting does not always gain enough attention and recognition from the general public. This is an issue which affects other larger dairy firms (Sun and Lange, 2023). One recent high-profile project Valio has participated in is the SIHTI Project initiated by the Ministry of Economic Affairs and Employment which assesses the human rights performance of Finnish companies (Tran-Nguyen et al., 2021).

Since dairy farmers own Valio, this stakeholder group is crucially important in terms of communication. Dairy farmers fear that sustainability requirements create an additional burden for farming without related economic benefits. The EVP - SSR pointed out the importance of convincing dairy farmers in committing to sustainability issues. A message focusing on 'doing right' is insufficient. There needs to be an understanding of factors such as economics, political pressure, consumer behaviour and competition. Dairy farmers need tailored communication and targeted communication channels which reflect farmers' specific needs. The SVP - GC also highlighted Valio's key role in safeguarding farmers' interests in Finland's agricultural policy. She highlighted the importance of making decision-makers aware of Valio's work in promoting sustainability in co-operation with farmers.

Communication for consumers is very important as well. Similar communication challenges exist with consumers as with decision-makers. The SVP - GC stated that Valio aimed to develop its communication to reach consumers as far as (grocery) store level with a focus on product-specific sustainability information. The EVP - SSR discussed the possibility of bringing sustainability issues more explicitly into financial reporting. One example that was mentioned was the carbon emissions adjusted Earnings per Share (EPS) figure.

7.1 UN Sustainable Development Goals

Valio has selected eight out of 17 SDGs as most relevant to its operations. These are:

- SDG 2 Zero hunger;
- SDG 3 Good health and well-being;
- SDG 8 Decent work and economic growth;

- SDG 10 Reduced inequalities;
- SDG 12 Responsible consumption and production;
- SDG 13 Climate action;
- SDG 14 Life below water; and
- SDG 15 Life on land.

Two SDGs discussed in more detail here (and particularly relevant to biodiversity) are:

- SDG 14 Life below water; and
- SDG 15 Life on land.

Valio's commitment to SDG 14 is focused to promotion of a 'Circular economy in primary production, reducing effect on water system (COD, Chemical Oxygen Demand)'. SDG 14 is an especially relevant goal because Finland has approximately 187 000 freshwater lakes within its borders. One of Finland´s nicknames is the 'Land of Thousand Lakes'. Turning to SDG 15 Valio focuses on 'Animal welfare, promoting biodiversity at dairy farms'. Valio has a biodiversity roadmap and targets. Animal welfare is promoted in part through a sustainability bonus scheme (2 cents per litre) which Valio wants all farmers to commit to by 2020, as well as planning a welfare index for dairy cattle.

Valio has also worked with other partners to promote sustainability. Initiatives include carbon farming training with the Baltic Sea Action Group (BSAG). Valio helped to develop BSAG´s E-College for regenerative agriculture. This is a free, science-based online course on regenerative farming. As mentioned earlier, Valio led the CARBO® project in collaboration with several preeminent research groups: Finnish Meteorological Institute, Natural Resources Institute (Luke), University of Eastern Finland, Yara and Atria Tuottajat. It has also supported R&D projects focused on carbon sequestration in grasslands and the efficiency of grassland farming.

The previous discussion of Valio's approach to sustainability also shows its choice of other SDGs. SDGs 2 and 3 are related to the availability of food. SDGs 8 and 10 relate to key Valio stakeholders: 8 to dairy farmers that are owners of Valio and 10 to employees and their equal treatment. SDG 12 strives for responsible consumption and production. Finally, SDGs 13, 14 and 15 goals relate closely to dairy farming. Farming should be done in a way that takes biodiversity into account. Use of fossil fuels can be reduced by using technology to convert waste and manure to energy, e.g. with biofuel replacing diesel in milk trucks.

8 Conclusion

The statement of SVP – GS 'sustainability in all actions and all markets' captures well Valio's intentions regarding sustainability. Valio's scientific roots are strong and deep and targets are set high. These roots go back to Professor A. I.

Virtanen who helped cows produce more nutritious milk year-round without imported fodder. Professor Virtanen's scientific heritage laid the foundation for scientific innovation to be valued at Valio.

Well-functioning CG has been a key element in Valio's approach to sustainable development throughout the production chain (from field to table). All three interviewees highlighted the importance of good CG and clear allocation of responsibilities. Teamwork in a matrix structure is also important. Expert work in teams is indispensable because sustainability issues in the dairy business are diverse and call for specialized solutions. Collaboration between Valio, universities and other expert organizations is intense, fruitful and productive. All this suggests that Finland provides a good base for the dairy industry to improve its sustainability and promote biodiversity. One indicator supporting this is Finland's number one position in the international sustainable development comparison of the 193 countries ranked (Sachs et al., 2021). The annual ranking is by the UN and the Bertelsmann Foundation. The ranking assesses countries' progress in implementing the 2030 Agenda for Sustainable Development and its SDGs.[6]

Valio's strong CG together with Finland's favourable sustainability regime offers a suitable framework facilitating sustainability developments. An important part of good CG is committed personnel and good communication. Valio follows an annual clock, has frequent forum meetings, has other frequent meetings and has various other communication channels to ensure adequate information flow within the organization.

Valio is an active player in its field. It listens carefully to markets, tries to pick up megatrends in consumption, is an active participant in relevant discussions and also communicates actively with government and other decision-makers. Stakeholder theory and legitimation theory help to explain Valios's performance (Deegan and Unerman, 2006). Valio's sustainability actions are in line with stakeholder theory. Legitimation theory helps to explain Valios's approach to communication with decision-makers.

Last but not least is Valio´s co-operative structure through which Valio is owned by dairy farmers. Thanks to this ownership structure Valio can avoid a more short-term approach. Valio's target is to be profitable and to pay all operating profits to owners. Dairy farmers are fundamental both in terms of ownership and rootstock of raw material. These roles make dairy farmers a special stakeholder group among all other stakeholders.

This chapter has some caveats. Even though the persons interviewed were highly experienced and had a thorough knowledge of Valio Group, a larger number of interviews would have given additional insight into some of the more specific areas discussed. One important theme to investigate further is

[6] https://valtioneuvosto.fi/en/-/10616/finland-ranks-first-in-international-sustainable-development-comparison.

the Carbon-Neutral Milk Chain Project. This project is led by the senior vice president, highlighting the importance of this flagship project, which was also confirmed by all three interviewees. It would also be particularly useful to analyse how financial reporting could be enriched by integrating carbon-related information. One such example, indicated by the EVP - SSR, could be emission-adjusted milk price per litre that Valio pays to farmers. Currently Valio publishes two separate reports (which separate sustainability and financial information):

- Sustainability Report
- Board of Directors' Report and Financial Statements

Conventional accounting and its communication could benefit by taking sustainability information more into account, e.g. as alternative performance measure figures. That would, in turn, assist stakeholders to recognize better the economic effects of sustainability actions.

Valio's selected eight SDGs would be a fruitful area for a deeper examination from an accounting integration perspective (Gunarathne et al., 2022). In addition, one future research area in accounting could focus on how complicated sustainability matters could be most effectively and most distinctly communicated to stakeholders. From a more economic perspective, it would be advantageous to stakeholders if sustainability and economic data could be more integrated. Enhanced integration, in turn, would help stakeholders to take the economic effects of, say, environmental resource allocations better into account. Overall, more research identifying natural capital and biodiversity accounting tools to support integrated reporting is called for.

9 Acknowledgements

The interviewees at Valio are acknowledged for their time and expertise facilitating the realization of this chapter. Professor Jill Atkins has offered useful comments in various phases of this chapter, for which I am most grateful. I would also like to acknowledge Joona Haavisto and Valtteri Toivanen for transcribing the interviews. I gratefully acknowledge the financial support of the Foundation for Economic Education. The author retains responsibility for the final interpretations.

10 Appendix 1 Regulations and recommendations related to sustainability for Valio

In Finland according to Accounting Act 3a§ requirement to report nonfinancial information does not apply to non-listed firms. The only exception is information given in the review of operations. Information should be given in the review of operations if it has significant economical relevance and is related to environmental and personnel matters.

On the voluntary sustainability reporting firms could have certain contracts they are committed to following. For example, those firms that have signed the Global Compact agreement have committed to report annually. That reporting can be done by following the Global Reporting Initiative (GRI) format or constructing a more tailored, firm-specific, format instead. There could be also other contracts such as energy-efficiency contracts. Those contracts could contain responsibilities to report about progress for contract parties and possible also more widely. These specific contracts do not include a requirement for sustainability reporting. The dairy industry as such does not require sustainability reporting. However, in Finland in the food industry, there exists agreed general principles and recommendations for reporting. For example, the Finnish Food and Drink Industries' Federation has published self-monitoring instructions (HACCP guidelines, Hazard Analysis Critical Control Point) tailored also for the dairy industry (link to Finnish Food and Drink Industries' Federation https://www.etl.fi/en/index.html).

11 Appendix 2 Interview guide for representatives of the Finnish company

The interview guide needs some adjustment to each interviewee background - this is a general guide only.

Preamble

- Introduce myself/ourselves.
- Our research: We are interested in firm sustainability. We try to figure out firm's sustainability orientation, sustainability systems and sustainability communication to firm stakeholders.
- Anonymity/confidentiality.
- General concept (definition) characterizing firm sustainability: definition of sustainability. The quality of being responsible to the economy, society and environment or not depleting economic, societal or natural resources, thereby supporting long-term economic, societal and ecological balance.
- The translation of sustainability in Finnish might work using the term responsibility 'vastuullisuus' among businesspeople.
- Is recording OK?

1. Demographic information

- What is your educational background?
- What is your professional background/career
- Position in the company
- Current job in the company

Prior experience

- Years of experience in the current position and in the company
- Do you have any experience in China?
- Have you worked in countries other than Finland? What kind of experiences have you had there?

2. Sustainability overview

- How would you characterize your company in terms of sustainability?
- How does sustainability affect your work?

3. Sustainability at the strategic level in the company

- What is the role of sustainability in the strategy of the company?
- How has sustainability at the strategic level emerged?
- How are these values communicated inside the company?
- How does the company ensure compliance with these values?
- Why has sustainability potentially been included in the strategy or is it even the core of the strategy?

Do you know any differences in sustainability at the strategic level between Finland and the China subsidiary?

- Are people, in general, in the company familiar and convinced with the strategy?

4. Sustainability at the operational level in the company

- To what extent is sustainability transposed into operational practices and in what areas does this occur?
- How is sustainability at the strategic level transposed into operational practices?
- Why sustainability practices at the operational level?
- What kind of sustainability practices does the company execute in its operations?
- To what extent do these operations manifest sustainability?

5. Sustainability in sustainability or in accounting systems (questions to an accounting person)

- What kind of sustainability systems the company have?
- How is sustainability reflected in sustainability systems? What kind of sustainability/accounting practices are there?
- How have these sustainability practices emerged? What kind of factors influenced their development?
- How are these sustainability-based accounting tools used inside the organization? How does accounting support sustainability operations?

- Are there differences between (sustainability-related) accounting practices in the Finnish company and Chinese subsidiary? What kind of?

6. Sustainability in communication/reporting

- How is sustainability communicated outside the company and to whom?
- Why is sustainability reporting done?
- Who/what departments are producing the reporting?
- Based on which data is the reporting produced?
- How is sustainability reporting linked to accounting practices and accounting department?
- How are the needs of stakeholders figured out?
- How do you know the degree of match between sustainability information supply and demand?
- Are there any differences between reporting of the Finnish company and the Chinese subsidiary?

7. Concluding questions

- What do you think about the future of sustainability in your company?

8. Other topics of interest for the interviewee regarding sustainability?

External influences on sustainability management: standards, laws, peer pressures, industrial guidance etc.

12 References

Allen, A., Lewis-Western, M. F. and Valentine, K. (2022). The innovation and reporting consequences of financial regulation for young life-cycle firms, *Journal of Accounting Research* 60(1), 45–95.

Cooper, S. M. and Owen, D. L. (2007). Corporate social reporting and stakeholder accountability: the missing link, *Accounting, Organizations and Society* 32(7–8), 649–667.

Deegan, C. and Unerman, J. (2006). *Financial Accounting Theory, European Edition*, Berkshire: McGraw-Hill.

Farooq, M. B. and de Villiers, C. (2019). Understanding how managers institutionalise sustainability reporting: evidence from Australia and New Zealand, *Accounting, Auditing and Accountability Journal* 32(5), 1240–1269.

Gunarathne, N., Lee, K.-H. and Kaluarachchilage, P. K. H. (2022). Tackling the integration challenge between environmental strategy and environmental management accounting, *Accounting, Auditing and Accountability Journal*. doi: 10.1108/AAAJ-03-2020-4452.

Hofman, P. S., Moon, J. and Wu, B. (2017). Corporate social responsibility under authoritarian capitalism: dynamics and prospects of state-led and society-driven CSR, *Business and Society* 56(5), 651–671.

Huppertz, T. and Vasiljevic, T. (Eds) (2022). *Understanding and Improving the Functional and Nutritional Properties of Milk*, Cambridge: Burleigh Dodds Science Publishing.

Kautto, M. (Eds) (2001). *Nordic Welfare States in the European Context*, Florence, KY: Routledge.

Kok, A., Oostvogels, V. J., de Olde, E. M. and Ripoll-Bosch, R. (2020). Balancing biodiversity and agriculture: conservation scenarios for the Dutch dairy sector, *Agriculture, Ecosystems and Environment* 302, 1-10.

Marquis, C., Zhang, J. and Zhou, Y. (2011). Regulatory uncertainty and corporate responses to environmental protection in China, *California Management Review* 54(1), 39-63.

McLaren, J. and Appleyard, T. (2020). Improving accountability for farm animal welfare: the performative role of a benchmark device, *Accounting, Auditing and Accountability Journal* 33(1), 32-58.

Röös, E., Patel, M. and Spångberg, J. (2016). Producing oat drink or cow's milk on a Swedish farm - Environmental impacts considering the service of grazing, the opportunity cost of land and the demand for beef and protein, *Agricultural Systems* 142, 23-32.

Sachs, J., Kroll, C., Lafortune, G., Fuller, G. and Woelm, F. (2021). *The Decade of Action for the Sustainable Development Goals: Sustainable Development Report 2021*. Cambridge University Press. doi: 10.1017/9781009106559.

Senn, J. and Giordano-Spring, S. (2020). The limits of environmental accounting disclosure: enforcement of regulations, standards and interpretative strategies, *Accounting, Auditing and Accountability Journal* 33(6), 1367-1393.

Sizemore, G. C. (2015). Accounting for biodiversity in the dairy industry, *Journal of Environmental Management* 155, 145-153.

Sun, Y. and Lange, Y. (2023). Implementing biodiversity reporting: insights from the case of the largest dairy company in China, *Sustainability Accounting, Management and Policy Journal* 14(1), 75-100.

Sustainability Accounting Standards Board (2018). *Meet, Poultry and Dairy Sustainability Accounting Standards*, San Francisco, CA: The SASB Foundation.

Thomson, I. (2014). Mapping the terrain of sustainability and accounting for sustainability. In: Bebbington, J., Unerman, J. and O'Dwyer, B. (Eds) *Sustainability Accounting and Accountability* (2nd edn., pp. 15-29), London: Routledge.

Tran-Nguyen, E., Halttula, S., Vormisto, J., Aho, L., Solitander, N., Rautio, S. and Villa, S. (2021). *Status of Human Rights Performance of Finnish Companies (SIHTI) Project: Report on the Status of Human Rights Performance in Finnish Companies*. Publications of the Ministry of Economic Affairs and Employment 2021:17.

Yang, H. H., Craig, R. and Farley, A. (2015). A review of Chinese and English language studies on corporate environmental reporting in China, *Critical Perspectives on Accounting* 28(May), 30-48.

Zhang, R., Han, Z., Lu, Q., Wang, K., Chen, Y., Cong, W.-F. and Zhang, F. (2022). Harnessing biodiversity for healthy dairy farms, *Frontiers of Agricultural Science and Engineering* 9(2), 238-244.

Chapter 11

Natural capital and biodiversity accounting in palm-oil production

Gunnar Rimmel, Aalborg University Business School, Denmark; and Maizatulakma Abdullah, The National University of Malaysia, Malaysia

1 Introduction

Over the past decade worldwide demand for palm oil has risen sharply, resulting in a significant increase in global palm-oil production and land use for plantations. In the 1990s, production of palm oil was relatively small scale with total global production of palm oil at around 11 million metric tonnes. In 2020, the United States Department of Agriculture (USDA) reported that global production had increased to 72.2 million metric tonnes. The USDA estimated in November 2020 that the total global area of palm-oil plantations had grown to 24.093 million ha, a little less than the size of New Zealand.

According to the USDA, oil palms are currently grown in 28 countries. Indonesia is responsible for 59% of world production of palm oil. It has the largest area under cultivation at 11.950 million ha in 2020 which is estimated to reach 17 million ha by 2025. Malaysia represents 24% of global palm-oil production with the second largest area of palm-oil cultivation (USDA, 2020; Abdullah and Rimmel, 2021). The Malaysian Palm Oil Certification Council (MPOCC) reported 5.9 million ha in cultivation in 2020, almost half of the size

http://dx.doi.org/10.19103/AS.2023.0128.09

of peninsular Malaysia. Shevade and Loboda (2019) estimated that between 1988 and 2012 about 651 757 ha of natural forest were converted to palm-oil production on the Malaysian peninsular.

Most of Malaysia's oil-palm plantations are located in the states of Sabah and Sarawak in the northern part of the island of Borneo. Sabah provides 25% of Malaysian palm-oil production, while Sarawak produces 19%. The other palm-oil-producing Malaysian regions are Johor with 16%, Pahang with 14%, and Perak with 7%. This growth in cultivation in Malaysia has been associated with significant deforestation, biodiversity loss, and the risk of the extinction of highly valued species. It has been estimated that palm-oil production was responsible for 60% of forest loss in the regions of Sabah and Sarawak between 1972 and 2015 (Meijaard et al., 2018). More generally, it has been estimated that 45% of oil-palm plantations in Southeast Asia came from areas that were forests in 1989 (Vijay et al., 2016).

The expansion of palm-oil production and its environmental effects are not confined to Southeast Asia. A study showed that Africa and Latin America have the greatest potential for oil-palm expansion after Southeast Asia (Vijay et al., 2016). In Africa, the greatest deforestation was observed between 1989 and 2013 in Cameroon and Ghana (Vijay et al., 2016). In Latin America, the majority of suitable land is located in the Amazon rainforest (Yui and Yeh, 2013). In Brazil alone, it has been estimated that 238 million ha are suitable for palm-oil production (Munasinghe et al., 2019). In Peru, 72% of new oil-palm plantations expanded into Amazon rainforest areas (Gutiérrez-Vélez et al., 2011). Between 2000 and 2015, oil-palm plantations destroyed about 84 500 ha of the Peruvian Amazon forest (Vijay et al., 2016). In Colombia, 60% of oil-palm plantations are located in areas of primary forest (Meijaard et al., 2018).

Despite its negative environmental impact, it has proved challenging to limit the growth of oil-palm cultivation. Potential alternatives such as rapeseed and sunflower require significantly more land to produce the same amount of oil (Grayson and Stampe, 2012; Wilcove and Koh, 2010). Replacing palm oil with alternative oil plants therefore risks more deforestation and biodiversity loss (Cattau et al., 2016; Meijaard et al., 2018). Attempts by non-governmental organisations (NGOs) to limit growth in oil-palm cultivation and its environmental impact, e.g. by encouraging a boycott of products containing palm oil, have had a limited impact (Potter, 2015). Official initiatives, such as the decision of the Parliament of the European Union (EU) to impose restrictions on palm-oil imports from Indonesia and Malaysia in 2018, have also only been partially successful. Since palm oil is used in so many products, global exports have continued to grow, driven by massive demand from the largest palm-oil-consuming countries such as India and the Republic of China (Voora et al., 2019). The palm-oil industry is also hugely important to the economies of countries such as Malaysia and Indonesia (Singagerda et al., 2018). According to Choiruzzad (2019), about 860 000 people in Malaysia made a living from the palm-oil industry in 2015, only a fraction of the 14 million people earning a living from the industry in Indonesia. As a result, growth has been actively promoted by e.g. the Malaysian government (Varkkey et al., 2018; Vijay et al., 2016). These factors suggest that, rather than trying to get rid of the industry, the most effective solution is to help the palm-oil industry to achieve sustainable production.

This chapter reviews natural capital and biodiversity accounting in Malaysian palm-oil production. It outlines the palm-oil industry's impact on biodiversity and natural capital. It then assesses the transparency of sustainability information in the Malaysian palm-oil industry. Five initiatives for improving the transparency of sustainability reporting and the implementation of best practice by palm-oil companies are highlighted. These five initiatives are Carbon Disclosure Project (CDP), Forest 500, No Deforestation, No Peat and No Exploitation Implementation Regulatory Framework (NDPE IRF), Sustainability Policy Transparency Toolkit (SPOTT), and Worldwide Fund for Nature (WWF) Palm Oil Buyers Scorecard. The chapter includes examples of SPOTT data from Malaysian palm-oil companies. This chapter ends with a discussion of weaknesses in current governance that justify the need for more governmental initiatives towards sustainable palm-oil production.

2 The impact of the palm-oil industry on biodiversity and natural capital

The palm-oil industry's impact on nature, biodiversity, and natural capital is easily visible by flying over the Malaysian regions of Sabah, Sarawak, Johor, Pahang, and Perak. In place of the once untouched Malaysian natural forests, with their irregular trees and plants, there are now neatly organised palm-oil plantations designed to maximise productivity. The palm-oil industry has caused widespread deforestation, leading to biodiversity loss, extinction risk, and contributing to climate change (Abdullah and Rimmel, 2021; Rimmel and Abdullah, 2022). As well as these ecological impacts, oil-palm cultivation has also been associated with air pollution (e.g. through slash-and-burn practices to clear land for cultivation), water and soil pollution, and soil erosion and contributing (through deforestation) to climate change (Savi, 2018).

Previous studies show that extensive conversion of natural forests to oil palm plantations increases biodiversity loss and increases the extinction risk of highly valued species, mainly through forest clearing and the use of slash-and-burn practices to prepare land for oil-palm plantation (Meijaard et al., 2018; Seaman et al., 2018; Abdullah and Rimmel, 2021; Rimmel and Abdullah, 2022). Abdullah et al. (2020) showed that the use of slash-and-burn practices causes the so-called Southeast Asian haze which is sufficiently severe to be a factor in assessing the environmental performance of palm-oil companies. The loss of Malaysian natural forest in particular has severely affected Bornean orangutan populations. The Bornean orangutan is classified as critically endangered according to the International Union for the Conservation of Nature and Natural Resources (IUCN) Red List of Threatened Species (Seaman et al., 2018). The massive growth in palm-oil production has e.g. led to the clearance of a million ha of forest in the eastern Sabah region. This has forced the Bornean orangutan to retreat from its natural habitat into surrounding fragmented natural forest areas. Seaman et al. (2018) found that orangutans struggle to move between fragmented natural forest areas and have become isolated populations, leading to overcrowded Bornean orangutan populations in remnant forest patches (Yue et al., 2015). The smaller the patch of remnant forest, the lower the survival chances because of increased competition for resources (Ancrenaz et al., 2010; Santika et al., 2017). This habitat loss has also increased human-orangutan contact, leading to increased killing of orangutans (Davis et al., 2013; Corlett, 2007).

As well as the orangutan, deforestation to boost oil-palm cultivation also threatens critically endangered species such as the Sumatran rhinoceros, the Sumatran elephant, the Sumatran tiger, white-winged wood ducks, and saltwater crocodiles (Meijaard et al., 2018). The slash-and-burn practices used to prepare the land for oil-palm plantation together forest clearance has also caused the loss of 46 species of forest birds (Koh et al., 2011). Fragmentation of natural forest areas into a few isolated areas means that surviving populations in these areas may not be viable and could become extinct over several generations (Rimmel and Abdullah, 2022).

Oil-palm plantations are encroaching into the Kalimantan Forest on Borneo Island which is home to 12 000 species of flowering plants, 3000 species of trees, 44 species of mammals such as rhinos and elephants, 100 species of endemic amphibian, 47 species of lizard, and 41 species of snake. This has resulted in a significant loss of butterfly and bird species (Ginoga et al., 2019; Yudea and Santosa, 2019). The excessive use of pesticides and fertilisers by oil-palm plantations has also had negative effects on over 160 species of fish in this area (Darras et al., 2019).

3 Sustainability initiatives in the Malaysian palm-oil industry and the issue of transparency

Although it is dominated by a handful of multinational companies at the refining stage, the palm oil value chain is complex. It comprises a wide range of stakeholders, from upstream growers of oil palm of all sizes to downstream processors (e.g. millers and refiners), traders, consumer goods manufacturers, and others producing final products (incorporating palm oil as an ingredient) through to retailers and, finally consumers (Pacheco et al., 2018). The Malaysian Government e.g. grants concessions to palm-oil producers through regulators such as the MPOCC and the Malaysian Palm Oil Board (MPOB), allowing producers to clear land and plant palm-oil trees. Palm-oil producers can be categorised into

- industrial plantations;
- medium-scale operators; and
- smallholder plantations.

Industrial plantations are typically managed by large national or multinational companies operating across thousands of hectares and typically with its own milling and processing operations (Meijaard et al., 2018). Medium-scale operators typically have plantations of between 50 and several thousand hectares. Smallholder plantations are usually family farms of less than 50 ha. In the Southeast Asia region, smallholders usually manage about 35–40% of planted oil palms (Gaveau al., 2019). Regulating and monitoring the activities of such a large number and diversity of producers is challenging.

The extent of the challenge can be seen in 2001, when representatives of the food processing and distribution sector launched a dialogue with NGOs such as the WWF and plantation companies, leading in 2004 to the establishment of the first voluntary sustainability standard in the sector: the Roundtable on Sustainable Palm Oil (RSPO). In Malaysia, this voluntary standard has also been supplemented by a national sustainability standard set up by the national government in 2013: Malaysian Sustainable Palm Oil (MSPO) (Djama, 2018). In 2021, MPOCC stated that over 88% of Malaysian oil-palm plantation were now certified under the MSPO certification scheme. Chain Reaction Research (2020) showed that more than 80% of large Malaysian palm-oil companies have implemented NDPE policies. However, as the previous section illustrates, initiatives such as RSPO have been only partially successful in practice in reducing deforestation and biodiversity loss in practice (Carlson et al., 2018; Cazolla Gatti et al., 2018).

A key issue is transparency. According to accountability theory, companies receive rewards if they provide information on how well they meet the expectations of stakeholders in the value chain (such as regulatory agencies, shareholders, potential investors, NGOs, and customers). Failing to meet these expectations may result in punishment (e.g. legal sanctions, withdrawal of investment, or reduced sales) (Afolabi et al., 2023; Rimmel, 2020). Transparent information about the positive changes enacted by palm-oil-producing companies and the degree to which their activities are sustainable are crucial to their accountability to stakeholders. International investors and other stakeholders have begun to intervene and play monitoring roles to ensure that palm-oil supply chains make progress towards meeting sustainability commitments (Grayson and Stampe, 2012). Initiatives such as the EU Platform on Sustainable Finance allow the global banking industry e.g. to play an important role in reducing deforestation if banks are able to use information on sustainable palm-oil production to make decisions on which companies in which to invest or e.g. issue green bonds to palm-oil companies to finance sustainable practices. Key to this process is access to reliable information (Luque-Vílchez et al., 2023; Afolabi et al., 2023; Rimmel and Abdullah, 2022).

However, an analysis of environmental reports of Malaysian palm-oil companies between 2013 and 2017 found that information about environmental policies, governance, and performance lacked transparency (Abdullah et al., 2020). As an example, fines and penalties for noncompliance with environmental laws were not disclosed. Malaysian regulators have not specifically defined the terms of what exactly needs to be disclosed and how, leaving this to the discretion of the companies. The voluntary nature of these reports helps explain the poor transparency of the palm-oil industry in Malaysia. To address this issue and make the palm-oil industry more accountable and sustainable, transparency initiatives have been developed by several non-profit organisations.

4 Transparency initiatives to monitor the palm-oil industry's impact on biodiversity and natural capital

Different stakeholder groups, especially environmental NGOs, have placed intense pressure on the Malaysian government and its regulatory institutions to make Malaysian palm-oil producers improve their disclosure of information about the sustainability of palm-oil production. The Accountability Framework Initiative defines disclosure as the public sharing of information by companies, including the reporting of company policies and commitments, business structures, financial interests, supplier lists, conflicts of interest, and political actions. Disclosure therefore is a tool for transparency and is key to stakeholders in making informed decisions (Accountability Framework, 2021).

In response, some NGOs have developed specific disclosure frameworks for palm-oil-producing companies that measure a number of indicators to show the impact of these companies on biodiversity, the environment, and ecosystems. Five initiatives are specifically dedicated to monitoring the palm-oil industry's impact on biodiversity and natural capital. A general overview is provided in Table 1, which is followed by a review of these five transparency initiatives for the Malaysian palm-oil industry.

4.1 Carbon Disclosure Project

CDP started as the Carbon Disclosure Project but is now known simply as CDP. CDP is a global non-profit charity, with regional offices e.g. in USA and Canada, Latin America, Europe, India, Australia and New Zealand, China, Japan, and Hong Kong, with local partners in over 50 countries. Established in 2000, CDP runs a global disclosure system for investors, companies, cities, states, and regions to demonstrate their environmental performance. CDP has assessed more than 11 400 companies across 14 industries, including upstream and downstream palm-oil companies (CDP, 2022).

As shown in Table 1, CDP uses primary data collected through open-ended questionnaires completed by companies. This method allows CDP to collect rich data because companies have more room to respond in this open-text format and provide more detailed explanations. Considering that CDP's questionnaires use a single source of information, they provide consistent and comparable data, which allow a more meaningful analysis (Luo et al., 2012). CDP's areas of focus are the climate, water, and forests. The questionnaires' scoring methodology is used to assess companies' progress towards environmental stewardship. These questionnaires can help companies to identify ways to manage their own environmental risks and opportunities. However, neither CDP nor its scoring partners verify the information they receive from individual companies (CDP, 2022). CDP also discloses only limited information on its website: www.cdo.net. Only the total band score for individual companies is publicly reported, without the detailed score for each indicator or question, making it difficult to track specific areas of performance. However, investors and the public can pay for full access to these detailed data. Eight Malaysian palm-oil companies have so far been assessed through this initiative.

4.2 Forest 500

Forest 500 was established in 2001 by an environmental NGO: Global Canopy. This organisation receives financial support from the UK government and the Climate and Land Use Alliance in the US. The main focus of this initiative is to accelerate progress towards a deforestation-free economy through improved transparency, innovative finance, and strategic communications. Forest 500 ranks the companies that are most responsible for driving tropical deforestation (including upstream and downstream palm-oil companies) and identifies the financial institutions that fund them. In order to facilitate ranking, the company uses multiple data sources, i.e. company websites, certification schemes (e.g. RSPO), and data from CDP. The Forest 500 assessment methodology is used annually to rank companies with the greatest influence on what they term

Table 1 General overview of five transparency initiatives towards the Malaysian palm-oil industry

	CDP	Forest 500	WWF Palm Oil Buyers Scorecard	SPOTT	NDPE IRF
ESTABLISHED	2000	2001	2009	2014	2018
Developer	CDP, non-profit charity	Global canopy, independent environmental organisation	WWF, independent conservation organisation	Zoological Society of London (ZSL), conservation charity	The collaborative group of 30 companies
Website	www.cdp.net	www.forest500.org	www.palmoilscorecard.panda.org	www.spott.org	www.ndpe-irf.net
Objective	To run the global disclosure system for investors, companies, cities, states, and regions to manage their environmental impact	To accelerate progress towards a deforestation-free global economy – through improved, transparency, innovative finance, and strategic communications	To ensure that is no destruction of nature including no deforestation ('conversion-free') along their supply chains and support a responsible and sustainable palm-oil industry beyond their own supply chain	To track transparency about companies' sustainability and to influence them to increase disclosures and improve their practices on the ground	Meeting NDPE commitments is an ongoing process to understand what is needed to achieve it, as well as a commitment to track progress, identify where action is needed and to implement the necessary changes and improvements
Industry coverage	14 industries including palm-oil industry (upstream and downstream)	Agriculture including palm oil, soy, beef, timber, leather, pulp, and paper (upstream and downstream)	Palm oil (downstream)	Palm oil, timber and pulp, natural rubber (upstream and downstream)	Palm oil (upstream and downstream)
Data collection methods	Questionnaire	Content analysis	Content analysis and questionnaire	Content analysis	Observations
Main source of data	Primary data – company's response	Secondary data – company websites, reporting portals such as CDP Forests and certification schemes	Primary and secondary data – RSPO records, company websites, company's response	Primary and secondary data – publicly available reports and publications	Primary data – checklist and template

Main areas of focus	Climate change, water and deforestation	Deforestation	Deforestation	Climate change, water and deforestation, labour rights, and governance	Deforestation, expansion on peat, labour rights, and land rights
Indicators	1 Governance 2 Policies 3 Commitments 4 Strategies 5 Certification 6 Traceability 7 Compliance 8 Supply chain 9 Initiatives 10 Ecosystem restoration and protection	1 RSPO certification 2 Deforestation 3 Traceability 4 Supply chain 5 Smallholders	1 RSPO certification 2 Deforestation 3 Palm-oil consumption 4 Traceability 5 Supply chain	1 Sustainability policy and leadership 2 Landbank, maps, and traceability 3 Deforestation and biodiversity 4 High Conservation Value (HCV), High Carbon Stock (HCS), and impact assessments 5 Peat, fire, and greenhouse gas (GHG) emissions 6 Water, chemical, and pest management 7 Community, land, and labour rights 8 Certification standards (RSPO, MSPO, ISPO) 9 Smallholders and suppliers 10 Governance and grievances	1 Supply chain 2 Certification standards (RSPO, MSPO, ISPO, International Sustainability and Carbon Certification, Palm Oil Innovation Group) 3 Traceability 4 Peat 5 Deforestation 6 Land rights 7 Smallholders 8 Grievance management
Disclosure of results	The full score is publicly reported but not the specific indicator scores	The full score and specific indicators are publicly reported	The full score and specific indicators are publicly reported	The full score and specific indicators are publicly reported	Not available yet
Malaysian companies palm oil assessed (based on the latest assessment)	8 companies	8 companies	3 companies	22 companies	Not available yet

the global forest risk commodity (FRC). Companies are assessed against four categories (Global Canopy, 2021):

- Overall approach;
- Content of commitments;
- Social considerations; and
- Reporting and implementation.

Companies are assessed on their general approach towards deforestation-free or conversion-free commitments, including policies and governance mechanisms. Under 'content of commitments', all companies are assessed for commitments that are relevant to their operations. Palm-oil companies are evaluated on the basis of four indicators:

- RSPO certification;
- Commitment to NDPE;
- Supply chain management; and
- Traceability.

Under 'social considerations', two main indicators are assessed, namely labour rights and smallholders. Under the 'reporting and implementation' category, companies are assessed on the basis of their commitment to delivering transparent information on their progress towards sustainable palm-oil production or consumption.

Table 1 shows that Forest 500 publicly discloses both the full score for individual companies, and a specific score for each indicator, which can help stakeholders better understand and track the progress of these companies towards their deforestation-free commitment. One limitation is that this initiative only focuses on deforestation-related key performance indicators (KPIs) and disregards other environmental KPIs, such as water- and air-related risks. Furthermore, Forest 500 also assesses companies on the basis of the data collected from different sources (using different types and methods of measurement), which can skew the data and make analysis of results difficult. To date eight Malaysian palm-oil companies have been assessed through this initiative.

4.3 The Worldwide Fund for Nature Palm Oil Buyers Scorecard

The WWF Palm Oil Buyers Scorecard was developed by WWF (World Wide Fund for Nature) in 2009. This scorecard is dedicated specifically to palm oil but only focuses on downstream palm-oil companies. As of 2021, 227 palm-oil-buying companies had been assessed by WWF, one of which is Malaysian. Unlike CDP and Forest 500, WWF uses secondary and primary data in its assessment. To evaluate palm-oil buyers, WWF uses data collected for the Palm Oil Buyers Scorecard through an extensive questionnaire, which includes information reported by RSPO members in their 2020 Annual Communication of Progress (ACOP). All the data collected relate to companies' palm-oil purchases and sustainability actions taken to date. Table 1 shows that for the WWF Palm Oil Buyers Scorecard, companies are assessed on five main indicators:

- RSPO certification;
- Deforestation-free commitment;
- Palm-oil consumption;
- Traceability; and
- 5. Supply chain management.

The 2021 WWF Palm Oil Buyers Scorecard assessed companies against a more comprehensive set of criteria aimed at evaluating the key actions they are taking to support ethical palm-oil supply chains, in alignment with the guidance outlined in the Accountability Framework. The WWF reports the full score, together with each indicator, on its website (www.palmoilscorecard.panda.org).

4.4 Sustainability Policy Transparency Toolkit

The SPOTT is a free online platform that is used to assess commodity producers, processors, and traders on their public disclosure regarding their organisation, policies, and practices related to environmental, social, and governance (ESG) issues (Oppenheimer et al., 2021). The Zoological Society of London (ZSL) established SPOTT in 2014 to develop a toolkit to score palm oil, tropical forestry, and natural rubber companies against over 100 sector-specific ESG indicators to benchmark their progress over time. SPOTT's main objective is to improve transparency in reporting and implementation of best practice in production of three commodities:

- Palm oil;
- Timber and wood pulp; and
- Natural rubber.

SPOTT currently covers 100 palm-oil production and processing companies, including 21 in Malaysia.

SPOTT has developed a specific scoring framework for each commodity. This includes requiring companies to provide 'practice indicators' such as evidence of how they monitor deforestation. SPOTT expects palm-oil companies to report three types of disclosure:

- Organisation;
- Policy; and
- Practice.

In the 'organisation' category, which features 39 disclosure items, a company should disclose information regarding its operations, assets, and management structure. In the 'policy' category, which includes 76 disclosure items, a company should disclose information regarding the policies, commitments, and processes it uses to guide its operations and practices on the ground. In the 'practice' category, which includes 76 disclosure items, a company should disclose information regarding the activities it undertakes in order to progress towards its targets and implement its policies and commitments on the ground. In the 'practice' category, disclosure items are broken down into three groups:

- Self-reported;
- Externally verified; and
- Certified.

These groups make it easier to assess the reliability of the information supplied. SPOTT also specifies how e.g. deforestation is to be monitored, including the extent of the area monitored and the timeframe involved. Similarly, for biodiversity protection, SPOTT requests companies to report whether they identify species whose conservation is a concern, referencing international or national lists of endangered species. The number of disclosures can be adjusted for individual companies, as they might not all include operations such as cultivation or processing. A number of indicators relate directly to the monitoring of natural capital and the prevention of biodiversity loss.

This framework is reviewed annually through feedback it gathers from investors and other stakeholders. This annual review ensures results are robust and that the framework reflects the latest developments in environmental, social, and governance transparency and reporting expectations (Rimmel and Abdullah, 2022). Like the WWF Palm Oil Buyers Scorecard, SPOTT assesses palm-oil companies by using secondary and primary data. In the first stage, SPOTT uses publicly available reports and publications to assess upstream and downstream palm-oil companies. After the first review, SPOTT submits the first draft of the assessment to each company, giving them the opportunity to review, provide feedback, and make further public disclosures if necessary. SPOTT then consolidates the first assessment draft with the feedback it receives from the company. The detailed results are eventually published publicly on its website.

SPOTT provides a unique assessment methodology to measure palm-oil companies' progress towards sustainable palm-oil targets (Rimmel and

Abdullah, 2022). It provides a different score for information disclosed based on the reliability of the data. The highest score is awarded to information that can be externally verified by a second or third party, whilst the lowest score is given to self-reported information. According to Rimmel and Abdullah (2022), investors, buyers, and other key influencers can use the SPOTT assessments to inform stakeholder engagement, manage ESG risk and increase transparency across multiple industries.

4.5 No Deforestation, No Peat, and No Exploitation Implementation Reporting Framework

NDPE IRE stands for the No Deforestation, No Peat, and No Exploitation Implementation Reporting Framework (NDPE IRF). NDPE IRF was initiated in 2018 by the Palm Oil Collaboration Group, a collaborative group of more than 30 companies from all stages of the palm-oil supply chain, including producers, refiners, traders, manufacturers, and retailers (www.ndpe-irf.net). The development of its framework is supported by environmental NGOs such as Proforest. NDPE IRF provides a checklist to be completed by each company. NDPE IRF is distinct because it was initiated by the companies themselves. Its development suggests that environmental and sustainability awareness amongst companies is increasing.

5 Transparency in practice: SPOTT framework examples from Malaysian palm-oil companies

Comparing these five transparency initiatives, it currently seems that SPOTT is the most comprehensive transparency initiative for the palm-oil industry. There is e.g. a lack of recognition of MSPO certification schemes in CDP, Forest 500, and the WWF Palm Oil Buyers Scorecard. In contrast SPOTT recognises these schemes in its scoring system. It also incorporates information missing in from other initiatives, such as plantation location/maps as well as water- and air-related risks. SPOTT indicators are reviewed regularly to reflect the latest developments in environmental, social, and governance transparency and reporting expectations. SPOTT indicator reviews also include the feedback received from investors and other stakeholders. SPOTT is therefore used in this section to illustrate some of the information available on Malaysian palm-oil-producing companies.

Out of the 100 assessed palm-oil companies (including producers, processors, and traders) from all over the world, 21 are based in Malaysia (see Table 2).

Table 2 shows that palm-oil companies based in Malaysia vary both in SPOTT rank, from 6th to 97th, as well as total score, ranging from over 862% to 0%. The company United Plantations Bhd is ranked 6th by SPOTT with a total

Table 2 Malaysian palm-oil companies in SPOTT ranking

	Company	SPOTT link	SPOTT rank	Total score	Landbank	Market cap	Media
1	United Plantations Bhd	https://www.spott.org/palm-oil/united-plantations-bhd/	6	86.2%	58.8 kha	US$1403.0M	9
2	Sime Darby Plantation Sdn Bhd	https://www.spott.org/palm-oil/sime-darby-plantation-sdn-bhd/	10	83.4%	718.2 kha	US$3674.0M	99
3	Hap Seng Plantations Holding Bhd	https://www.spott.org/palm-oil/hap-seng-plantations-holdings-bhd/	13	79.4%	39.7 kha	US$390.0M	4
4	Kuala Lumpur Kepong Bhd	https://www.spott.org/palm-oil/kuala-lumpur-kepong-bhd/	15	77.9%	Not disclosed	US$5376.0M	76
5	IOI Corporation Bhd	https://www.spott.org/palm-oil/ioi-corporation-bhd/	17	76.4%	206.5 kha	US$5711.0M	95
6	FGV Holdings Bhd	https://www.spott.org/palm-oil/fgv-holdings-bhd/	19	74.2%	432.8 kha	US$1313.0M	65
7	Genting Plantations Bhd	https://www.spott.org/palm-oil/genting-plantations-bhd/	23	72.2%	243.2 kha	US$4769.0M	34
8	Kulim (Malaysia) Bhd	https://www.spott.org/palm-oil/genting-plantations-bhd/	26	67.0%	76.2 kha	Private co.	1
9	IJM Plantations Bhd	https://www.spott.org/palm-oil/ijm-plantations-bhd/	33	62.2%	76.2 kha	US$656.0M	6
10	TSH Resources Bhd	https://www.spott.org/palm-oil/tsh-resources-bhd/	34	61.2%	103.9 kha	US$371.0M	10
11	Boustead Plantations Bhd	https://www.spott.org/palm-oil/boustead-plantations-bhd/	45	52.9%	93.9 kha	US$366.0M	4

12	Glenealy Plantations Sdn Bhd	https://www.spott.org/palm-oil/glenealy-plantations-sdn-bhd/	46	52.3%	51.3 kha	Private co.	10
13	Sarawak Oil Palms Bhd	https://www.spott.org/palm-oil/sarawak-oil-palms-bhd/	52	44.8%	122.0 kha	US$497.0M	13
14	Carotino Group	https://www.spott.org/palm-oil/carotino-group/	53	44.6%	40.7 kha	Private co.	1
15	BLD Plantation (Bintulu Lumber Development Bld plantation)	https://www.spott.org/palm-oil/bld-plantation-bhd-bintulu-lumber-development-bld-plantation/	67	26.6%	51.4 kha	US$208.0M	14
16	QL Resources Bhd	https://www.spott.org/palm-oil/ql-resources-bhd/	72	14.3%	Not disclosed	US$2.8M	2
17	FGV IFFCO	https://www.spott.org/palm-oil/fgv-iffco/	76	11.2%	N/A	Private co.	6
18	FELCRA Bhd	https://www.spott.org/palm-oil/felcra-bhd/	78	7.8%	257.3 kha	Private co.	2
19	Rimbunan Hijau Group	https://www.spott.org/palm-oil/rimbunan-hijau-group/	79	7.5%	N/A	Private co.	2
20	Tradewinds Plantation Bhd	https://www.spott.org/palm-oil/tradewinds-plantation-bhd/	82	5.8%	Not disclosed	Private co.	8
21	Sazean Holdings	https://www.spott.org/palm-oil/sazean-holdings/	97	0%	Not disclosed	Private co.	0

Table 3 Three types of SPOTT disclosures by Malaysian palm-oil companies

	Company	Total score	Organisation	Policy	Practice	Self-reported	External	Certified
1	United Plantations Bhd	**86.2%** 158.62/184	34/37	68.5/76	56.1/71	7.2/71	18.5/71	30.4/71
2	Sime Darby Plantation Sdn Bhd	**83.4%** 159.2/191	32.5/39	67.5/76	59.2/76	16/76	12.8/76	29.4/76
3	Hap Seng Plantations Holding Bhd	**79.6%** 122.65/154	26/30	59/71	37.6/53	7.2/53	4/53	26.5/53
4	Kuala Lumpur Kepong Bhd	**77.9%** 148.01/190	28/38	64.5/76	55.5/76	13.5/76	12.3/76	29.7/76
5	IOI Corporation Bhd	**76.4%** 145.9/191	27.5/39	62.5/76	55.9/76	13.7/76	12/76	29.2/76
6	FGV Holdings Bhd	**74.2%** 140.24/189	29/37	62/76	49.2/76	8.8/76	12.5/76	28/76
7	Genting Plantations Bhd	**72.2%** 136.54/189	29/38	62/76	45.5/75	7.5/75	10/75	28/75
8	Kulim (Malaysia) Bhd	**67.0%** 104.47/156	20/30	43.5/71	41/55	6.4/55	7.8/55	26.9/55
9	IJM Plantations Bhd	**62.2%** 110.75/178	24/33	59/75	27.8/70	19.5/70	2/70	6.3/70
10	TSH Resources Bhd	**61.2%** 99.77/163	19.5/32	50.5/72	29.8/59	3.7/59	4/59	22.1/59
11	Boustead Plantations Bhd	**52.92** 83.07/157	17.5/31	36.5/71	29.1/55	3.7/55	2/55	23.4/55
12	Glenealy Plantations Sdn Bhd	**52.3%** 86.75/166	22/31	43.5/72	21.3/63	5/63	9.8/63	6.5/63
13	Sarawak Oil Palms Bhd	**44.8%** 82.5/184	21.5/35	39.5/75	21.5/74	9.5/74	4.8/74	7.3/74
14	Carotino Group	**44.6%** 81.7/183	10/35	38/75	33.7/73	4.1/73	3/73	26.6/73
15	BLD Plantation (Bintulu Lumber Development Bhd plantation)	**26.6%** 46.75/176	7/32	30/74	9.8/70	5.5/70	1/70	3.3/70
16	QL Resources Bhd	**14.3%** 25.25/170	11/34	8.5/72	4.8/64	3.3/64	0.8/64	0.8/64
17	FGV IFFCO	**11.2%** 12.03/107	1.5/21	4.5/48	6/38	0/38	4/38	0/38
18	FELCRA Bhd	**7.8%** 13.5/174	6/34	5/74	2.5/66	1.5/66	1/66	0/66
19	Rimbunan Hijau Group	**7.5%** 12.25/163	5.5/32	4.5/71	2.3/60	1.5/60	0.8/60	0/60
20	Tradewinds Plantation Bhd	**5.8%** 11/190	1.5/39	7/76	2.5/75	0/75	1/75	1.5/75
21	Sazean Holdings	**0%** 0/191	0/39	0/76	0/76	0/76	0/76	0/76

Bold values indicate the percentage of the SPOTT total score, which SPOTT.org provides.

score of 86.2%, making it the highest-ranked Malaysian company. The company provides land-holding information that shows that it grows oil palm on 58.8 kha and boasts a market capitalisation of RM 1403.0. SPOTT also includes nine media reports that have been published on the company's activities. However, United Plantations Bhd is small in comparison to the company Sime Darby Plantation, which grows oil palm on 718.2 kha and achieved a total score of 83.4% with a SPOTT rank at No 10.

As Table 3 shows, some companies fail to disclose key information e.g. on land holdings or market capitalisation. The depth of disclosure varies across Malaysian palm-oil companies, with some providing detailed information on total landbank, planted areas, unplanted areas, and areas set aside for conservation, whilst information from other companies is lacking, making it hard to assess their degree of commitment to and performance in protecting natural capital.

The table shows that Malaysian palm-oil companies tend to disclose more information about organisation and policy than information about practice and performance. This result is consistent with prior studies (Clarkson et al., 2013; Clarkson et al., 2008), which show that companies are likely to disclose soft information which is relatively uncontroversial, unverifiable, and subjective. This is because such information is often already available, designed to show a company in the best light and less likely to attract criticism, and is consistent with what other companies are typically prepared to disclose. SPOTT addresses this by providing a detailed explanation of how a company has performed against a particular indicator, including any sources of external verification or third-part certification. To cite one example:

> *... [Indicator] 55. Commitment to zero deforestation? The company makes this commitment through the RSPO Principles and Criteria 2018. Full points have been therefore awarded on the basis of the company's RSPO-certified landbank (≥75%) and a further commitment to be 100% certified within 5 years. The company also makes this commitment separately. 1/1 Score.*

The level and reliability of information supplied by a company is used in the scoring system. Ranking and comparing the performance of individual companies both makes it easier for investors and others to identify the best-performing companies and to encourage companies to improve their commitment to and performance in reducing deforestation and biodiversity loss.

6 Conclusion

The Malaysian economy has benefited greatly from the growth of the palm-oil industry. However, this growth is a double-edged sword, as unsustainable palm-oil production has had serious negative effects on the environment and ecosystems. The slash-and-burn practices used to prepare the land for monocultural oil-palm plantations have led to the significant fragmentation of natural habitat, biodiversity loss, extinction risk, and climate change (Abdullah and Rimmel, 2021; Rimmel and Abdullah, 2022).

This chapter has compared five transparency initiatives, CDP, Forest 500, WWF Palm Oil Buyers Scorecard, SPOTT, and NDPE IRF, in terms of their methodology and indicators and their disclosure of results. These five international transparency initiatives were established to encourage palm-oil companies to disclose more reliable, transparent and verifiable information regarding their environmental performance. Despite the fact that these five initiatives have gained support from many global stakeholders, only a few Malaysian companies seem to have engaged with them to date. The example of SPOTT showed that there is still significant room for improvement for Malaysian palm-oil companies to disclose reliable information to global stakeholders in order to show their commitment towards sustainable palm oil. Otherwise sustainable palm oil may not be so sustainable, as it could be (Gatti et al., 2019).

While the Malaysian government has made progress in improving the governance of the palm-oil industry by developing national certification schemes, palm-oil companies still e.g. do not always publish their concession maps, reflecting a continuing lack of transparency and accountability. One way forward would be to make disclosures regarding palm-oil production, processing, and trade, which are currently voluntary, mandatory. Another would be to make it necessary for these disclosures to be externally verified, which would reduce the scope for self-reported information.

Certification can also play an important role (Carlson et al., 2018). For example, the MSPO certification scheme could tighten its mandatory disclosure requirements. This would require a high level of organisational governance. It is important to monitor company policies, practices and performance to ensure they align with the interests of all stakeholders. Palm-oil companies should shift their perspective from that of short-term shareholder value towards the creation of long-term and sustainable value, which includes financial value aligned with the environment and communities.

7 Where to look for further information

- www.accountabilityframework.org.
- www.cdn.cdp.net.
- www.chainreactionresearch.com.
- www.globalcanopy.org.
- www.ipad.fas.usda.org.
- www.spott.org.
- www.panda.org.

8 References

Abdullah, M., Hamzah, N., Ali, M. H., Tseng, M. L. and Brander, M. (2020). The Southeast Asian haze: The quality of environmental disclosures and firm performance. *Journal of Cleaner Production* 246, 118958.

Abdullah, M. and Rimmel, G. (2021). Promoting sustainable palm oil through corporate transparency initiatives, Chapter 7. In: Makhbul, Z. M., Khalid, N. and Sarmidi, T. (Eds.), *Management of Sustainable Palm Oil - Policies and Environmental Considerations*. Malaysia: Penerbit UKM Press, 88–89.

Accountability Framework (2021). Report: Disclosure for a deforestation-free supply chain. Available at: https://accountabilityframework.org/fileadmin/uploads/afi/Documents/Resources/Disclosure_For_Deforestation_Free_Supply_Chain_AFi_CDP_2020-11.pdf.

Afolabi, H., Ram, R. and Rimmel, G. (2023). Harmonization of sustainability reporting regulation: Analysis of a contested arena. *Sustainability* 14(9), 5517.

Afolabi, H., Ram, R. and Rimmel, G. (2023). Influence and behaviour of the new standard setters in the sustainability reporting arena: Implications for the Global Reporting Initiative's current position. *Sustainability Accounting, Management and Policy Journal* 14(4), 743–775.

Ancrenaz, M., Ambu, L., Sunjoto, I., Ahmad, E., Manokaran, K., Meijaard, E. and Lackman, I. (2010). Recent surveys in the forests of Ulu Segama Malua, Sabah, Malaysia, show that orang-utans (P. p. morio) can be maintained in slightly logged forests. *PLoS ONE* 5(7), e11510.

Carlson, K. M., Heilmayr, R., Gibbs, H. K., Noojipady, P., Burns, D. N., Morton, D. C., Walker, N. F., Paoli, G. D. and Kremen, C. (2018). Effect of oil palm sustainability certification on deforestation and fire in Indonesia. *Proceedings of the National Academy of Sciences of the United States of America* 115(1), 121–126.

Cattau, M. E., Marlier, M. E. and DeFries, R. (2016). Effectiveness of Roundtable on Sustainable Palm Oil (RSPO) for reducing fires on oil palm concessions in Indonesia from 2012 to 2015. *Environmental Research Letters* 11(10), 105007.

CDP (2022). Measuring progress towards a sustainable palm oil supply chain a company's journey. Available at: https://cdn.cdp.net/cdp-production/cms/reports/documents/000/006/522/original/CDP_Palm_Oil_Report_2022_Final.pdf?1660821343.

Chain Reaction Research (2020). *The Chain: Deforestation's Public Health Impact in the Spotlight Amid COVID-19 Outbreak*. Available at: https://chainreactionresearch.com/the-chain-deforestations-public-health-impact-in-the-spotlight-amid-covid-19-outbreak/.

Choiruzzad, S. A. B. (2019). Save palm oil, save the nation: Palm oil companies and the shaping of Indonesia's national interest. *Asian Politics and Policy* 11(1), 8–26.

Clarkson, P. M., Fang, X., Li, Y. and Richardson, G. (2013). The relevance of environmental disclosures: Are such disclosures incrementally informative? *Journal of Accounting and Public Policy* 32(5), 410–431.

Clarkson, P. M., Li, Y., Richardson, G. D. and Vasvari, F. P. (2008). Revisiting the relation between environmental performance and environmental disclosure: An empirical analysis. *Accounting, Organizations and Society* 33(4–5), 303–327.

Corlett, R. T. (2007). The impact of hunting on the mammalian fauna of tropical Asian forests. *Biotropica* 39(3), 292–303.

Darras, K. F. A., Corre, M. D., Formaglio, G., Tjoa, A., Potapo, A., Brambach, F., Sibhatu, K. T., Grass, I., Rubiano, A. A., Buchori, D., Drescher, J., Fardiansah, R., Hölscher, D., Irawan, B., Kneib, T., Krashevska, V., Krause, A., Kreft, H., Li, K., Maraun, M., Polle, A., Ryadin, A. R., Rembold, K., Stiegler, C., Scheu, S., Tarigan, S., Valdés-Uribe, A., Yadi, S., Tscharntke, T. and Veldkamp, E. (2019). Reducing fertilizer and avoiding herbicides in oil palm plantations–ecological and economic valuations. *Frontier for Global Change, Section Tropical Forests*. doi: 10.3389/ffgc.2019.00065.

Davis, J. T., Mengersen, K., Abram, N. K., Ancrenaz, M., Wells, J. A. and Meijaard, E. (2013). It's not just conflict that motivates killing of orangutans. *PLoS ONE* 8(10), e75373.

Djama, M. (2018). Certifying sustainability in oil palm cultivation. In: Rival, A. (Ed.), *Achieving Sustainable Cultivation of Oil Palm Volume 2: Diseases, Pests, Quality and Sustainability*. Cambridge: Burleigh Dodds Science Publishing, 315-330.

Gatti, R. C., Liang, J., Velichevskaya, A. and Zhou, M. (2019). Sustainable palm oil may not be so sustainable. *Science of the Total Environment* 652, 48-51.

Gaveau, D. L. A., Locatelli, B., Salim, M. A., Yaen, H., Pacheco, P. and Sheil, D. (2019). Rise and fall of forest loss and industrial plantations in Borneo (2000-2017). *Conservation Letters* 12(3), e12622.

Ginoga, L. N., Santosa, Y. and Mutmainnah, A. R. (2019). The loss, gain, and diversity of butterfly species due to the development of PT PKWE oil palm plantation, West Kalimantan Province. In *IOP Conference Series: Earth and Environmental Science* 336(1): 012025. IOP Publishing.

Global Canopy (2021). *The Little Book of Investing in Nature - A Simple Guide to Finding Life on Earth*. Available at: https://globalcanopy.org/wp-content/uploads/2021/07/LBIN_2020_RGB_ENG.pdf.

Grayson, J. and Stampe, J. (2012). *Palm Oil Investor Review: Investor Guidance on Palm Oil. The Role of Investors in Supporting the Development of a Sustainable Palm Oil Industry*. Switzerland: World Wide Fund for Nature (WWF).

Gutiérrez-Vélez, V. H., DeFries, R., Pinedo-Vásquez, M., Uriarte, M., Padoch, C., Baethgen, W., Fernandes, K. and Lim, Y. (2011). High-yield oil palm expansion spares land at the expense of forests in the Peruvian Amazon. *Environmental Research Letters* 6(4), 044029.

Koh, L. P., Miettinen, J., Liew, S. C. and Ghazoul, J. (2011). Remotely sensed evidence of tropical peatland conversion to oil palm. *Proceedings of the National Academy of Sciences of the United States of America* 108(12), 5127-5132.

Luo, L., Lan, Y. C. and Tang, Q. (2012). Corporate incentives to disclose carbon information: Evidence from the CDP Global 500 report. *Journal of International Financial Management and Accounting* 23(2), 93-120.

Luque-Vílchez, M., Cordazzo, M., Rimmel, G. and Tilt, C. A. (2023). Key aspects of sustainability reporting quality and the future of GRI. *Sustainability Accounting, Management and Policy Journal* 14(4), 637-659.

Meijaard, E., Garcia-Ulloa, J., Sheil, D., Wich, S. A., Carlson, K. M., Juffe-Bignoli, D. and Brooks, T. M. (2018). *Oil Palm and Biodiversity. A Situation Analysis by the IUCN Oil Palm Task Force*. IUCN: Switzerland.

Munasinghe, M., Jayasinghe, P., Deraniyagala, Y., Matlaba, V. J., dos Santos, J. F., Maneschy, M. C. and Mota, J. A. (2019). Value-Supply Chain Analysis (VSCA) of crude palm oil

production in Brazil, focusing on economic, environmental and social sustainability. *Sustainable Production and Consumption* 17, 161–175.
Oppenheimer, P., Clarke, E., Cupit, O., Delabre, I., Dodson, A., Guindon, M., Hatto, A., Lam, J., Lawrence, L., Melot, C., Spencer, E., Wainwright-Déri, E., Freeman, R. and Hoffmann, M. (2021). The SPOTT index: A proof-of-concept measure for tracking public disclosure in the palm oil industry. *Current Research in Environmental Sustainability* 3, 100042.
Pacheco, P., Levang, P., Dermawan, A. and Schoneveld, G. (2018). The palm oil governance complex: Progress, problems and gaps. In: Rival, A. (Ed.), *Achieving Sustainable Cultivation of Oil Palm Volume 1: Introduction, Breeding and Cultivation Techniques*. Cambridge: Burleigh Dodds Science Publishing, 49–72.
Potter, L. (2015). *Managing Oil Palm Landscapes: A Seven-Country Survey of the Modern Palm Oil Industry in Southeast Asia, Latin America and West Africa*. Bogor: CIFOR, 154.
Rimmel, G. (2020). Chapter 1. Accounting for sustainability – historical development of the field. In: Rimmel, G. (Ed.), *Accounting for Sustainability*. Abingdon: Routledge.
Rimmel, G. and Abdullah, M. (2022). Oil palm industry and biodiversity loss: Enhancing governance through the sustainability policy transparency toolkit (SPOTT), Chapter 6. In: Macpherson, M. and Atkins, J. (Eds.), *Extinction Governance: Implementing a Species Protection Action Plan for the Financial Markets*. London: Routledge. pp. 1115-138, 610 pages.
Santika, T., Meijaard, E., Budiharta, S., Law, E. A., Kusworo, A., Hutabarat, J. A., Indrawan, T. P., Struebig, M., Raharjo, S., Huda, I., Sulhani, Ekaputri, A. D., Trison, S., Stigner, M. and Wilson, K. A. (2017). Community forest management in Indonesia: Avoided deforestation in the context of anthropogenic and climate complexities. *Global Environmental Change* 46, 60–71.
Savi, S. (2018). The palm oil market: Growth and trends. In: Rival, A. (Ed.), *Achieving Sustainable Cultivation of Oil Palm (Volume 1): Introduction, Breeding and Cultivation Techniques*. Cambridge: Burleigh Dodds Science Publishing, 3-16.
Seaman, D., Randriahaingo, H. N. T., Randrianarimanana, H. L. L., Ravaloharimanitra, M., Humle, T. and King, T. (2018). Edge effects on indri and black-and-white ruffed lemur distribution in the Andriantantely lowland rainforest, Eastern Madagascar. *Lemur News* 21, 51–56.
Shevade, V. S. and Loboda, T. V. (2019). Oil palm plantations in Peninsular Malaysia: Determinants and constraints on expansion. *PLoS ONE* 14(2), e0210628. doi: 10.1371/journal.pone.0210628.
Singagerda, F. S., Hendrowati, T. Y. and Sanusi, A. (2018). Indonesia growth of economics and the industrialization biodiesel based CPO. *International Journal of Energy Economics and Policy* 8(5), 319–334.
USDA (2020). U.S. Department of Agriculture, palm oil explorer 2022. Available at: https://ipad.fas.usda.gov/cropexplorer/cropview/commodityView.aspx?cropid=4243000.
Varkkey, H., Tyson, A. and Choiruzzad, S. A. B. (2018). Palm oil intensification and expansion in Indonesia and Malaysia: Environmental and socio-political factors influencing policy. *Forest Policy and Economics* 92, 148.e159.
Vijay, V., Pimm, S. L., Jenkins, C. N. and Smith, S. J. (2016). The impacts of oil palm on recent deforestation and biodiversity loss. *PLoS ONE* 11(7), e0159668.
Voora, V., Larrea, C., Bermudez, S. and Baliño, S. (2019). *Global Market Report: Palm Oil* (p. 16). Manitoba: International Institute for Sustainable Development.

Wilcove, D. S. and Koh, L. P. (2010). Addressing the threats to biodiversity from oil-palm agriculture. *Biodiversity and Conservation* 19(4), 999-1007.

Yudea, C. and Santosa, Y. (2019). How does oil palm plantation impact bird species diversity? A case study from PKWE Estate, West Kalimantan. In *IOP Conference Series: Earth and Environmental Science* 336(1): 012026. IOP Publishing.

Yue, S., Brodie, J. F., Zipkin, E. F. and Bernard, H. (2015). Oil palm plantations fail to support mammal diversity. *Ecological Applications* 25(8), 2285-2292.

Yui, S. and Yeh, S. (2013). Land use change emissions from oil palm expansion in Pará, Brazil depend on proper policy enforcement on deforested lands. *Environmental Research Letters* 8(4), 044031.

Chapter 12

Natural capital accounting and biodiversity in the Italian winemaking industry

Federica Doni, University of Milano-Bicocca, Italy; Antonio Corvino, University of Foggia, Italy; and Silvio Bianchi Martini, University of Pisa, Italy

1 Introduction

Biodiversity is becoming a critical aspect of the 'environmental' dimension of environmental, social and governance (ESG) standards governing corporate behaviour. Biodiversity is a precondition for the effective functioning of ecosystems and ultimately for sustaining life on Earth. Biodiversity loss leads to simplification, decline and the progressive disappearance of life (Corvino et al., 2021; Martini et al., 2018). The United Nations highlighted safeguarding biodiversity as Sustainable Development Goal (SDG) No. 15 of the UN's 2030 Agenda for Sustainable Development. The importance of biodiversity has been recognised at both global (e.g. the impact of the: Final Report - the Economics of Biodiversity. The Dasgupta Review (https://www.gov.uk/government/publications/final-report-the-economics-of-biodiversity-the-dasgupta-review)) and European levels (e.g. the EU Biodiversity Strategy for 2030 (https://environment.ec.europa.eu/strategy/biodiversity-strategy-2030_en)).

This chapter first introduces the Italian winemaking sector. It then discusses how biodiversity is defined in the winemaking sector. The chapter goes on to analyse a range of reports on sustainability and biodiversity produced by a

http://dx.doi.org/10.19103/AS.2023.0128.10

sample of leading Italian winemaking companies. It then discusses a particular case study of Cantine Teanum, a family-run small-medium enterprise (SME) situated in the Apulia Region, using detailed qualitative analysis.

2 The Italian winemaking sector

Italy has 690 000 ha of vineyards with 650 000 ha grown for wine grapes. According to The International Organization of Vine and Wine (OIV), with 80 major varieties cultivated across 75% of the area for winemaking, Italy has the greatest diversity of varieties for winemaking. Other countries, such as Portugal and Romania, grow 40 varieties whilst major wine producers such as Spain and France grow only 15 principal varieties across the majority of the area used for viticulture. The main variety cultivated in Italy is Sangiovese, representing 8% of the total area for winemaking, followed by Montepulciano, Glera and Pinot Grigio (4% each) and Merlot (3%), leaving other varieties accounting for the remaining 77%.

One hundred fifteen major Italian wineries represent 55.6% of the sector's total turnover, 65.36% of exports and 46.2% of domestic sales (https://www.areastudimediobanca.com/). Overall, the largest 73 private wineries and the 42 cooperatives in 2021 had a turnover of just over €8 billion, exports of €4.6 billion and €3.4 billion in domestic sales, representing an increase of 18.31% of total turnover, export growth of 20.86% and domestic growth of 15.02%.

In this sector, we need to distinguish two different types of business:

- Private companies; and
- Cooperatives (bring together individual winegrowers).

The private sector (accounting for 48.5% of overall turnover from the top 115 wineries) has been particularly successful, as demonstrated by the strong results of many companies, such as

- Tenute Piccini: growth of 50% in 2021;
- Villa Sandi: growth of 32% in 2021; and
- Mionetto: growth of 22% in 2021.

These companies now have a €100 million turnover. Together with the Ermes Cooperative Group, they are now members of the exclusive club of 24 'over 100' wineries. Alongside other players, such as the Fantini group founded by Valentino Sciotti, with a turnover of €93 million, the market has shown great dynamism. For instance, Spumanti Valdo, with a turnover of €64 million, is investing €16 million to double production capacity. Tenute del Leone Alato, the winemaking arm of Genagricola (in the Assicurazioni Generali Group), has launched exclusive distribution partnerships in Salento with Cantine Fiorentino and in Sardinia with Tenute Dettori to strengthen its domestic sales. Piccini has expanded into the Langhe region by purchasing the Porta Rossa brand. Hyle Capital Partners has invested in Contri Spumanti, led by Paolo Contri, to support its development plans.

Cooperatives account for 41.5% of overall turnover from the top 115 wineries, 33.1% of exports and 52.9% of domestic sales. For the first time, cooperatives have a reduced share amongst the leading wineries. Cantine Riunite & Civ is still in first place with a turnover of €635.2 million, with two industrial groups, Iwb and Polo Botter-Mdv, in the second and third place. Amongst the top 20 cooperative wineries, ten have a turnover exceeding €100 million. The immense extent of vineyards owned by individual members of cooperatives should also be underlined.

3 Defining biodiversity in the winemaking sector

Biodiversity broadly refers to the variety of plant and animal species living in a specific area (Vandermeer and Perfecto, 1995). The OIV provides grape and wine-producing and consuming countries with guidance to develop regulations, minimise trade barriers, promote sustainable production and common quality standards for consumers (https://www.oiv.int/who-we-are/presentation). A key objective is to promote international standardisation of different national standards and practices. The OIV defines biodiversity as 'all species of plants, animals and microorganisms existing and interacting within an ecosystem' (OIV, 2018). In accordance with Boller et al. (2004), it is possible to identify three levels of biodiversity:

1. Genetic diversity (varieties);
2. Taxonomic diversity (species); and
3. Diversity of communities of organisms (ecosystems).

On this basis, we can identify three different types of biodiversity:

1. Intraspecific diversity (related to the genetic varieties within the individuals of the same species);
2. Interspecific diversity (related to the differences between species); and
3. Ecosystem level diversity (related to the variety of ecosystems and ecological niches in a specific area).

Protecting biodiversity means conserving the natural habitats within which these various types of biodiversity have evolved. However, farming has led and continues to lead to a significant loss of natural habitat. The Food and Agricultural

Organization (FAO) e.g. estimates deforestation rates of approximately 50 000 km² of tropical forest a year with tropical forests now reduced to approximately half of the area (7 500 000 km²) that they occupied in prehistoric times, primarily due to the expansion of agriculture (FAO - Global Forest Resources Assessment 2020 (https://www.fao.org/forest-resources-assessment/en/). The 2005 Millennium Ecosystem Assessment largely confirmed the predictions made by E.O. Wilson, a famous American naturalist, that deforestation would lead to the extinction of about 30 000–40 000 species per year (https://www.millenniumassessment.org/).

As well as halting habitat loss by not expanding agriculture further, there is also a need to explore how existing farming practices can promote biodiversity on land used for agriculture. Optimising yields in vineyards e.g. has usually meant creating a monoculture and reducing biodiversity (e.g. by eliminating weeds and other plants which might compete for resources with vines, with a resulting loss of floral biodiversity that then threatens the insect and animal species that depend on these plants). However, this loss of biodiversity can also threaten agricultural production. A monoculture can create an ideal environment for pests to thrive whilst reducing the natural predator species that might normally help to control pest numbers. This forces the farmer to rely on increasing use of pesticides which can both create a problem of resistance building amongst target pest species and further harming non-target beneficial species, such as pest predators and pollinators.

In 2010, in the 'International Year of Biodiversity', the World Biodiversity Association (WBA) announced 'Biodiversity Friend®', a new voluntary certification scheme to promote biodiversity practices in farming based on the principle that 'the more life there is in a territory, the less the impact of human activities on it has been' (https://biodiversityassociation.org/en/what-we-do/biodiversity-friend-2/). Assessment of the company's performance in promoting biodiversity involves a series of checklists, including establishing and monitoring biodiversity indices for soil, water and air. As part of the scheme, a company signs the 'Biodiversity Friend Declaration' which commits it to pursue ten actions that make up 'Decalogue of sustainability':

1 Developing a sustainable cultivation model;
2 Protection of soil fertility;
3 Rational management of water resources;
4 Protection of hedges, woods, meadows and other nectariferous plant species;
5 Protection of agricultural and structural biodiversity;
6 Energy saving and use of renewable energy sources;
7 Protection of rural landscapes;
8 Social sustainability;

9 Economic sustainability; and
10 Enhancing soil, water and air quality.

Each action requires various activities which can be scored in terms of performance. Because of its importance, Action 4 relating to the protection of hedges and woods scores highest. Companies must achieve a minimum score of 60 points out of 100 to obtain certification. The certification body checks performance annually, requiring an improvement by at least one point compared to the previous year. Once a company achieves 80 points, an audit moves from being carried out annually to every 2 years.

Following these principles, winemakers can undertake a wide range of measures to promote biodiversity. These include

- planting hedges or building dry stone walls (which both provide potential homes for a range of flora and fauna);
- planting trees;
- avoiding tillage whilst grassing areas beneath vines; together with alternating mowing of rows to allow a variety of grass and flower species to establish;
- promoting nectariferous plant species;
- installing bat boxes for bats and/or artificial nests for birds
- promoting wild bees and other pollinators; and
- retaining wild areas of vegetation bordering vineyards.

Developing a more diverse landscape and range of plant species provides habitat and food resources for a diverse range of birds, small mammals, amphibians, reptiles, insects, bees and molluscs. These can help manage pests, reducing reliance on pesticides. Such practices also further soil physical, chemical and biological health. Promoting grass cover e.g. improves soil physical structure, retains soil moisture, contributes to soil organic matter and encourages microorganisms in the soil. The safeguard of soil health then promotes vine growth with potentially less reliance on fertilisers as well as pesticides and less need for irrigation. A summary of the key OIV documents relating to good biodiversity practices is provided in Table 1.

There are also three voluntary certification systems specifically for Italian winemakers focused on sustainability:

- National Integrated Production Quality System (SQNPI) - a national scheme for integrated crop management developed by the Italian Ministry of Agriculture (MPAAF) (https://www.ccpb.it/en/blog/certificazione/integrated-crop-management-national-quality-system-sqnpi/);
- Equalitas (https://www.equalitas.it/en); and
- Viva (https://viticolturasostenibile.org/en/).

Table 1 Summary of OIV documents on biodiversity

Title	Aim	Technical details
OIV Collective Expertise Document 2018 Functional biodiversity in the vineyard	The 'Functional Biodiversity (FB) approach', according to Boller (2004), can be defined as 'the utilitarian part of biodiversity'.	The functional Biodiversity approach aims to 'integrate ecological infrastructures (hedgerows, woodlands, dry-stone walls, ground covers, etc.) supporting and enhancing biodiversity into the vineyard and to improve their management increasing, simultaneously, the quality of the production, while maintaining the quality of the landscapes' (OIV, 2018, p.10).
OIV-VITI 640-2020 Multi-criteria assessment of environmental impact in vitivinicultural sector-Life cycle assessment (LCA): General principles for conduction and communication of results	Recognition of the importance of assessing the overall environmental footprint of vitivinicultural production in order to develop appropriate action plans. Promotion in the vitivinicultural sector of using the Life Cycle Assessment methodology (LCA) as the most successful tool in the global and multi-criteria assessment of environmental impacts of product systems.	Biodiversity loss is one of the most relevant issues as the potential environmental impacts, like climate change, ozone depletion etc.
OIV-VITI 641-2020 OIV Guide for the implementation of principles of sustainable vitiviniculture	Updated guide for implementing the general principles of sustainable vitiviniculture throughout an organisation, identifying and hierarchising relevant stakeholders, and prioritising relevant action areas. General principles of sustainability and related actions.	*Five principles of sustainable vitiviniculture* Principle 1: Sustainable approach integrates environmental, social and economic aspects Principle 2: Sustainable vitiviniculture respects the environment Principle 3: Sustainable vitiviniculture is sensitive to social and cultural aspects Principle 4: Sustainable vitiviniculture seeks to maintain economic viability Principle 5: Sustainable initiatives require planning and assessment

(Continued)

Table 1 (*Continued*)

Title	Aim	Technical details
		Biodiversity is part of Principle 2, Area for action 2.3 'Preserving biodiversity'. Moreover, the protection of biodiversity is cited as an important issue in strategic business decision-making, particularly before planting new vineyards or establishing other vitivinicultural facilities. For preserving biodiversity, some indicators are recommended, such as: 1 % of the surface of ecological interest (terraces, stone walls, groves of trees, hedges, buffer zones near watercourses, etc.) 2 Number of species protected by specific actions of the organisation. 3 Number of varieties and genotypes in use. To evaluate biodiversity on site, the guide recommends: 1 Survey of sensitive and protected species of flora and fauna situated in the proximity of its site (through local associations, authorities, scientific institutions, etc.). 2 Keep an inventory of its sites' habitats/ecological niches and their evolution yearly.
OIV-VITI 654-2021 OIV Guidelines for the sustainable production of grape juice, concentrated juice and for processing	General sustainability principles applied to grape juice, concentrated juice, and processing methods.	Preservation of biodiversity by promoting intra and inter-varietal diversity in commercial vineyards. Concerning the natural environment, schemes should be established to preserve and promote biodiversity where vitivinicultural activities are developed, encouraging the development of local microorganisms, flora, and fauna. A new approach to functional biodiversity management is recommended concerning the planning of vineyards.

(*Continued*)

Table 1 (*Continued*)

Title	Aim	Technical details
OIV-VITI 655-2021 OIV Recommendations about valuation and importance of microbial biodiversity in a sustainable vitiviniculture context	Focus on developing policies for quantitative and qualitative valuation of microbial abundance and its biodiversity in the vineyard, its use as an indicator of functional biodiversity, biological sustainability and as a measurable value of the effect and impact of vitiviniculture practices and management systems. (OIV, 2021, p.2)	Study of the biogeographical patterns, the microbial populations and the use of harmonised monitoring methodologies for the study of microbial diversity concerning different types of soil, management techniques and age of the vineyards. Promotion of technical-scientific interactions for bringing together the skills between microbiologists, ecologists and physiologists of the vine to enhance the role and importance of the microbial biodiversity of the vineyard system.
OIV Collective expertise document 2021 Sustainable use of water in winegrape vineyards	After the adoption 2018 of Resolution OIV-VITI 569-2018, 'OIV protocol for the sustainable use of water in viticulture', OIV decided to prepare a reference and guidance tool for the sector regarding the sustainable use of water.	This document seeks to meet the objectives of the OIV's Strategic Plan for the period 2020-2024 – particularly axis 1, 'Promote environmentally friendly vitiviniculture', through points B and C: 'Improvement of environmental performance' and 'Preservation of natural resources' – and the recently approved Resolution OIV-VITI 641-2020 'OIV guide for the implementation of principles of sustainable vitiviniculture'.
OIV-VITI 652-2021 OIV recommendations concerning the selection and breeding of grapevine varieties for their adaptation to the effects of climate change	Improvement of knowledge on adapting the existing varietal assortment and the biodiversity potential in genus Vitis concerning climate change and its response to changes in growing conditions and related issues.	The use of the genetic and phenotypic diversity and variability of the grapevine and rootstock varieties to respond to the issues of adaptation to the environmental and biological constraints (abiotic and biotic stress) accentuated by climate change.

Source: https://www.oiv.int/what-we-do/standards?search_title_body=&type%5Bresolutions%5D=resolutions&items_per_page=2525

These certification schemes include similar requirements for protecting biodiversity as the WBA scheme.

There have been a number of studies reviewing e.g. how business models can incorporate sustainability in sectors, such as viticulture (Broccardo and Zicari, 2020; Ferrer et al., 2022). Other studies have discussed potential constraints to adopting more sustainable practices (Signori et al., 2017) as well as potential advantages, in terms of marketing to consumers (Tait et al., 2019). A particularly interesting field is environmental certification (Casolani et al., 2023) and the need for an indicators system (Merli et al., 2018) or framework (Flores, 2018) that can support wineries in implementing sustainable and biodiversity practices. The deployment of sustainable management practices can foster the competitiveness of businesses located in tourist destinations, as demonstrated e.g. in South Brazilian companies by Barbosa et al. (2018). Some studies have focused on Italian wineries (Corbo et al., 2014; Broccardo and Zicari, 2020; de Steur et al., 2020). However, there is a lack of an in-depth analysis of biodiversity practices and indicators amongst Italian winemakers, particularly on how wineries can improve disclosure about biodiversity practices e.g. by adopting particular practices or by complying with particular certification schemes.

4 Sustainability and biodiversity reporting in the Italian winemaking sector

As a context for the case study in the next section, the authors sampled a number of leading Italian winemaking companies to see how they reported on biodiversity (Table 2).

4.1 Caviro Group

Caviro Group operates 27 cellars in 7 regions in Italy (totalling 36 300 ha of vineyards), producing 615 000 tons of grapes (9% of all Italian grapes). It published its first sustainability report in 2019. The latest sustainability report covers the 2022 fiscal year. The protection of biodiversity is included in a list of 'material projects' with 'adoption of sustainable agricultural practices' as an

Table 2 Sample of companies

≠	Company	Revenues (millions of Euros)	Sustainability report - year
1	Gruppo Caviro	417.4	2022
2	Berlucchi Franciacorta	47.5	2021
3	Cantina Produttori Valdobbiadene	56.16	2021
4	Cecchetto	10.9	2021
5	Tasca Conti D'Almerita	17.8	2022

action to take. The company is working with renowned international oenologists to create management protocols relating to sustainability e.g. to reduce sulfites and shift to organic production. This issue is linked to SDG No. 13 to limit and adapt to climate change.

The report refers to 'the environment as a point of strength: the circularity of the productive cycle'. The main commitment is to create a circular economy, focused on waste recovery. An example of this circularity regards the process of crop residues and waste products using anaerobic digestion to produce biofertilisers which can then be used to fertilise vineyards with new organic matter where it all began. The company also wants to produce renewable electricity, heat and biofuels from almost complete recovery of supply chain waste, in a circular systemic perspective. Other issues include reducing water consumption by lessening groundwater extraction and the optimisation of transport and packaging systems to decrease non-renewable material, energy use and emissions.

4.2 Berlucchi Franciacorta

Guido Berlucchi is the founder of Berlucchi Franciacorta. It cultivates 115 ha of vineyards, has 15 000 square metres of cellars to store wine and produces 4 million bottles per year. Its 2021 sustainability report (third edition) has been prepared by following global reporting initiative (GRI) standards. The company lists biodiversity as a material issue ('protection of the soil and biodiversity') with a high economic, social and environmental impact and a medium relevance for stakeholders (Materiality matrix at p. 39). The company has committed to use environmentally friendly agronomic and vineyard management practices to protect soil health and biodiversity. These practices include integrated pest management, grassing, green manure, organic fertilisation (see e.g. the F.A.Re. Su.BIO project) and protecting natural and protected areas near vineyards. Biodiversity is linked to the SDG No. 2 (relating to zero hunger - target 2.5) and No. 15 (conserving natural habitats and stopping biodiversity loss - targets 15.5 and 15.9).

The company is carrying out different projects to preserve biodiversity. Since 2014 the company has conducted the Biopass project to measure and safeguard biodiversity in viticulture. The project is based on measuring organic matter in the soil over consecutive years to identify targeted solutions to promote soil health and thus healthy vines with high-quality grapes. Since 2019 the company has collaborated with Treedom to fund tree planting. In 2021 the Guido Berlucchi 'forest' reached 2500 trees. Another project launched in 2016 is LifeVitisom for the innovative distribution of organic fertilisers. Thanks to this project, there has been a 5% increase in organic matter in vineyard soils, a 5% increase in soil biodiversity, a 10% reduction of equivalent CO_2 emissions due to reduced use of chemical fertilisers and a 20% reduction in costs associated with fertiliser distribution. In 2017 the company launched the F.A.Re. Su.BIO Project which involves 19 companies. The project aims to improve levels of organic matter and identify the most suitable cultivation techniques for enhancing both below-ground and above-ground biodiversity.

4.3 Cantina Produttori Valdobbiadene

Cantina Produttori Valdobbiadene cultivates 10 009 780 ha of vineyards, producing 15 861 189 bottles. Its 2021 sustainability report (i.e. the second edition) has been prepared using GRI standards. The company also has Equalitas and VIVA certifications with a strong focus on minimising the impacts of its viticulture operations on the environment.

According to VIVA certification, the company measures its commitment to biodiversity through the Territory indicator which assesses the impact of corporate activities on a particular area or region, understood as the biodiversity, landscape and local community in that territory. The indicator is divided into three macrosections:

- Landscape and Biodiversity;
- Society and Culture; and
- Economy and Ethics.

The indicator shows whether the company meets environmental, social, ethical and economic requirements and helps, through a self-assessment, to identify areas for improvement. The company launched the Bo.Di.Ca research project to protect biodiversity between 2019 and 2020. The project focuses on the winegrowing area of Valdobbiadene which is protected by the denomination of origin scheme (DOCG) developed by the Italian government. The project covers the Conegliano Valdobbiadene Prosecco Superiore DOCG vineyards to define 'environmentally compatible' development strategies in the DOCG area. The Cantina Produttori di Valdobbiadene has e.g. set up a large research laboratory in collaboration with other regional winemakers to develop more sustainable practices as well as improve training and technological innovation. The company believes it is important to combine wine-growing needs with the preservation of the biodiversity and aesthetic quality of the territory. Key areas of research are as follows:

1 The value of terroir ('the set of two specific characteristics of the soil, of the topography, of the climate, landscape and biodiversity');
2 The soil microbiome: an invisible resource for the vineyard;

3 The microbial biodiversity of the vineyard and its contribution to the terroir; and
4 The genetic biodiversity of vines (e.g. the range of varieties cultivated).

The company has to take particular care because its vineyards are close to a site of community importance (SCI) (i.e. Grave del Piave - River Soligo - Fosso di Negrisia) and three special protection areas (SPAs) (i.e. Pre-Alpine ridge between Valdobbiadene and Serravalle, Garzaia di Pederobba and Settolo Basso).

4.4 Cecchetto

Cecchetto is a family-owned winery located in the Veneto region. It produces 68 850 hectolitres of wine from 224 ha of vineyards. The 2021 report (i.e. the first edition) has been prepared according to GRI standards. The company identifies the protection of biodiversity as a material issue (No. 3) linked to SDG No. 15 (on protecting natural habitat and halting biodiversity loss). To achieve this goal, the company declares a commitment to respecting the terrestrial ecosystem, favouring pollinating insects, managing a forest certified under the Forest Stewardship Council (FSC) scheme, choosing sustainable raw materials and facilitating the consumer in reuse and recycling. The company declares it wants to become 'climate positive' by 2026. In 2017, the company achieved VIVA certification which implies the assessment of firm's sustainability performance every 2 years, according to the following indicators:

- Air;
- Water;
- Territory; and
- Vineyard.

Assessment allows the company to identify areas of improvement for each sustainability indicator with the aim of upgrading the sustainability of operations year on year.

The company has a 'Biodiversity in the vineyard' approach e.g. by grassing of the vineyard during all vegetative phases to promote beneficial insects such as pollinators (such as bees) and pest predators. Cecchetto manages two proprietary FSC-certified forests of about 6 ha. After an initial process of restoration in 2018, the company is organising educational activities and providing forest roads to encourage tourists. The forests both promote overall biodiversity and support biodiversity in vineyards. The Cecchetto report highlights how biodiversity has been evaluated using GRI Standards 304-1, 304-2 and 304-3 (p. 81).

4.5 Tasca Conti d'Almerita

Tasca Conti d'Almerita is situated in Sicily and was set up in 1830 by a well-known local family, founder of one of the oldest Italian wineries. Since 2009, such firm partakes in the 'SOStain' programme, which is focused to continuously improving the sustainability of winemaking in Sicily (https://www.fondazionesostainsicilia.it/en/sostain-programme/). The commitment to sustainability helped Tasca Conti d'Almerita win the 'European Winery of the Year' award from the 'Wine Enthusiast' magazine and the prestigious Robert Parker Green Emblem for sustainable practices (https://www.robertparker.com/). In 2023, the B Corp certification has been gained, so that the firm is involved in an international network into which the attainment of the hugest social and environmental performance is deemed a crucial goal.

The word 'biodiversity' is quoted 47 times in the company's 2022 sustainability report (i.e. the 12th edition, in Italian and English version). Examples of practices to promote biodiversity include phasing out chemical weeding as damaging to soil biodiversity, setting aside uncultivated areas representing at least 5% of a vineyard's total area (as recommended by the SOStain Programme) and promoting crop diversification to support biodiversity of flora and fauna. Meeting this goal has also facilitated VIVA certification. Another important initiative is supporting bees, because the 'Tasca Conti d'Almerita Estates are home to bees' (p. 114). Other practices include grassing, a ban on hunting in territory owned by the company, as well as bug hotels to provide habitat for insects. There is a section focused on 'Projections on 2023' where the company reports progress in increasing biodiversity (i.e. creating bug hotels, planting a vegetable garden, installing a Sicilian black bee hive). The sustainability report also includes a very interesting in-depth analysis of biodiversity named 'Changes in biodiversity: Where do we stand?'.

4.6 Summary

This analysis provides a brief overview of leading sustainability practices and reporting in the Italian winemaking sector. All companies discuss protecting biodiversity although with varying degrees of detail. All companies adopt GRI standards, but only three companies add the GRI content index. Most companies have a specific section on biodiversity in their sustainability reports and consider it a material issue. The most important sections relate to projects and initiatives, in some cases educational activities, to protect and promote awareness of biodiversity in viticulture. Three companies out of five have VIVA certification. All companies link biodiversity to some SDGs e.g Nos. 2, 13 and 15. There is, however, a lack of standardisation of reporting practices and key performance indicators (KPIs) to track progress are limited. One company provides an interesting overview of some indicators and sets out targets and how they are being met in a very clear way.

5 Case study: Cantine Teanum

This case study uses qualitative analysis to explore 'why' and 'how' biodiversity is beginning to play a significant role in the Italian winemaking industry. It complements and adds depth to quantitative approaches (Eisenhardt, 1989; Ryan et al., 2022; Yin, 2009). The analysis is based on a case study: Cantine Teanum, a family-run SME situated in the Apulia Region, one of the leading wine-producing regions at national and international levels. The study was undertaken over 3 years from 2019 to 2022. Drawing upon prior studies (Yin, 2009; Creswell and Plano Clark, 2011), the research group consisting of three members (i.e. one supervisor and two seniors) followed the protocol shown in Table 3 to ensure the robustness of the findings.

Our qualitative study is based on both secondary and primary data. The former was from the following sources:

- Annual reports;
- Management correspondence; and
- The company website.

To address the lack of formal sustainability reporting in Cantine Teanum, we also selected a number of Italian winemaking companies that have published sustainability reports (see previous section). These helped to benchmark the performance of Cantine Teanum.

The primary data was gathered through a questionnaire given to the sales director (one of the founder's sons). The list of questions - arranged in two sections - is shown in Table 4. The research group carried out three interviews partly in person and partly online (via Google Meet). Each interview lasted around 2½ h. Data were collected during the first half of 2023.

Table 3 Research protocol

Type of validity	Steps
Internal validity	• Literature review • Selection of the relevant theoretical frameworks
External validity	• Selection of the case study • In-depth details of the case company chosen • Setting of the case study database
Reliability	• Description of the data collection procedure • Triangulation of secondary data sources • Assessment and exploration of primary data between the research group and management team

Table 4 List of questions included in the questionnaire

Sections	Questions
1	1. What do you think about the commitment of the agricultural sector and, particularly, the winemaking sector to biodiversity and species extinction?
	2. What do you think about the recent EU's Biodiversity Strategy?
	3. What are the primary tools used by the agricultural industry to preserve biodiversity?
	4. Focused comments on the Italian agricultural sectors and winemaking industry?
	5. Is there a specific responsibility amenable to winemaking companies?
2	1. What is the Cantine Teanum's interpretation of the concept of biodiversity?
	2. Which initiatives does Cantine Teanum have in favour of biodiversity?
	3. What are the main actions on risk management?
	4. What are the main actions on business strategies and (eventually) corporate reporting?
	5. What are the main actions on wine production to preserve biodiversity?
	6. What is the main impact of wine production on the natural capital?
	7. Is there a national plan or action promoted by the association of winemaking producers at the national and local (south of Italy) levels?

Cantine Teanum was founded in 2000 (in San Paolo di Civitate, a small city close to Foggia in the Apulia region). The initial idea was to restore a 1970s winery, focusing solely on bulk wine production. However, the company soon decided to focus on making its wines more recognisable, so it decided to invest in bottling. For this purpose, it acquired a second winery in the province of Foggia, in the town of San Severo in the Predicatella district.

Cantine Teanum's brand identity is closely based on Apulian history. The company name is inspired by Teanum Apulum, the Roman *municipium* built on the ruins of Tiati which eventually became the city of San Paolo di Civitate. Tiati's ancient coins are shown on the labels of Cantine Teanum's best wines: Gran Tiati, Metodo Classico and Gold Vintage PDO Reserve. The message is also reinforced by the logo design in a Romanesque style with serif typefaces and inclusion of motifs, such as *pila* (Roman javelins). Teanum wines are available in most of the key markets in the wine sector (https://teanum.com/en/winery/). The company cultivates 150 ha of vineyards in five estates, producing 1.5 million bottles with an annual turnover of €4.6 million.

The company's two wineries are located in a hilly area, first known as Daunia and then as Capitanata in the so-called 'head' of Apulia. Set between Molise, Campania and Basilicata, this distinctive area is surrounded by two coastal lakes (Lesina and Varano), a mountain (the Gargano), a mountain massif (the Monti Dauni) and a large flat area (the Tavoliere delle Puglie). In 1968, San Severo was the first Apulian city to obtain a protected designation of origin (PDO) developed by the EU to protect local producers highlighting the specific local origins of their productions. The clayey or clayey-silty soils that characterise the San Severo PDO derive from the unique rock formations in the region which are rich in potassium but relatively poor in organic matter. The altitude at which vines are grown varies between 90 m and 250 m in height, while the climate is warm, with rainfall in the autumn and winter. This particular combination of soil and climatic conditions and the wise use of modern oenological practices makes it possible to obtain high-quality and regionally distinctive wines. After ageing in barriques for 24 months, in barrels for 12 months, in steel containers for 3 months and in bottles for 24 months, Tiati Gold Vintage is the only PDO San Severo Reserve in the market.

Teanum is part of a group of companies supporting the local economy. These include DUE D, a mill producing olive oil, a product celebrated as *Apulian green gold*. The two extra virgin olive brands (Filo d'Olio and Denocciolato) are entirely produced from Apulian Peranzana variety olives. Another company is Rosso Gargano which processes long tomatoes (primarily grown in Apulia, particularly in the province of Foggia). Founded in 2008 by four local farming families, Rosso Gargano turns 900 000 quintals of tomatoes into products that are shipped to restaurants and supermarkets all over the world.

In March 2023, the questionnaire was sent to the sales manager (one of the founder's sons) and to the company agronomist after introducing the main aspects of the research. Two online meetings via Teams were organised over the period April–May 2023. The content of these meetings was transcribed, analysed and discussed with the interviewees. The main findings are summarised next.

The main interpretation of biodiversity by Cantine Teanum can be summarised as the adoption of a 'short supply chain'. The founder highlights that Teanum is part of a short supply chain in which it controls both raw material production (grape cultivation) and its processing into the final product (bottled wine) for sale:

> *The glue for the birth of the cellar was the short supply chain or not selling the raw material to other processors but 'closing the circle'. This concept also extends not only from the point of view of the transformation of the oenological product but we also to other associated companies ranging from the resale of agro-pharmaceutical products to irrigation systems; therefore, the concept of the supply chain starts from the beginning and in particular in the area of Gargano we can control the process from the beginning of the seed to obtaining the raw material. Few entrepreneurs have the raw material and decide to work it, enhance it in the Apulian territory, and develop their brand linked to the territory. This represents a distinctive character of the company as most of the producers of raw materials give to the cooperative and then take and sell from the cooperative to the Neapolitan or Tuscan entrepreneur. The idea of the short supply chain comes from the fact that I produce and follow the*

process; for example, in this case, I follow the theme of biodiversity and enhance it in the final product. So, the raw material is valorised from the land up to the sales counter to the final consumer. In this case, the entrepreneurial vision, which unfortunately is not very present in the Apulian territory, should be emphasised.

The primary driver for Cantine Teanum is the market. As the founder says:

The market drives biodiversity as we produce different varieties of wine grapes, Lambrusco, Merlot, Falanghina, all international vines, and then Apulian wines, Primitivo, Negroamaro, Nero di Troia. This choice of what to grow, if it suits me to produce wine grapes in a given area, is mainly determined by the market, or what is the market asking for? However, we must monitor what the market is asking for. We have to travel on these two tracks: territoriality and the market. Territorial work is also market work. Commercial decisions are important as we export to 30 countries worldwide and are forced to travel on these two 'tracks'. As the Cantine Teanum is strongly linked to the Apulian territory, biodiversity is interpreted as the wide variety of vines. Moreover, the company's history is linked to the territory of the Roman Empire; the wine is connected to other products, which amplifies even more the contribution that the company can offer for biodiversity purposes.

The founder and the agronomist confirmed that they had no specific requirements to meet biodiversity goals or belonged to environmental certification schemes with a biodiversity component. The most important certification scheme they need to comply with is GLOBALG.A.P (https://www.globalgap.org/uk_en/). However, this focuses primarily on good agricultural practices (GAPs) related to quality assurance. These relate e.g. to safety from potential contaminants (due e.g. to improper use of pesticides, fertilisers and other chemicals) and practices to ensure production meets product specifications and sensory quality standards. GLOBALG.A.P has an additional focus on social and employee matters (e.g. the rights of labourers) but limited coverage of environmental issues. Meeting GLOBALG.A.P standards therefore encourages careful control of raw materials and the supply chain. As the founder said:

For example, if we have a new supplier, this requires strict control over the grapes and the products he used to produce them, if you have used products you should not have used, etc. If he does not handle the grapes properly, he is banned as our supplier.

Reinforcing this point, the agronomist said:

We ask for a 'historical' check that ensures a certain continuity in the product we want to obtain. Therefore, we ask suppliers to respect certain analytical parameters for the grapes when a wrong treatment has been made due to the negligence of a farmer who goes to pour everything on the vine, on the must and then on the wine. We are 'rigorous' on this aspect; therefore, our suppliers must respect the chemical and analytical parameters because otherwise, we can decide not to collect the grapes.

> *We invest: if they respect the parameters, we pay them more. This is the rewarding aspect. In this sense, it is again the market that rewards certain behaviours.*

As noted in earlier comments, the agronomist highlighted the Cantine Teanum commitment to biodiversity, in terms of the range of vine varieties cultivated (i.e. in relatively narrow terms of intraspecific diversity), arguing that:

> *If it is true that in Italy we have 80 different varieties of vines and it is, therefore, one of the nations with the greatest diversification from the point of view of oenological material in this sense, in our work we have tried to do a great job from the point of view of the selection of best varieties that responded to the characteristics of our territory, we understand well that having a company of 150 hectares we have managed to diversify our territory as much as possible therefore by planting the best vines for our best geographical areas. This has allowed us to thoroughly study the vine's responses concerning zoning, i.e. the area where it was planted.*

Climate change is another key issue highlighted by the agronomist, reflecting damage caused by recent hailstorms, severe rain storms and floods in the region:

> *The other aspect that I would like to underline is climate change, which in the last two years has been making us think seriously because we can take all possible precautions, but if we get water bombs, that phenomenon that we have never seen in Pulia, it is true that it gives us great problems because it risks compromising an entire year's work in the production of wine grapes. So let us say that we try to intervene by preserving the characteristics of our vineyards as much as possible; we try to use chemistry as little as possible in agriculture to create a sort of vegetative-productive balance. This because? I connect to the commercial point of view; today, we have an infinite variety of wines, and each territory has its peculiarities, but I believe that a trump card in the wine market is precisely the diversification concerning the market, therefore creating an identity and respecting it as much as possible. Puglia was born as the 'reservoir of Italy': first, many grapes were produced, destined for northern or central Italy or northern Europe. However, let us say that thanks to the will of many entrepreneurs who have decided to enhance the area and invest resources and energy to transform the product, in fact, one of our goals is to transform and sell the fruit of our land. As far as biodiversity is concerned, we try as much as possible to preserve nature, or what has been left to us by our ancestors and at the same time, we try to enhance it as much as possible.*

Biodiversity is linked to tourism which has been a significant factor in sales of the company's Primitivo wine brand. Teanum managers also highlighted the effect of biodiversity issues on firm's image and reputation. The founder stated that e.g. they had plans to diversify into organic wine in a few years.

The interviewees also commented on the issue of greenhouse gas (GHG) emissions contributing to climate change. The agronomist said:

> *Fortunately, our sector is not very energy-intensive compared to the tomato sector. We are independent from an energy point of view for ten months of the year. Your production causes emissions to the outside. No, we do not have burners or machinery that can cause emissions outside. In the vinification company, we work extensively with solar panels and try to mitigate energy consumption.*

The founder and the agronomist also highlighted sustainability issues relating to water consumption:

> *Water is a commodity that costs in Puglia, so all farms must make the most of it. Here, we have drought problems; therefore, we have to get it from wells, so the costs of getting water from wells are very high. So respect for the consumption of natural resources is imposed in this case by costs and therefore by the market.*
> *Water and cold temperature are the most critical elements from an environmental point of view. However, water is used for washing; it is not a question of polluted water. However, we see it as a matter of conscience and responsibility to have environments that are always clean. From a technical point of view, the main residues are water and sugar which can easily be attacked by non-yeast organisms. Hence, we try to control the environment as much as possible; I do not mean sterile but 'sanitised'. So we have sufficient consumption of water to guarantee a healthy final product. However, the water is purified and can be drinkable and usable in agriculture. We have a large basin and can clean it and return it to the environment after 7.8 days. The water is reusable and can be used for irrigation.*

The last issue discussed is the usage of potential pollutants or pesticides in winemaking. The agronomist stated: *There is a team that works on monitoring to respect the environment. Four agronomists monitor the vineyards and 150 hectares of vineyards.* On the biodiversity side, the research team asked: *Are there any animal or plant species that the production of wine can endanger?* They replied: *We have the opposite problem. Every year different pathogens arrive on the vineyard. We are the ones who have to defend the vineyard from the environment.*

At the end of the interview, the authors put a question on the need to establish common biodiversity and sustainability indicators in the winemaking sector. The interviewees argued that they do not have a framework to support them: *There is an urgent need for an action by legislator*. The founder also reiterated his understanding of biodiversity in terms of the range of local varieties: *For example, we produce the best typology of grape, which is the Montepulciano's grape doc San Severo, but we cannot use this grape for our production of wine because a specific geographical area in Abruzzo protects this kind of grape and I cannot put this denomination on my wine. In this perspective, I have to delete this aspect of 'biodiversity' and I limit the biodiversity in my production of grapes because the market does not use the denomination.*

Finally, we reported a general comment about the difficult balance between sustainability and production performance in the winemaking sector: *We know very well the problem of pesticides: nowadays one must keep an eye on biodiversity and eco-sustainability. However, as a company that has to work I have to guarantee production because I cannot lose economically from problems with disease attacks on the vineyards. There are times to exercise control and when you have to wait before you can control. There are times of 'shortage' that the farmer, in that case, must respect before harvesting the grapes and before taking them to the cellar. Unfortunately, I cannot guarantee a farmer uses only copper and sulfur as was done 40 years ago to assure a good harvest. Why is this? Because the weather conditions have changed and, today, if I get a hailstorm pre-harvest, or torrential rain, you must treat the vineyards; otherwise, you will lose all the grapes. It takes the right balance between these two crucial issues.*

The agronomist concluded: *There is a strict respect for pesticide use. We use some permitted products, i.e. that, after seven days, lose their effectiveness and do not penetrate the plant; therefore, the product cannot be found in the grapes. However, it is also true that if you do a treatment before harvesting, you will find it in the cellar. Hence, supply chain control is very relevant for us. If you do not know the farmer, the risk is significant. You cannot trust them 100% because that slight grape discharge may contaminate the final product. Also, the analyses that the large groups to which we sell if they are buying organic wine require us to have zero parameters, i.e. we are talking about traces micrograms per litre, which compromise the batch of wine.*

In summary, the founder and managers highlight two critical issues:

- The need for more support by legislators to incentivise winery companies to promote biodiversity; and
- The need to balance financial performance and sustainable practices in winemaking.

Addressing these issues will be essential to promoting biodiversity in the sector.

6 Conclusion

Biodiversity is sometimes seen as an obstacle to productivity based on an increasingly unsustainable monoculture. However, on closer inspection, biodiversity turns out to be an 'ally' of the vineyard. Indeed, as for sustainability, biodiversity must be seen as an instrument of productive development in promoting more resilient and successful winemaking. A change of vision is needed to rethink the viticulture of the future in all its 'intrinsic biodiverse' complexity.

There is a need to establish clear principles and effective methods that can prevent the degradation of a habitat/ecosystem (Atkins and Maroun, 2018). The primary goal of winemaking that respects biodiversity is to harmonise wine production and environmental conservation. The introduction of wooded and hedged areas, improving soil, water and air quality and the recovery of ancient varieties of crops, such as traditional vines, leads to an improvement of the agricultural landscape and enables farmers to become real 'guardians' of the territory, in addition to biodiversity.

7 Acknowledgements

The authors would like to thank Cantine Teanums' Founder and managers that have been interviewed and who also dedicated their time to this research. Moreover, the authors would like to acknowledge the Editor for her invaluable support during the review process.

8 References

Barbosa, F. S., Scavarda, A. J., Sellitto, M. A. and Marques, D. I. L. (2018). Sustainability in the winemaking industry: An analysis of Southern Brazilian companies based on a literature review. *Journal of Cleaner Production* 192, 80-87.

Berlucchi Franciacorta (2021). *Report di sostenibilità*, available at https://online.flippingbook.com/view/90930129/ Last Access on 30th August 2023.

Broccardo, L. and Zicari, A. (2020). Sustainability as a driver for value creation: A business model analysis of small and medium enterprises in the Italian wine sector. *Journal of Cleaner Production* 259, 120852.

Cantina Produttori Valdobbiadene (2021). *Bilancio di sostenibilità 2021*, available at https://www.valdoca.com/wp-content/uploads/2023/03/Bilancio-Sostenibilita-2021-PDF.pdf. Last accessed on 30th August 2023.

Cantine Teanum (2023). Available at https://teanum.com/. Last accessed on 30th August 2023.

Casolani, N., Chiodo, E. and Liberatore, L. (2023). Continuous improvement of VIVA-certified wines: Analysis and perspective of greenhouse gas emissions. *Sustainability* 15(3), 2349.

Caviro (2022). *Bilancio di sostenibilità 2022*, available at https://www.caviro.com/wp-content/uploads/2023/03/BdS_CAVIRO_2022.pdf. Last accessed on 30th August 2023.

Cecchetto (2021). *Bilancio di sostenibilità 2021*, available at https://www.rabosopiave.com/wp-content/uploads/2022/11/Cecchetto_BilancioSostenibilita_2021.pdf. Last accessed on 30th August 2023.

Corbo, C., Lamastra, L. and Capri, E. (2014). From environmental to sustainability programs: A review of sustainability initiatives in the Italian wine sector. *Sustainability* 6(4), 2133-2159.

Creswell, J. W. and Plano Clark, V. L. (2011). *Designing and Conducting Mixed Methods Research*. Thousand Oaks: Sage.

Corvino, A., Bianchi Martini, S. and Doni, F. (2021). Extinction accounting and accountability: Empirical evidence from the west European tissue industry. *Business Strategy and the Environment*, 30(5), 2556–2570.

Dasgupta, P. (2021). The economics of biodiversity: The Dasgupta review. London: HM Treasury, available at https://www.gov.uk/government/publications/final-report-the-economics-of-biodiversity-the-dasgupta-review. Last accessed on 30th August 2023.

De Steur, H., Temmerman, H., Gellynck, X. and Canavari, M. (2020). Drivers, adoption, and evaluation of sustainability practices in Italian wine SMEs. *Business Strategy and the Environment* 29(2), 744–762.

Eisenhardt, K. M. (1989). Building theories from case study research. *Academy of Management* 14(4), 532–550.

EU (2020). *Biodiversity Strategy for 2030*, available at https://environment.ec.europa.eu/strategy/biodiversity-strategy-2030_en Last Access on 30th August 2023.

Ferrer, J. R., García-Cortijo, M. C., Pinilla, V. and Castillo-Valero, J. S. (2022). The business model and sustainability in the Spanish wine sector. *Journal of Cleaner Production* 330, 129810.

Flores, S. S. (2018). What is sustainability in the wine world? A cross-country analysis of wine sustainability frameworks. *Journal of Cleaner Production* 172, 2301–2312.

Martini, S. B., Doni, F. and Corvino, A. (2018). Deforestation risk and the tissue industry in Italy. In Atkins, J. and Atkins, B. *Around the World in 80 Species* (pp. 273–285). Routledge

Ryan, B., Scapens, R. and Theobald, M. (2002). *Research Method and Methodology in Finance and Accounting*. London: Thomson.

Signori, P., Flint, D. J. and Golicic, S. L. (2017). Constrained innovation on sustainability in the global wine industry. *Journal of Wine Research* 28(2), 71–90.

Tasca Conti D'Almerita (2022). *12° Report di sostenibilità*, available at https://cdn.adoratorio.app/tasca/SOSTAIN_REPORT_TDA_2022-1.pdf. Last accessed on 30th August 2023.

Tait, P., Saunders, C., Dalziel, P., Rutherford, P., Driver, T. and Guenther, M. (2019). Estimating wine consumer preferences for sustainability attributes: A discrete choice experiment of Californian Sauvignon blanc purchasers. *Journal of Cleaner Production* 233, 412–420.

Vandermeer, J. and Perfecto, I. (1995). *Breakfast of Biodiversity: The Truth about Rainforest Destruction*. Oakland: Food First Books.

WBA World Biodiversity Association Onlus (2023). Biodiversity friend, available at https://biodiversityassociation.org/en/what-we-do/projects/conservation/biodiversity-friend/. Last accessed on 30th August 2023.

Yin, R. (2009). *Case Study Research: Design and Methods*, 4th edition. Thousand Oaks: Sage.

Chapter 13

The role of global investors in protecting natural capital and delivering ecosystem services within agriculture, food and other land use value chains

Martina Macpherson, Henley Business School, UK/University of Zurich, Switzerland/Future of Sustainable Data Alliance, UK/All-Party-Parliamentary-Group (APPG) on ESG – Houses of Parliament, UK

1 Introduction

This chapter summarises key challenges facing agriculture, food and other land use (AFOLU) systems. It outlines the concept of 'five pillars of a sustainable AFOLU system'. It points to the role of financial markets products and investor engagement initiatives as potential enablers for medium- to longer-term transformative solutions in the sector. After first discussing the interconnected challenges facing AFOLU systems and its negative impacts on the environment, the chapter discusses ways AFOLU systems can become more sustainable. It

http://dx.doi.org/10.19103/AS.2023.0128.05

focuses particularly on the ways the investment sector can and is contributing to enabling this transition. The chapter reviews regulatory frameworks and standards relating to financial reporting on sustainability issues as well as ways of benchmarking the impact of sustainability initiatives. After looking at financial instruments such as biodiversity credits, it focuses particularly on the growing range of multi-stakeholder initiatives involving the investment sector promoting environmental, social and governance (ESG) values and practices in the agri-food sector.

2 Challenges facing agriculture, food and other land use systems

The world's population is currently expanding by over 67 million people per year, and it is expected to grow by over a third, or 2.3 billion people, between the last two decades and 2050 (UN: https://www.un.org/en/global-issues/population). Per capita incomes in 2050 are also projected to increase significantly (World Bank, 2006). Relative inequality in per capita incomes could be reduced considerably by 2050. However, absolute differences are likely to remain pronounced, and they could even increase further, given the current huge gaps in absolute per capita incomes and the lessons learned from the recent financial and pandemic crises and their (socio)-economic impacts.

As per capita income increases, the developing world's middle class is likely to continue to upgrade its dietary requirements and thus consume more protein and grains. Dietary changes might override population growth as the major driver of global demand for land use in the near future. This could have major environmental consequences as meat-based food production requires about five times as much land per unit of nutritional value as its plant-based equivalent, and also has a higher water footprint which is around 20 times higher for beef than for cereals (Water Footprint Network, 2010). Since the 1960s, global average meat consumption has almost doubled, from 23 kg per person to 42 kg, with the highest consumption in the US and Europe, while China and Brazil have recorded significant increases in the last 20–30 years. Estimates suggest that global annual demand for meat products may increase by a further 76% relative to 2005 to 455 million tonnes in 2050 (Ritchie et al., 2017).

In the face of continuous population growth and increasing demand for food products and services, especially linked to (meat-based) proteins, as well as in the context of a limited land base, there is likely to be continued pressure on prices for food-producing land, on food processing, manufacturing and distribution, as well as on global food supply chain logistics. The Intergovernmental Panel on Climate Change has categorised these and related activities as: 'Agriculture', 'Food' and 'Other Land Use' (AFOLU) (Smith et al., 2014).

There are multiple layers of complexity involved in the production, processing, marketing, retail, consumption and disposal of goods that originate from AFOLU systems, including food and non-food products, livestock, pastoralism, fisheries including aquaculture and forestry. AFOLU systems and challenges also involve a wide range of stakeholders, people and institutions, as well as the socio-political, socio-economic, technological and environmental areas and activities.

As noted, by 2050 agricultural producers will have to support a population of more than 9.7 billion people. This increase will require a 60% boost in agricultural productivity (UN FAO, 2009). The Global Harvest Initiative 'GAP Report' (2019) (Virginia Tech College of Agriculture and Life Sciences (CALS) Global Programs, 2019) estimates that, to meet global demand by 2050, agricultural producers would have to double their output from 2010 levels. This will require an annual average growth of at least 1.73% in total factor productivity (TFP) or the output per unit of total resources employed in production (Fig. 1).

This challenge is increasing pressure on global agriculture and food sector actors as well as global supply chains. At the same time, industrialisation and urbanisation continue to encroach upon the world's finite farmland resources and products.

According to the United Nations Farmland and Agriculture Organization (UNFAO, 2009) (Fig. 2), 'agriculture [*as well as farming, and other land use AFOLU*] in the 21st century faces multiple challenges: it has to produce more food and fibre to feed a growing population with a smaller rural labour force, more feedstocks for a potentially huge bioenergy market, contribute to overall development in the many agriculture-dependent developing countries, adopt more efficient and sustainable production methods and adapt to climate change'.

3 Challenges facing the environment: the role of agriculture, food and other land use systems

It has been estimated ('Our World in Data', 2019) (Ritchie, 2019) that 10% of the Earth's surface is covered by glaciers; a further 19% is made up of barren land, i.e. deserts, dry salt flats, beaches, sand dunes and exposed rocks. This leaves the rest classified as 'habitable land'. Humans have claimed half of all this habitable land on Earth for agriculture (IPBES, 2019), with livestock accounting for 77% of global farming (producing 18% of the world's calories and 37% of total proteins (Poore and Nemecek, 2018). This leaves just 37% for forests, 11% as shrubs and grasslands and 1% as freshwater. The remaining 1% encompasses

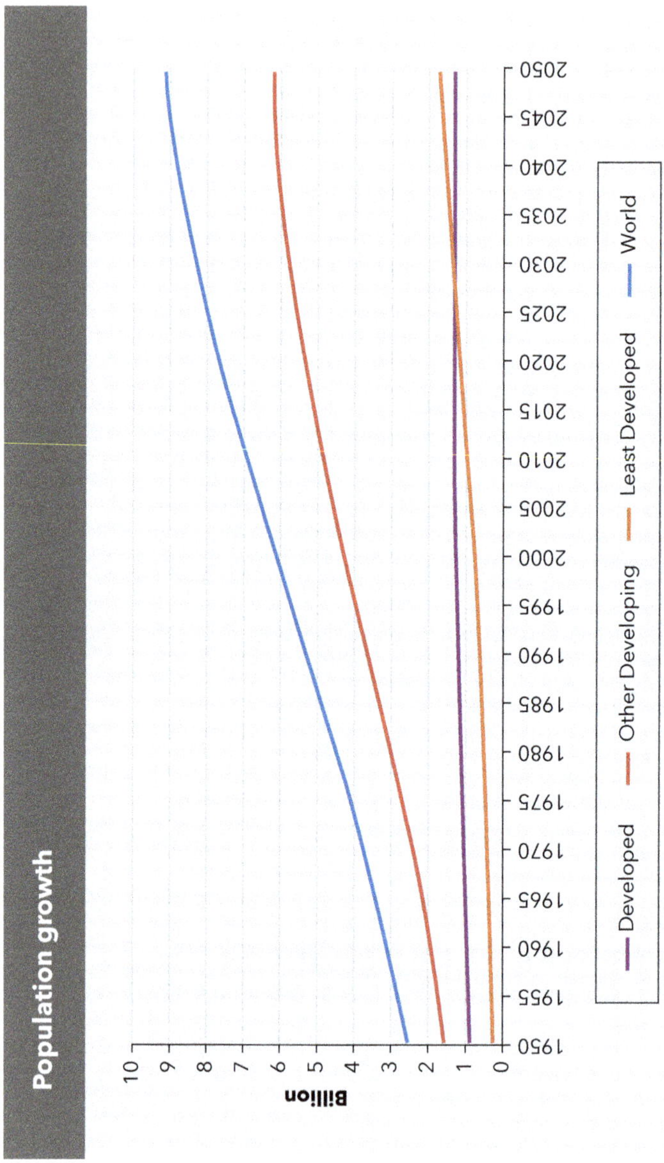

Source: UN Population Division, from van der Mensbrugghe et al.2009

Figure 1 Global population growth, 1950–2050.

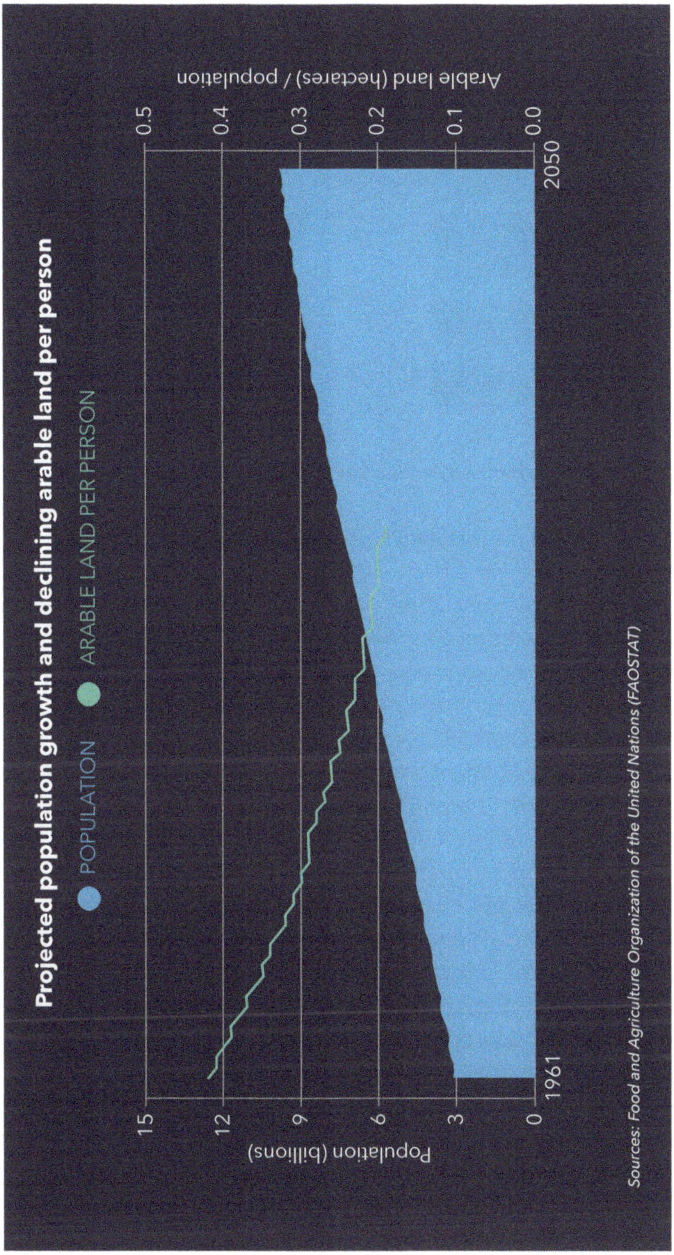

Figure 2 Projected population growth and declining arable land, 1961–2050.

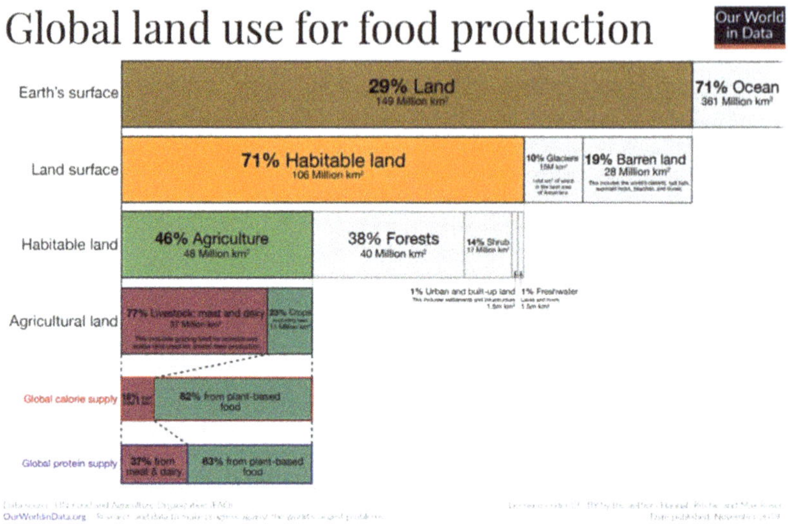

Figure 3 Global land use for food production. Sources: UN FAO/IUCN (2019).

urban areas including cities, towns, villages, roads and other areas of human infrastructure (Fig. 3).

The expansion of agriculture, land use, human-driven pollution and overconsumption have been classified as some of humanity's largest impacts on the Earth and the environment. The AFOLU system has e.g. transformed habitats and has had a major impact on biodiversity and on ecosystem services. According to the often-cited IPBES Report (2019), approximately 75% of the land-based environment and 66% of the marine environment has been significantly altered by human activity. The IPBES Report highlights that natural ecosystems have declined by 47% while the global biomass of wild mammals has fallen by 82%. Global wildlife populations have declined by 60% over the last 40 years. Of the 28 000 species evaluated as to be threatened with extinction on the 'IUCN Red List' (2019), the AFOLU system was listed as a threat for 24 000 of them (see IUCN's 'Red List' database: https://www.iucnredlist.org/search). According to the United Nations 'UNCCD Report' (2019), as much as 40% of Earth's land surfaces are now considered as degraded, and more than 1 million species are currently threatened with extinction, vanishing at a rate not seen in 10 million years.

Looking more broadly, scientists have identified nine planetary boundaries which define the key physical, chemical and biological process boundaries that must not be exceeded if the Earth is to maintain the ecological balance it has retained for around 12 000 years. These boundaries are being defined as (Richardson et al., 2023) follows:

- Climate change;
- Novel entities;

- Stratospheric ozone depletion;
- Atmospheric ozone loading;
- Ocean acidification;
- Biogeochemical flows;
- Freshwater change;
- Land-system change; and
- Biosphere integrity.

Human-led activity is currently seen as breaching six of these nine boundaries. The AFOLU system has been identified as 'the single biggest transgressor' (Campbell et al., 2017) of Earth's 'nine planetary boundaries', ultimately overtaking the fossil-fuel industry. The AFOLU system is directly responsible for 90% of global deforestation, 70% of species threatened with loss and 65% of agrochemical pollution (Bhavnagri, 2023). The agriculture is responsible, e.g. for (Bhavnagri, 2023)

- 80% of biodiversity loss;
- 80% of land-system change; and
- 70% of freshwater withdrawals.

The AFOLU sector has also become a major contributor to climate change: around one quarter of all greenhouse gas emissions stem from AFOLU sector outputs (Bhavnagri, 2023).

4 Defining solutions to create a sustainable agriculture, food and other land use system

Biodiversity and ecosystem services are vital for a sustainable AFOLU system. These ecosystem services include crop pollination, water purification, flood protection and carbon sequestration. Globally these services are worth an estimated US$125–140 trillion per year, i.e. more than one and a half times the size of global GDP (OECD, 2019). For all these reasons, managing the impact of AFOLU on biodiversity and ecosystem services has also become a key area of concern for both AFOLU businesses and investors alike.

The scale of change is considerable. ''Protecting 30% of land and sea by 2030' ('Campaign for Nature', 2022)[1] will require an end to agriculture and fishery expansion, the incorporation of more biodiversity on farms and a reduction in the current footprint of conventional agriculture to create space for the restoration of nature. In creating sustainable food systems, it has been

[1] For more information, see (homepage): https://www.campaignfornature.org/more-than-100-countries-commit-to-protect-at-least-30-of-land-and-oceans-by-2030; more than 100 countries have committed to protect at least 30% of land and oceans by 2030 at the United Nations (UN) Ocean Conference on 30 June 2022.

estimated that human would ultimately have to return around 1 billion hectares of agricultural land to nature, an area roughly the size of China (Odier, 2023).

BloombergNEF research (2023) (Bhavnagri, 2023) suggest 'five broad solutions' to address AFOLU system challenges which are listed here (with additions suggested by the author which are of particular relevance for financial markets, systems and instruments). These solutions include:

- **1 'Sustainable proteins and fats:** Producing protein and fat in a more sustainable way is one of the key challenges for global food systems and supply chains. A key challenge is how to produce protein using less land and with (near)-zero greenhouse gas emissions. A radical change to protein production is ultimately necessary to address the other environmental challenges. Plant-based proteins cultured or lab-grown meats and precision fermentation are options but ultimately consumer adoption is critical to determine how much of an impact these technologies will have'.
- **2 'Green agrochemicals:** The future food system will undoubtedly continue to need agrochemicals like fertilizers, pesticides and herbicides to enable high-yield farming. The challenges related to these chemicals and pollinator extinction issues have been widely assessed (Biehl and Macpherson, 2022) but still these chemicals will need to be made and used with fewer side effects. A new generation of fertilizers and pesticides that are less environmentally hazardous and produce fewer greenhouse gas emissions is under development and include hydrogen, bio stimulants that help plants better absorb nutrients and improve plant health, and biopesticides and other genetic innovations that improve the targeting of pesticides. More advanced agricultural machinery and the application of artificial intelligence (AI) will also have a key role to play to maintain and scale global food systems'.
- **3 'Sustainable intensification technologies:** A variety of technologies and innovations are under development to increase farm productivity and yields, while using less resources. These will be essential to scaling up production to feed billions of increasingly affluent people. Perennial crops are a notable example. These are a new generation of plants that grow and yield grain for several years, rather than dying after one season, making them more drought and stress resilient, more tolerant of pests, able to store more carbon and better able to access nutrients. Gene editing technologies will almost certainly continue to support productivity improvements, and precision agriculture technologies and approaches will also help to maximize yields, while reducing inputs'.
- **4 'Carbon-smart farming:** There are a variety of land management practices, such as regenerative agriculture and agroforestry, that can help sequester carbon in the soil, improve soil health, boost on-farm

biodiversity and increase crop resilience. However, the challenge is in scaling implementation and global adoption and applications. Farmers need to get paid to overcome the risk of changing practices and meet upfront costs. Carbon markets are the most obvious way to do this. Business model innovations, new methodologies and rising prices for carbon units through regulated and voluntary Carbon markets concepts will play a vital role to create more opportunities for farmers and to align AFOLU sector innovation with financial markets systems and new market instruments'.

- **5 'Biodiversity markets and finance:** The key to achieving the AFOLU system transformation will ultimately need to be found in finance and accounting and nature's services must be valued as core assets if they are to be protected and sustainably managed for the long-term. A variety of long-term value creation and system change approaches are being developed to help enabling this. These include biodiversity credit markets, disclosure and reporting frameworks like the Taskforce on Nature-related Financial Disclosures, target-setting regimes, natural capital accounting frameworks and changes to banks' valuation methods to recognize the essential services provided by nature'.

These options provide a variety of impact-focussed solutions to address AFOLU system challenges.

However, to create a transformative (global) food and supply system, there is also a need to involve the world's smallholder farmers which, in turn, face a multitude of additional challenges. However, the role of these smallholders cannot be underestimated in the context of global food (and supply) systems: 80% of Asia and Africa are fed by these smallholdings, this means that one-third of the entire population of the world being fed by smallholder farmers (Lowder et al., 2021).

5 The role of sustainable investment in improving agriculture, food and other land use systems

Sustainable investment in AFOLU systems refers to the creation of productive assets and capital formation comprising physical, human or intangible capital which is oriented to support the realisation of food security, nutrition and sustainable development, including increased production and productivity.

The potential role of investors needs to be seen in the context of the significant increase in direct investment in farmland and agriculture in the last two decades. Demand for farmland among institutional investors has exploded, according to research by Valoral Advisors (2020).

In 2005, there were less than 20 farmland-focussed investment funds globally. This number began to change exponentially, following the 2008-9 financial crisis, as more assets begin to flow into safe-haven investments like farmland funds. By 2020, the number of funds specifically focussed on farmland had grown to 166+.

In addition to protecting biodiversity and ecosystem services, sustainable investment in AFOLU systems can make a significant contribution to enhancing sustainable livelihoods for smallholders and members of marginalised and vulnerable groups. Sustainable investment can help to create decent work for all agricultural and food workers, eradicate poverty, foster social and gender equality, eliminate the worst forms of child labour, promote social participation and inclusiveness, increase economic growth, helping to achieve sustainable development. It is paramount that the capacity of smallholders to invest is strengthened. Responsible investment includes hence also priority investments in, by, and with smallholders, including those that are small-scale producers and processors, pastoralists, artisans, fishers, communities closely dependent on forests, indigenous peoples and agricultural workers.

Investing in sustainable AFOLU systems can produce multiplier effects for complementary sectors, such as service or manufacturing industries, thus further contributing to food security and nutrition and overall economic development. At the same time, the value of safety and health in generating productive agriculture and food systems is important and investing successfully means taking a holistic approach in terms of human, animal, environmental and overall public health.

The viability of investments in AFOLU systems is dependent on well-functioning biodiversity- and eco-systems, and on the sustainable use of natural resources. Depleting biodiversity and natural capital can create significant near-term operational costs, threatens the supply of critical resources, and puts global GDP at risk. It is estimated that US$44 trillion of economic value generation, over half the world's total GDP, is potentially at risk because of the dependence of the global economy on nature and its services (WEF, 2020). Biodiversity loss e.g. can lead to a variety of micro-economic risks, within the corporate value and supply chains, from reputational to operational and supply chain, to regulatory and even litigation risks. This, in turn, can create risks for investors.

The importance of biodiversity and nature themes as strategic key ESG financial and capital markets issues is rooted in academic research spanning analysis that dates back almost 30 years. According to the 'World Resource Institute, IUCN and the UN Environment Programme Report' (1992) the destruction of the natural world can result in loan defaults, lower returns and increasing insurance liabilities. In 2007, the Stern Report identified the potential economic costs of climate change and the economic case for a transition to a low-carbon economy (Stern, 2007). More recently, the issues have become better understood in the business and financial community: for example, the 'Organisation for Economic Cooperation and Development' Report (2019)[2] outlined that, if natural capital is not protected, businesses, banks and investors may also face higher costs of capital, and losses and or additional costs related to investment, lending and underwriting activities.

In one of the first studies of its kind, the Dutch Central Bank (De Nederlandsche Bank NV, 2020) calculated that the country's financial intuitions held EUR 510 billion (US$ 604 billion) of investments that were highly or very highly dependent on one or more ecosystem services. The Central Bank, in alignment with the 'Network for Greening the Financial System – Central Banks' (Network for Greening the Financial System, 2020), has been focussing its efforts on measuring the interconnected issues and impacts related to climate and nature risks.

More recently, biodiversity risk has also been classified as the top three risk for humanity and our natural systems in the coming decade, by the World Economic Forum (2023). However, addressing better protection of natural capital, as well as the ''four dimensions of food security and nutrition', including 'availability', 'access', 'stability' and 'utilisation', requires a significant increase in sustainable investment in AFOLU systems and in sustainable finance-centred project financing (UN FAO, 2014).[3]

6 Supporting sustainable investment in agriculture, food and other land use systems: regulations, standards and benchmarking

Over the last 5 years, there have been a variety of new normative and regulatory initiatives, new policies, classification systems and disclosure frameworks.[4] Some of them have been launched to address a wide range of ecosystem, biodiversity, natural capital and species loss and extinction challenges (Biehl and Macpherson, 2022) (such as Green/Conservation Bonds).

On the policy and regulatory side, the new EU Taxonomy (2020),[5] supported by 'Guidelines on Non-Financial Reporting' (European Commission, 2017) and 'Guidelines on Non-Financial Reporting: Supplement on Reporting

[2] OECD, 'Biodiversity: Finance and the Economic and Business Case for Action Annexes to the Report, Prepared by the OECD for the French G7 Presidency and the G7 Environment Ministers', Meeting Report, 5-6 May 2019.

[3] UN FAO, 'Principles for Responsible Investment in Agriculture and Food Systems', 2014; The Committee on World Food Security (CFS) endorsed the Principles for Responsible Investment in Agriculture and Food Systems on October 15, 2014.

[4] For a summary on key normative and regulatory frameworks in the ESG domain, see e.g. Roncalli, Thierry, *Handbook of Sustainable Finance*, Universite Paris-Saclay, revised version 26 January 2023.

[5] EU Commission, 'EU Taxonomy for Sustainable Activities' (homepage): https://ec.europa.eu/info/business-economy-euro/banking-and-finance/sustainable-finance/eu-taxonomy-sustainable-activities_en; in June 2020, the European Parliament formally adopted the EU Taxonomy Regulation, and it has now been published in the Official Journal of the European Union. The Taxonomy creates a classification system for sustainable economic activities that will help guide investments towards more sustainable technologies and business.

Climate-Related Information' (European Commission, 2019), seeks to create a classification framework for policy and industry best practices and actions. The EU Taxonomy includes six environmental ('green') economic activities and categories, including 'Protection and Restoration of Ecosystems and Biodiversity' (Official Journal of the European Union, 2020), which apply to products developed by actors in the financial and insurance industry.

The EU Taxonomy complements the EU's corporate and investor disclosure frameworks and has been established in accordance with the EU Non-Financial Reporting Directive (NFRD) (EU 2014/95),[6] and the upcoming EU Corporate Social Responsibility Directive (CSRD) (EU 2022/2464).[7] The taxonomy is also being used to identify green economic activities of financial and capital market participants (Lehmann et al., 2022). A key financial investment product regulation, the EU Sustainable Finance Disclosure Regulation (SFDR) (EU 2019/2088) (European Commission, 2019), which came into full effect on 1 January 2023, requires investment funds registered for sales in Europe to disclose and report to their investors if and how their funds address sustainability risks and 'Principle Adverse Impacts', in pre-contractual information, on websites and in periodic reports in line with EU SFDR Article 6, 8 and 9. These risks can include biodiversity risks.

Alongside the work of the EU's classification framework (EU Taxonomy), a suite of new disclosure regimes have been launched, including EU CSRD (see earlier, to be implemented by the European Financial Reporting Advisory Group (EFRAG)[8]), as well as normative frameworks by international standard setters such as the international Financial Reporting Standards - International Sustainability Standards Board (IFRS/ISSB) (framework expected in June 2023[9]), and by the International Standards Organisation (ISO) (ISO/TC 322 was launched in early 2023[10]).

[6] European Commission, Directive 2014/95/EU of the European Parliament and of the Council of 22 October 2014 amending Directive 2013/34/EU as regards to Disclosures of Non-Financial and Diversity Information by certain Large Undertakings and Groups, 15 November 2014; the EU NFRD amends the accounting directive 2013/34/EU, companies are required to include non-financial statements in their annual reports from 2018 onwards. EU NFRD is currently under review, the 2nd edition in process following a public consultation process in 2020.

[7] European Commission; 'Regulation Directive (EU) 2022/2464 of the European Parliament and of the Council of 14 December 2022 amending Regulation (EU) No 537/2014, Directive 2004/109/EC, Directive 2006/43/EC and Directive 2013/34/EU, as regards Corporate Sustainability Reporting', 14 December 2022; on 5 January 2023, the Corporate Sustainability Reporting Directive (CSRD) entered into force. This new directive modernises and strengthens the rules about the social and environmental information that companies have to report. A broader set of large companies, as well as listed SMEs, will now be required to report on sustainability, approximately 50 000 companies in total.

[8] For more information, see (homepage): https://www.efrag.org; EFRAG is a private association established in 2001 with the encouragement of the European Commission to serve the public interest. EFRAG extended its mission in 2022 following the new role assigned to EFRAG in the CSRD, providing Technical Advice to the European Commission in the form of fully prepared draft EU Sustainability Reporting Standards and/or draft amendments to these Standards. Its Member Organisations are European stakeholders and National Organisations and Civil Society Organisations.

[9] For more information, see (homepage): https://www.ifrs.org/groups/international-sustainability-standards-board/; the Trustees of the IFRS Foundation announced the formation of the ISSB on 3 November 2021 at COP26 in Glasgow, following strong market demand for its establishment. The ISSB is developing standards that aim to result in a high-quality, comprehensive global baseline of sustainability disclosures focussed on the needs of investors and the financial markets.

[10] For more information, see (homepage): https://www.iso.org/committee/7203746.html; the ISO TC 322 aims to support the alignment of the global financial and capital markets (i.e. investor, banks) system with the UN's Sustainable Development Goals.

The EU Commission has also launched a dedicated corporate dialogue workstream for multi-stakeholder engagement in the context of biodiversity: the 'EU Business @ Biodiversity Platform' (2020).[11] It provides a unique EU forum for dialogue to explore the links between business, biodiversity and natural capital and works towards fulfilling the objectives of the 'EU Biodiversity Strategy to 2030'. International efforts include the International Platform for Sustainable Finance (IPSF).[12]

Alongside developments in financial reporting standards relating to sustainability, there have been initiatives to establish benchmarks to measure and reduce impacts. A recent example is 'Science Based Targets for Nature'.[13] This is a five-step model which aims to provide corporations with an 'Action Framework' ('AR3T') to avoid and reduce their impacts on biodiversity and nature loss, with assessments to better understand the needs for restoration and regeneration activities, and with guidance around the transformation of underlying systems to address the drivers of nature loss. This new science-based model could also become a framework for the up-and-coming biodiversity credit market models, ultimately copying the success of its sister program in the context of carbon off-sets.

There are also initiatives to provide the kind of data that such models need. An example is the Future of Sustainable Data Alliance (FoSDA), established after the 2020 World Economic Forum at Davos in Switzerland (https://futureofsustainabledata.com). FoSDA seeks to map data needs in identifying and costing nature-related related financial risks, isolate and help to fill data gaps as well improve ways of measuring ESG performance.

Other initiatives include providing guidance to finance professionals on how to identify risks so they can make more informed investment decisions. An example is the Cambridge Institute for Sustainability Leadership (CISL), established by the University of Cambridge as a collaboration between academics working on conservation issues, banks and asset managers. CISL has developed a *Handbook for Nature-Related Financial Risks* designed for financial practitioners with limited prior knowledge of issues such as biodiversity loss and the potential financial risks such losses pose (CISL, 2021). The handbook outlines types of nature-related business risks (e.g. physical, transitional and liability risks) and the ways these feed through to credit, market, liquidity and

[11] For more information, see EU Commission, Business @ Biodiversity, launched in 2020 (homepage): https://ec.europa.eu/environment/biodiversity/business/index_en.htm.

[12] For more information, see (homepage): https://finance.ec.europa.eu/sustainable-finance/international-platform-sustainable-finance_en. The platform is a forum for dialogue between policymakers, with the overall aim of increasing the amount of private capital being invested in environmentally sustainable investments.

[13] For more information, see (homepage): https://sciencebasedtargetsnetwork.org/resources; nature-focussed SBTs will allow companies to take action that aligns with a subset of the goals set out in the UN Conventions on Biodiversity (UNCBD), Climate Change (UNFCCC), Land Degradation (UNCCD), and the 2030 Agenda for Sustainable Development (SDGs).

business risks. It also provides a detailed framework for identifying these kinds of risks in sectors such as agriculture.

In 2022, the World Benchmarking Alliance (WBA) also launched its 'Nature Benchmark' (https://www.worldbenchmarkingalliance.org/nature-benchmark) to support investor decision-making. The 'Nature Benchmark' will ultimately measure 1000 companies across 22 industries between two initial research cycles, in 2022 and 2023. During the first quarter of 2022, WBA published a draft methodology and conducted a public consultation to map existing behaviours and assess the best practice actions for the sector.

7 Supporting sustainable investment in agriculture, food and other land use systems: new investment instruments

Although biodiversity and ecosystem losses pose great challenges, and potentially significant economic and financial risks to all actors, they also provide new investment opportunities. Sustainable/impact investment and sustainable finance in capital markets has a vital role to play to support, finance and guide ecosystem governance, protection and the 'ecological transition'. Further efforts and new products in sustainable finance might also encourage new (regulatory) disclosure frameworks for biodiversity risks in the global banking sector.

One example is biodiversity credit markets which are developing rapidly, side by side with voluntary carbon markets, creating new potential revenue streams for farmers. These markets are intended to be improved on biodiversity offset schemes. These are well-established and numerous. However, because they are used to offset negative environmental impacts in one location (e.g. a new housing development) by promoting environmental benefits in another location (e.g. through a conservation project), they have been accused of managing rather than reversing biodiversity decline (Bull et al., 2013). Biodiversity credit schemes need to incentivise the restoration and conservation of biodiversity above baseline levels. How to create appropriate incentives and measure progress in restoring biodiversity is a subject of active debate (Taskforce on Nature Markets, 2023).

The financial market has also been showing new interest in debt-for-nature swaps (Georgieva et al., 2022). The concept was established in the late 1980s. Organisations such as Conservation International, The Nature Conservancy and the World Wildlife Fund have participated in international debt-for-nature swaps whereby the NGO transfers the debt title to the debtor country and, in exchange, the country agrees to either enact certain environmental policies or endow a government bond in the name of a conservation organisation, with the aim of funding conservation programs. However, despite these initiatives, 'prudential regulatory frameworks' to disclose biodiversity and nature risks and their impacts in the same way as climate risks[14] are still not available at scale in the banking sector.

8 Multi-stakeholder initiatives involving the investment sector to promote sustainability in agriculture, food and other land use systems

The last few years have seen a range of (collective) active ownership and investor engagement initiatives[15] that aim to address interrelated AFOLU and biodiversity/ecosystem challenges. Multi-stakeholder initiatives are targeting collective AFOLU system and ecosystem services protection, restoration, and/or regeneration actions. Key frameworks to underpin these engagement initiatives include the following.

In 2015, the United Nations published its 2030 Agenda for Sustainable Development. Protecting 'Life Below Water' (U.N. SDG Goal 14) and 'Life on Land' (U.N. SDG Goal 15), which are part of the United Nations' 'Agenda 2030' (United Nations (UN SDG), 2015), have highlighted key issues linked to biodiversity, nature and species protection across the investment value chain. The United Nations also hosts an annual 'United Nations Convention on Biological Diversity' (United Nations Convention on Biological Diversity, Sustainable Development, 2020), which brings together political, financial and business leaders and other leading nature-linked coalitions and taskforces such as

- the 'Capitals Coalition' (formerly the 'Natural Capital Coalition');[16]
- the 'Task Force on Nature-related Financial Disclosures' (TNFD, launched in 2021[17]); and
- the 'Business for Nature's Call for Action' (launched in 2020[18]).

Building on the developing framework of sustainable finance reporting standards (see TCFD[19]), the TNFD consists of 40 individual taskforce members

[14] New disclosure regulation for banks, the 'Green Asset Ratio' (GAR), which is part of banks compliance efforts with CRR Pillar 3 risk disclosures, provides insights about a bank's EU Taxonomy alignment and the carbon footprint of loans and investments on a bank's balance sheet. The idea behind this is that carbon-intensive industries carry a higher transition risk to move to a low carbon economy. The more exposure a bank has to companies with high carbon emissions, the more it could be impacted by transition risk.

[15] A part of their commitments to implement the Principles for Responsible Investment (PRI), investor signatories are being asked and tasked to consider ESG issues, including biodiversity risks, in their active ownership policies and practices ('Principle 2') and collaborate in their efforts to build a dialogue with investee companies ('Principle 5') for more than a decade.

[16] For more information, see Capitals Coalition (homepage): https://capitalscoalition.org/.

[17] For more information, see Task Force on Nature-related Financial Disclosures (homepage): https://tnfd.info/.

[18] For more information, see Business for Nature Call to Action (homepage): https://www.businessfornature.org/call-to-action.

[19] For more information, see Financial Stability Board, 'Task Force on Climate-Related Financial Disclosures' (TCFD, launched in 2017) (homepage): https://www.fsb-tcfd.org/.

(data as at March 2023) representing financial institutions, corporates and market service providers with over US$20 trillion in assets.[20] If the TNFD becomes as successful as its sister Taskforce for Climate-related Financial Disclosures (TCFD), then it has the potential to be a significant actor in supporting conservation, restoration or regeneration activities. It provides a promising new framework which distinguishes between two types of nature-related risks for investors:

- Physical risks: The risk resulting directly from reduced ecosystem services such as the lack of pollination for the agricultural industry or reduced flood control for the real estate market. Certain industries are more exposed to physical risk from biodiversity loss than others, for instance, tourism, agriculture, fishery or forestry.
- Transitional risks: The risk resulting from societal and political pressure, including reputational, regulatory scrutiny and litigation risks. As future potential losses of ecosystem services start to have an impact, pressure on societies is expected to grow and transitional risk will steadily increase. Industries that will be especially affected include agriculture, forestry, oil & gas, mining and construction.

Another initiative is the 'Partnership for Biodiversity Accounting Financials' (PBAF)[21] which consists of 40 financial institutions from 11 countries representing US$11 trillion in assets. PBAF enables financial institutions to assess and disclose impacts and dependencies on biodiversity in loans and investments they make, raise more awareness about the financial risks and impacts of biodiversity and ecosystem loss, and provide tools to measure corporate, investor and capital markets impacts and dependencies on ecosystem services.

Under the auspices of the UN, Principles for Responsible Investment (PRI - https://www.unpri.org/)[22] were launched in April 2006 at the New York Stock Exchange (NYSE) by a group of investors wishing to promote ESG values and practices. PRI was one of the first ESG coalitions providing a governance framework and multi-stakeholder forum for collaborative engagement. Over the years, they have initiated a range of nature and biodiversity-centred engagement activities. In January 2023, they announced a concentrated effort

[20] For more information, see (homepage): https://tnfd.global/about/; in November 2022, the TNFD released the third version of its beta framework for market consultation, building on v0.1 released in March 2022 and v0.2 released in June 2022. The next version of the beta framework will be released in March 2023 (v0.4) before the release of version v1.0 of the full framework for market adoption in September 2023.

[21] For more information, see (homepage): https://pbafglobal.com/; PBAF's primary aim is to develop the 'PBAF Standard which enables financial institutions to assess and disclose impact and dependencies on biodiversity of loans and investments.

[22] For more information, see Principles of Responsible Investment (PRI - homepage: https://www.unpri.org/). In early 2005, the UN Secretary-General Kofi Annan invited a group of the world's largest institutional investors to join a process to develop the 'Six Principles for Responsible Investment'. A 20-person investor group drawn from institutions in 12 countries was supported by a 70-person group of experts from the investment industry, intergovernmental organisations and civil society. The PRI were launched in April 2006 at the New York Stock Exchange.

and new recruitment for 'a new collaborative nature stewardship initiative' (Gambetta, 2023).[23] The new initiative aims to support investors in nature-positive engagement efforts to address forest loss, land degradation and other interrelated environmental issues and potentially will include support for new public policy initiatives.

'The FAIRR Initiative'[24] was founded by Jeremy Coller in 2016, with the intention of providing key data on risks associated with intensive animal agriculture. It now represents US$70 trillion in assets, and has become the world's fastest-growing ESG network and food sector engagement initiative, working closely with investors to change the conversation around animal agriculture and the way that food is produced.

'Finance for Biodiversity'[25] is a leading investor engagement initiative focussed on biodiversity. In 2023, 126 financial member institutions representing 21 countries and over EUR 18.8 trillion in assets signed the Finance for 'Biodiversity Pledge'. The Pledge was initiated by a group of 26 financial institutions to commit to protecting and restoring biodiversity through their finance activities and investments. The initiative was launched during the 'Nature for Life Hub' (25 September 2020) and the 'Biodiversity Summit of the United Nations General Assembly' (30 September 2020).

A recent global investor engagement initiative is 'Nature Action 100'.[26] It focuses on driving greater corporate ambition and action to reduce nature and biodiversity loss. The initiative engages companies in key sectors that are deemed to be particularly important in reversing nature and biodiversity loss by 2030.

Other initiatives include the Network for Sustainable Financial Markets, which has launched 'NextGen' to bring together young finance professionals interested in sustainability issues (https://www.nsfmnextgen.org/), as well as two groups working with CISL: the Investments Leaders Group (ILG) which seeks to ensure that the over $9 trillion in assets managed by its members meets ESG goals; and the Banking Environment Initiative, a group of global banks committed to practical action to achieve sustainability in sectors such as the agri-food sector (including e.g. HSBC, NatWest Group, Lloyds, Santander, Deutsche Bank and Standard Chartered) (https://www.cisl.cam.ac.uk/convening/centre-for-sustainable-finance).

Individually and collectively, these various multi-stakeholder/investor engagement efforts aim to address the complex challenges, up and downstream, within the AFOLU sectors, and/or the challenges linked to the depletion or loss

[23] Gambetta, Gina, 'PRI to Launch Collaborative Nature Stewardship Initiative', in Responsible Investor, 17 January 2023, available online: https://www.responsible-investor.com/pri-to-launch-collaborative-nature-stewardship-initiative.
[24] For more information, see (homepage): https://www.fairr.org/about-fairr/making-the-argument.
[25] For more information, see (homepage): https://www.financeforbiodiversity.org.
[26] For more information, see (homepage): https://www.natureaction100.org; the initiative's Secretariat and Corporate Engagement Working Group is co-led by Ceres and the Institutional Investors Group on Climate Change (IIGCC), and the initiative's Technical Advisory Group is co-led by the Finance for Biodiversity Foundation and Planet Tracker.

of biodiversity and ecosystem services. Neglecting the interconnectedness of these issues can increase losses of natural habitats, biodiversity and ecosystem services (Nabuurs et al., 2022).

9 Conclusion and trajectories for nature-smart concepts

Confronted with a need to address carbon emissions, biodiversity and ecosystem loss, changing consumer preferences and potentially disruptive new technology, AFOLU systems and their supply chains will undergo vast transformations over the coming decades. With those changes come significant threats to the existing products, services and business models of 'sluggish incumbents', but also opportunities for more innovative players to create new products and value-add initiatives.

Key to a transformation in the AFOLU sector is policy enablement. It is estimated that governments around the world allocate more than US$500 billion per year in agricultural subsidies that can be harmful to the environment.[27]

As with the journey to net-zero emissions, reorienting a core part of the global economy will require overcoming vested interests, addressing environmentally harmful subsidies, and ensuring a just transition for affected communities. New policies on fertilisers, pesticides, water use and land use practices will also be essential to developing, deploying and scaling up these efforts.

Meanwhile, market actors, such as The World Bank, are making increasingly the 'Economic Case for Nature' (Johnson et al., 2021, World Bank Report, 2021). The World Bank's Report is based on a series of papers that recognise how and how much global economies rely on nature for services that are currently largely under-valued and under-priced. The report presents a first-of-its-kind global integrated ecosystem-economy modelling blueprint between nature's services and the global economy to 2030. It includes a range and a combination of policy scenarios available to reduce the impact of nature's loss on global economies. This modelling framework represents an important milestone towards 'nature-smart' decision-making, as it seeks to support policymakers who face complex trade-offs when establishing concepts for the management of natural capital, while achieving growth that is 'resilient and inclusive'.

All in all, biodiversity-focussed (sustainable and impact) investment, banking and lending (sustainable finance) activities as well as (collective) investor engagement efforts, e.g. led by 'FAIRR', the 'Biodiversity Pledge' by 'Finance for Biodiversity', or by the 'Nature Action 100' coalition, will be fundamental in achieving a meaningful change in the medium to longer term.

[27] For further references, see UN FAO, 'NSP: Policies that Save and Grow', year of publication unknown, available online: https://www.fao.org/agriculture/crops/thematic-sitemap/theme/spi/scpi-home/framework/policies-for-scpi/policies-that-save-and-grow/en/.

10 References

Bhavnagri, K. 2023. "Sustainable Agriculture: The New Green Revolution", in *BloombergNEF Research Paper*, February 21, available online: https://assets.bbhub.io/professional/sites/24/Introduction-to-BNEF-Sustinable-Agriculture.pdf (accessed on 10/03/2023).

Biehl, C. and Macpherson, M. 2022. "Extinction Bonds: Addressing Biodiversity Issues through Sustainable Finance", in Atkins, J. and Macpherson (Eds), *Martina: "Extinction Governance, Finance and Accounting: Implementing a Species Protection Action Plan for the Financial Markets"*, Routledge, London. p. 229.

Bull, J. W., Suttle, K. B., Gordon, A., Singh, N. J. and Milner-Gulland, E. J. 2013. "Biodiversity Offsets in Theory and Practice", *Oryx: Flora and Fauna International*, 47(3), 369–380, available online: https://www.cambridge.org/core/journals/oryx/article/biodiversity-offsets-in-theory-and-practice/EDBF70717C273662B6D8EE0876370095 (accessed on 09/12/2022).

Cambridge Institute for Sustainability Leadership (CISL). 2021. *Handbook for Nature-Related Financial Risks: Key Concepts and a Framework for Identification*, University of Cambridge, Cambridge.

Campbell, B. M., Beare, D. J., Bennett, E. M., Hall-Spencer, J. M., Ingram, J. S. I., Jaramillo, F., Ortiz, R., Ramankutty, N., Sayer, J. A. and Shindell, D. 2017. "Agriculture Production as a Major Driver of the Earth System Exceeding Planetary Boundaries," *Ecology and Society*, 22(4), Resilience Alliance, available online: https://ora.ox.ac.uk/objects/uuid:25c20aa7-88b3-4457-9e4d-8ffbf57fc943/download_file?file_format=application%2Fpdf&safe_filename=Campbell%2Bet%2Bal%2B2017%2BAgriculture%2Bproduction%2Bas%2Ba%2Bmajor%2Bdriver%2Bof%2Bthe%2BEarth%2Bsystem.pdf&type_of_work=Journal+article (accessed on 08/12/2022).

Dutch Central Bank. 2020. "Indebted to Nature Exploring Biodiversity Risks for the Dutch Financial Sector", available online: https://www.dnb.nl/en/binaries/Indebted%20to%20nature%20_tcm47-389172.pdf (accessed on 10/03/2023).

European Commission. 2014. "Directive 2014/95/EU of the European Parliament and of the Council of 22 October 2014 Amending Directive 2013/34/, EU as Regards to Disclosures of Non-Financial and Diversity Information by certain Large Undertakings and Groups", 15 November, available online: https://eur-lex.europa.eu/legal-content/EN/TXT/?uri=CELEX%3A32014L0095 (accessed on 12/03/2023).

European Commission. 2017. "Guidelines on Non-financial Reporting - Methodology for Reporting Non-financial Information (2017/C 215/01)", 5 July, available online: https://eur-lex.europa.eu/legal-content/EN/TXT/?uri=CELEX:52017XC0705(01) (accessed on 12/03/2023).

European Commission. 2019. "Guidelines on Non-financial Reporting: Supplement on Reporting Climate-Related Information (2019/C 209/01)", 20 June, available online: https://eur-lex.europa.eu/legal-content/EN/TXT/?uri=CELEX:52019XC0620(01) (accessed on 12/03/2023).

European Commission. 2019. "Regulation (EU) 2019/2088 of the European Parliament and of the Council of 27 November 2019 on Sustainability-Related Disclosures in the Financial Services Sector", 9 December, available online: https://eur-lex.europa.eu/eli/reg/2019/2088/oj (accessed on 12/03/2023).

European Commission. 2022. "Regulation Directive (EU) 2022/2464 of the European Parliament and of the Council of 14 December 2022 amending Regulation (EU) No

537/2014, Directive 2004/109/EC, Directive 2006/43/EC and Directive 2013/34/EU, as regards Corporate Sustainability Reporting", 14 December, available online: https://eur-lex.europa.eu/legal-content/EN/TXT/?uri=CELEX:32022L2464 (accessed on 12/03/2023).

Gambetta, G. 2023. "PRI to Launch Collaborative Nature Stewardship Initiative", in *Responsible Investor*, January 17, available online: https://www.responsible-investor.com/pri-to-launch-collaborative-nature-stewardship-initiative (accessed on 15/03/2023).

Georgieva, K., Chamon, M. and Thakoor, V. 2022. "Swapping Debt for Climate or Nature Pledges Can Help Fund Resilience", *International Monetary Fund Blog*, December 14, available online: https://www.imf.org/en/Blogs/Articles/2022/12/14/swapping-debt-for-climate-or-nature-pledges-can-help-fund-resilience#:~:text=Debt%2Dfor%2Dclimate%20swaps%20and,spending%20on%20other%20development%20priorities (accessed on 06/03/2023).

Intergovernmental Science-Policy Platform on Biodiversity and Ecosystem Services (IPBES). 2019. "Global Assessment Report on Biodiversity and Ecosystem Services of the Intergovernmental Science-Policy Platform on Biodiversity and Ecosystem Services", available online: https://zenodo.org/record/6417333 (accessed on 20/01/2023).

Johnson, J. A., Ruta, G., Baldos, U., Cervigni, R., Chonabayashi, S., Corong, E., Gavryliuk, O., Gerber, J., Hertel, T., Nootenboom, C., Polasky, S., Gerber, J., Ruta, G. and Polasky, S. 2021. *The Economic Case for Nature: A Global Earth-Economy Model to Assess Development Policy Pathways*, World Bank, Washington, DC, available online: https://openknowledge.worldbank.org/entities/publication/fcc11682-c752-51c4-a59f-0ab5cd40dc6f (accessed on 24/10/2023).

Lehmann, M., Macpherson, M. and Ung, D. 2022. *ESG Investing and Analysis: A Practitioner's Guide*, RiskBooks, London.

Lowder, S. K., Sánchez, M. V. and Bertini, R. 2021. "Which Farms Feed the World and Has Farmland Become More Concentrated?", *World Development*, 142, available online: https://www.sciencedirect.com/science/article/pii/S0305750X2100067X?via%3Dihub (accessed on 10/12/2022).

Nabuurs, G.-J., Mrabet, R., Abu Hatab, A., Bustamante, M., Clark, H., Havlík, P., House, J., Mbow, C., Ninan, K. N., Popp, A., Roe, S., Sohngen, B. and Towprayoon, S. 2022. "Agriculture, Forestry and Other Land Uses (AFOLU)", Chapter 7, in *IPCC Report, 2022: "Climate Change 2022: Mitigation of Climate Change," Contribution of Working Group III to the Sixth Assessment Report of the Intergovernmental Panel on Climate Change*, available online: https://www.ipcc.ch/report/ar6/wg3/downloads/report/IPCC_AR6_WGIII_Chapter07.pdf (accessed on 12/03/2023).

Network for Greening the Financial System. 2020. "Guide for Supervisors Integrating Climate-Related and Environmental Risks into Prudential Supervision", May, available online: https://www.ngfs.net/sites/default/files/medias/documents/ngfs_guide_for_supervisors.pdf (accessed on 12/03/2023).

Odier, L. 2023. "Rethinking Food Systems at Davos 2023: Shifting a Ten Trillion Dollar Industry", January 20, available online: https://www.lombardodier.com/contents/corporate-news/corporate/2023/january/rethinking-food-systems-at-davos.html (accessed on 06/03/2023).

Official Journal of the European Union. 2020. "Taxonomy Regulation", 12 July, available online: https://eur-lex.europa.eu/legal-content/EN/TXT/?uri=CELEX:32020R0852 (accessed on 12/03/2023).

Organisation for Economic Co-operation and Development (OECD). 2019. "Biodiversity: Finance and the Economic and Business Case for Action", G7 Meeting Report, 5-6 May, available online: https://www.oecd.org/environment/resources/biodiversity/G7-report-Biodiversity-Finance-and-the-Economic-and-Business-Case-for-Action.pdf (accessed on 02/01/2023).

Poore, J. and Nemecek, T. 2018. "Reducing Food's Environmental Impacts through Producers and Consumers", *Science*, 360(6392), 987-992, available online: https://www.science.org/doi/10.1126/science.aaq0216 (accessed on 10/12/2022).

Richardson, K., Steffen, W., Lucht, W., Bendtsen, J., Cornell, S. E., Donges, J. F., Drüke, M., Fetzer, I., Bala, G., von Bloh, W., Feulner, G., Fiedler, S., Gerten, D., Gleeson, T., Hofmann, M., Huiskamp, W., Kummu, M., Mohan, C., Nogués-Bravo, D., Petri, S., Porkka, M., Rahmstorf, S., Schaphoff, S., Thonicke, K., Tobian, A., Virkki, V., Wang-Erlandsson, L., Weber, L. and Rockström, J. 2023. "Earth beyond Six of the Nine Planetary Boundaries", *Science Advances*, 9, 37, available online: https://www.science.org/doi/10.1126/sciadv.adh2458 (accessed on 15/02/2023).

Ritchie, H. 2019. "Half of the World's Habitable Land is Used for Agriculture: Our World in Data", available online: https://ourworldindata.org/global-land-for-agriculture (accessed on 11/02/2023).

Ritchie, H., Rosado, P. and Roser, M. 2017. "Meat and Dairy Production", *Our World in Data* (revised 2019), available online: https://ourworldindata.org/meat-production (accessed on 10/12/2022).

Roncalli, T. 2022. "Handbook of Sustainable Finance", *Universite Paris-Saclay*, November 15 (revised version, accessed on 26/01/2023), available online: https://papers.ssrn.com/sol3/papers.cfm?abstract_id=4277875 (accessed on 15/03/2023).

Roncalli, T. 2023. *Handbook of Sustainable Finance*, Universite Paris-Saclay (revised version) 26 January, available online: thierry-roncalli.com/download/HSF.pdf (accessed on 10/12/2022).

Smith, P., Bustamante, M., Ahammad, H., Clark, H., Dong, H., Elsiddig, E. A., Haberl, H., Harper, R., House, J., Jafari, M., Masera, O., Mbow, C., Ravindranath, N. H., Rice, C. W., Robledo Abad, C., Romanovskaya, A., Sperling, F. and Tubiello, F. 2014. "Agriculture, Food and Other Land Use (AFOLU)", in Edenhofer, O. et al. (Eds), *Climate Change: Mitigation of Climate Change. Contribution of Working Group III to the Fifth Assessment Report of the International Panel on Climate Change*, Cambridge University Press, Cambridge.

Stern, N. 2007. *The Economics of Climate Change: The Stern Review*, Cambridge University Press, Cambridge.

Taskforce on Nature Markets. "Biodiversity Credit Markets: The Role of Law, Regulation and Policy", available online: www.naturemarkets.com (accessed on 24/10/2023).

Taskforce on Nature Markets. "The Future of Biodiversity Credit Markets: Governing High-Performance Biodiversity Credit Markets", available online: www.naturemarkets.com (accessed on 24/10/2023).

United Nations (UN SDG). 2015. "Transforming Our World: The 2030 Agenda for Sustainable Development", available online: https://sdgs.un.org/2030agenda.

United Nations Convention on Biological Diversity, Sustainable Development. 2020. *Convention on Biological Diversity*, New York, 12 October, available online: https://www.cbd.int/doc/speech/2020/sp-2020-10-12-unga-en.pdf (accessed on 10/12/2022).

United Nations Convention to Combat Desertification (UNCCD). 2019. "Land Degradation Neutrality for Biodiversity Conservation", *Briefing Note*, 23 December, available online: https://www.unccd.int/resources/publications/land-degradation-neutrality-biodiversity-conservation-briefing-note (accessed on 06/03/2023).

United Nations Food and Agriculture Organisation (UN FAO). 2009. "Global Agriculture Towards 2050", High Level Expert Forum Briefing Paper, Rome, 12–13 October, available online: https://www.fao.org/fileadmin/templates/wsfs/docs/Issues_papers/HLEF2050_Global_Agriculture.pdf (accessed on 02/01/2023).

United Nations Food and Agriculture Organisation (UN FAO). 2014. "Principles for Responsible Investment in Agriculture and Food Systems", available online: https://www.fao.org/3/au866e/au866e.pdf (accessed on 10/12/2022).

United Nations Food and Agriculture Organisation (UN FAO). "NSP: Policies That Save and Grow", year of publication unknown, available online: https://www.fao.org/agriculture/crops/thematic-sitemap/theme/spi/scpi-home/framework/policies-for-scpi/policies-that-save-and-grow/en/ (accessed on 02/01/2023).

Valoral Advisors. 2020. "Mapping the Global Opportunities in the Food and Agriculture Investment Space Post Covid-19", Report, April, available online: https://www.valoral.com/wp-content/uploads/Valoral-Advisors-Food-And-Agriculture-Investments-Post-COVID-19.pdf (accessed on 15/03/2023).

Virginia Tech College of Agriculture and Life Sciences (CALS) Global Programs. 2019. "Global Agricultural Productivity (GAP) Report: Productivity Growth for Sustainable Diets and More", 7 November, available online: https://www.harvestrcn.org/post/global-agricultural-productivity-gap-report (accessed on 02/01/2023).

Water Footprint Network. 2010. "Water Footprint of Crop and Animal Products: A Comparison", available online: https://waterfootprint.org/en/water-footprint/product-water-footprint/water-footprint-crop-and-animal-products/ (accessed on 08/12/2022).

World Bank. 2006. *The Road to 2050: Sustainable Development for the 21st Century*, World Bank, Washington, DC.

World Economic Forum (WEF). 2020. "New Nature Economy Report II the Future of Nature and Business", available online: https://www3.weforum.org/docs/WEF_The_Future_Of_Nature_And_Business_2020.pdf (accessed on 20/02/2023).

World Economic Forum (WEF). 2023. "Global Risks Report", January, available online: https://www3.weforum.org/docs/WEF_Global_Risks_Report_2023.pdf (accessed on 12/03/2023).

World Resources Institute. 1992. "IUCN: The World Conservation Union, and the United Nations Environment Programme Global Biodiversity, Status of the Earth's Living Resources", (digitized in 2010), available online: https://www.unenvironment.org/resources/report/global-biodiversity-status-earths-living-resources (accessed on 12/03/2023).

Chapter 14

Exploring corporate weather accounting by the UK food retail industry

Elisabetta A. V. Barone, Northeastern University London and University of Galway, UK; Jill Atkins, Cardiff Business School – Cardiff University, UK; and Warren Maroun, University of the Witwatersrand, South Africa

1 Introduction

The UK is notorious for its variable and unpredictable weather. The main reason for the 'not-so-great' British weather is the country's geographic location. As an island between continental Europe and the Atlantic, the UK is located at the meeting point of five air masses (Winterman, 2013).[1] They collide with each other with the dominant stream deciding the weather at that point in time. Consequently, for those who live in the UK, the daily weather is always a topic of discussion and a good way to start a conversation when people meet for the first time.

An acknowledged outcome of climate change is weather variability with more frequent extremes. The UK is forecast to receive heavier rainfall with increased risk of flooding, particularly in winter, while summers may be hotter and drier. Unusual events, such as 'Ex-Hurricane' Ophelia or Storm Otto or Storm Ciarán, are expected to become more common, highlighting the risks posed by changes in climate. This is especially true for the agricultural sector.

As populations continue to grow and food security becomes an increasing concern for governments and policy-makers, the ways in which the food production and retail sectors respond to weather-related risks is becoming

[1] An air mass is defined as a large body of air that has similar temperature and moisture properties throughout (Winterman, 2013).

http://dx.doi.org/10.19103/AS.2023.0128.20

ever more important. UK agriculture has evolved to cope with the traditionally unpredictable British weather but the magnitude of the changes being experienced, largely due to global warming, is leaving farmers struggling to adjust. This, in turn, has implications for the reliability of food supplies and the cost at which these goods can be made available to consumers.

In this chapter, we explore the extent to which companies in the UK food industry are disclosing information on weather-related risk in their annual, integrated and sustainability reports as well as on their websites. While the primary focus is on UK food retailers, we also consider the impact of climate change on non-food areas of their business and in other countries due to supply chain effects. The aim is not to quantify risk disclosures or test potential reporting drivers. The research is grounded in a social constructivist stance which sees language as a tool used to construct and convey meaning. Using interpretative content analysis, we examine the content, tone and consistency of disclosures and draw conclusions on whether or not companies are reporting proactively on the risks posed by climate change and the boundaries of their weather accounts. We are interested in how companies understand and internalise the effects of erratic weather; if information on risk management strategies is being provided and the focus of their stakeholder engagement.

Our study makes an important contribution to the environmental accounting and accountability literature. First, we offer additional evidence on the construction of alternate 'charts of account' which have the potential to widen the scope of the conventional accounting system. This is done using an innovative case. While different types of environmental reporting and accounting techniques have been considered, we are not aware of prior research dealing specially with the British weather. Illustrations of how meaning is constructed and conveyed using environmental disclosures and how this contributes to the broader 'risk society framework are rare (see Giddens, 1990; Gallhofer and Haslam, 1996; Solomon et al., 2011). We also complement the prior research on climate change and carbon reporting. This has dealt with issues such as impression management/symbolic reporting (Borghei et al., 2016); how carbon accounting is framed as part of the broder sustainability reporting and accounting project (Ascui and Lovell, 2011) and the assurance of carbon disclosures (Green et al., 2017). How climate change forms part of an emerging risk discourse is not explicitly considered. The relationship between reporting on climate change and potentially emancipatory forms of accounting such as extinction (Atkins and Maroun, 2018) and ecological accounting (Russell et al., 2017) has not been addressed. As a result, an examination of reporting on the risks posed by weather variability, in the context of what Giddens (1990) describes as a risk society, offers an excellent opportunity to examine the accounting's reflexive change potential and contribution to the emancipatory environmental accountability agenda.

2 The discourse of risk: climate change, farming and the Great British weather

2.1 Modernity and the discourse of risk

Risk and the way in which risk is internalised, acted on and reported has been a topic of interest for social scientists studying periods of late modernity for several years. In one of the primary references on the emergence of a 'risk society', Giddens (1990) deals specifically with environmental disasters as a consequence of modernity and the increased prominence of environmental risk in modern business discourse. There is a growing awareness of the impact humanity is having on the planet and a realisation that we are on an unsustainable trajectory which ends in environmental disaster. Anxiety about environmental degradation is clearly evident in the scientific literature where we are warned about the catastrophic biological implications of climate change, habitat destruction and loss of species (Klein et al., 2007; Ceballos et al., 2017). Civil society, inter-governmental organisations and environmental research groups have also been ringing the alarm bells (see, e.g. Intergovernmental Panel on Climate Change, 2013). Closer to home, the environmental and sustainability accounting community have long challenged the myopic focus of conventional business practice (Gray et al., 1995b; Spence, 2009) and call for a proactive approach to accounting which works to avert environmental disasters (Milne et al., 2009; Russell et al., 2017; Atkins and Maroun, 2018).

From a risk society perspective, the proliferation of natural and social science research dealing with the state of the environment points to a shift from classical economics, institutions and political theory to a period of modernity 'characterised by [environmental] dangers...which are unmanageable and uninsurable' (Solomon et al., 2011, p. 1121). Ecological concerns are no longer seen as the 'side effect' of industrial activity but a serious threat to the legitimacy of existing institutional orders with the potential to destabilise the capital system. In agrarian and industrial society, behaviours and outcomes are capable of being ordered, ranked and controlled (Gordon, 1980; Foucault, 1983). The 'calculable infrastructure' of the accounting system can be mobilised both to quantity and to hold accountable (Mennicken and Miller, 2012). In periods of late modernity - where meaning is coordinated, managed and ordered according to risk discourse - taken-for-granted assumptions about the relationship between the environment and business practice which have safeguarded the status quo are shaken (Lash et al., 1996).

> In the phase of risk society, recognition of the incalculability of the hazards produced by technical-industrial development compels self-reflection on the foundations of the social context and a review of prevailing conventions and principles of 'rationality'. In the self-conception of risk society, society becomes *reflexive* (in the

> narrow sense of the word) - that is, becomes an issue and a problem to itself (Beck, 1996, p. 32, emphasis in the original).

It is possible for accounting to be intertwined in this process of reflexive awareness and change. Accounting's role in promoting the centrality of finance and economics to the exclusion of social and environmental imperatives is well documented (Gray et al., 1995b; Solomon and Thomson, 2009; Atkins et al., 2015a). How conventional accounting logics have constrained efforts at promoting more transparent reporting and a sustainable approach to doing business has also been discussed in detail (see, e.g. Tregidga et al., 2014; Flower, 2015). Accounting is, however, a 'fluent and emerging craft' (Hopwood, 1987, p. 207) capable of reflecting current organisational arrangements and providing us with a glimpse of alternate (and more desirable) realities (Dillard and Reynolds, 2008; Atkins et al., 2015a; Gallhofer and Haslam, 2019).

> By its routinisation of information flows and the ways in which it imposes a spatialisation on time, [accounting] can change conceptions of the past, the present and the future, contributing different saliences to each which can, in turn, moderate temporal preferences and emphases, and thereby, organisational actions (Hopwood, 1987, pp. 207-209).

We see evidence of this in the emergence of new forms of what can be referred to broadly as environmental accounting designed to broaden the scope of conventional financial reporting models and "emancipate" organisations from the myopic trappings of financial discourses (see Gallhofer and Haslam, 1996).

2.2 On the self-reflective awareness of the need to broaden accounting

A detailed review of the origin and history of environmental accounting is beyond the scope of this chapter (for details, see Gray et al., 1995a; Parker, 2005; Fifka, 2013). What is relevant are the most recent developments in environmental accountability aimed at constructing new fields of organisational visibility and promoting positive changes to business practice. Perhaps most notable is emergence of integrated reporting which provides:

> a concise communication about how an organization's strategy, governance, performance and prospects, in the context of its external environment, lead to the creation of value over the short, medium and long term (International Integrated Reporting Council [IIRC], 2013).

The IIRC has been criticised for placing too much emphasis on providers of financial capital and need to ensure market efficiency (Flower, 2015; Thomson, 2015). Like sustainability reporting, the possibility of integrated reporting forming part of a coordinated impression management exercise, which does little to promote genuine environmental responsibility, has also been considered (Brown and Dillard, 2014; Stubbs and Higgins, 2014; McNally et al., 2017). It may, however, be premature to dismiss the potential of this new reporting trend.

The IIRC take a broad approach to risk, requiring organisations to identify economic, environmental and social factors which could undermine value creation over the short-, medium- and long-term. These risks should be clearly identified and explained to stakeholders, including the steps being taken to mitigate their impact (IIRC, 2013). Early indicators suggest that the multi-capital model espoused by the IIRC may heighten awareness of the importance of environmental protection and provide a framework for organisations to internalise environmental risk as part of their strategic development and management processes (consider Atkins and Maroun, 2015; Al-Htaybat and von Alberti-Alhtaybat, 2018; McNally and Maroun, 2018). It is also possible for integrated reporting to be used in conjunction with other emerging forms of accounting as part of a more comprehensive approach to accounting for risk. Biodiversity, ecological and extinction accounting are excellent examples.

Jones and Solomon (2013) situate the accounting for biodiversity as part of a broader cultural, philosophical and accountability framework. The challenges of biodiversity reporting are outlined including for example, the difficulty of defining and accounting for changes in biodiversity mass (for details, see Cuckston, 2013; Freeman and Groom, 2013; Tregidga, 2013). The potential for biodiversity accounting and reporting to heighten awareness of environmental issues and lead a reflexive change in how companies protect biodiversity is also present (Jones, 2003). This is because the act of reporting on biodiversity loss can make the anthropocentric and deep ecological case for conserving species more apparent, a key objective of ecological and extinction accounting.

In anthropocentric terms, these two forms of environmental accounting and accountability place "natural capital' at the heart of organisations' business models (Atkins and Atkins, 2016, 2019). Risks posed by climate change and environmental degradation are key business risks which organisations ought to manage in order to preserve the 'inventory' of natural capital and, in turn, ensure business continuity (see also Jones, 2003; Siddiqui, 2013). At the deep ecological level, nature has an inherent or intrinsic value which is independent of the direct benefits it provides to people (Khisty, 2006). The risk of extinction or biological degradation must, therefore, be framed from a moral, cultural, religious and ecological perspective (see Russell et al., 2017; Gray and Milne, 2018).

Atkins and Maroun (2018) and Maroun and Atkins (2018) provide an illustration of an extinction account grounded in an integrated thinking and risk management discourse. This blends the anthropocentric and deep ecological arguments for internalising environmental risks and advances an emancipatory framework which sees accounting mobilised as an agent of positive change. Risk is understood in both financial and biological terms. The extinction account includes a clear explanation of how business activities affect different species and ecosystems, the steps taken to mitigate these ecological risks and the different capitals or resources which the organisation must mobilise to reverse or prevent environmental damage (see also Atkins and Atkins, 2016, 2019). The partnerships formed with NGOs, efforts to educate and engage with stakeholders and how well the organisation has performed in redressing adverse environmental consequences form a key part of an emancipatory extinction accounting model (Atkins et al., 2019, 2018).

This type of extinction accounting needs not be limited to a single organisation. Gaia and Jones (2017) and Buchling and Maroun (2019), for example consider the importance of ecological or extinction accounts covering natural capitals under the control of the public sector. For Atkins et al. (2018), the unit of account is a specific environmental issue addressed by multiple organisations in their integrated and sustainability reports. Similarly, Cuckston (2017, 2018) proposes accounting at the ecological level something which could, conceivably, entail a type of 'consolidated biological report' which aggregates the extinction or ecological accounts of multiple organisations, environmental groups and government agencies.

As part of the biodiversity, ecological and extinction accounting project, researchers have considered a range of species and environmental concerns to contextualise environmental risks. While not referring specifically to 'risk society' logic, Atkins et al. (2018) and Sibanda and Mulama (2019) deal with the adverse consequences of the possible extinction of the rhinoceros from both an economic and ecological perspective. Nicolov (2019) and Solomon and Clappison (2019) deal with accounting for whales. Butterflies (Lanka, 2019), bees (Atkins et al., 2017) polar bears (Jonäll and Sabelfeld, 2019) and pandas have also received recent attention. What has not been considered are the possible implications of changing weather patterns and how the unpredictability of periods of drought and excessive rain is incorporated as part of the broader risk discourse.

2.3 The (not-so) Great British weather

The traditionally changeable British weather is becoming even more unpredictable. The general consensus is that this is primarily the result of global warming. Researchers have found a general shift towards more frequent and heavy rainfall during winter and hotter and drier summers[2] (Maraun et al., 2008; Burt and Ferranti, 2012). In a speech at Lloyds of London about the risks to the insurance and financial sectors of climate change, Mark Carney, Governor of the Bank of England and the Head of the Financial Stability Board, stated:

> Potential increases in the frequency or severity of extreme weather events driven by climate change could mean longer and stronger heat waves; the intensification of droughts; and a greater number of severe storms. Despite winter 2014 being England's wettest since the time of King George III; forecasts suggest we can expect at least a further 10% increase in rainfall during future winters (Carney, 2015).

These changes pose significant operating and financial risks for the UK farming industry (Blackmore, 2014). Climate change and unpredictable weather patterns can have direct implications for the areas in which crops can be grown or livestock can be reared. Examples include soil nutrition, the spread of invasive species, incidence of disease and the risk of crops being lost due to adverse weather conditions (see Rial-Lovera et al., 2017).

While climate change is often framed as a long-term consideration, there are already indicators of extreme weather beginning to have an effect on the yields of essential crops such as wheat and soy (Gömann, 2015). In turn, the insurance industry has been hard-hit reporting that:

> Since the 1980s the number of registered weather-related loss events has tripled; and Inflation-adjusted insurance losses from these events have increased from an annual average of around \$10bn in the 1980s to around \$50bn over the past decade (Carney, 2015).

As a result, governments and regulators have called for climate change to be addressed as a systemic risk in the financial system which is more appropriately managed and disclosed (Deloitte, 2017). More detailed industry-level efforts to mitigate the risks arising from changing weather patterns and extreme weather events are also under way.

In a risk society context, resilience is a useful strategy for preparing for and mitigating the effects of the unexpected. This can include adopting a systems thinking approach, incorporating resistance as a core element of a system and improving system diversity, connectivity, redundancy and recovery capacity. Forward looking approaches which focus on additivity and transformation are also relevant (de Brujn et al., 2017). The emphasis is on 'learning to change farming and water management practices ... to respond appropriately to challenges of sustainability and climate change' (Blackmore, 2014, p. 177).

[2] For example, the UK Meteorological Office (Met Office) reported December 2015 as the warmest and wettest year since records began (Rial-Lovera et al., 2017).

One approach used by companies to manage weather-related risks is to trade in weather derivatives. These are a relatively new form of financial instrument introduced in the mid-1990s[3] (Wang et al., 2010). The derivatives allow companies to hedge 'moderate departures' from expected weather conditions as opposed to traditional indemnity insurance which covers 'large departures and catastrophes' (Dischel, 2002; Brockett et al., 2005; Bates and Goodale, 2017). They provide farmers with a means of hedging the financial effects of weather risks defined as:

> ...the uncertainty in cash flow and earnings caused by non-catastrophic weather events including temperature, humidity, rainfall, snowfall, stream flow and wind (Wang et al., 2010, p. 358).

Indeed, many food production and retail companies express concern over poor sales because of unfavourable weather. For example, temperature and precipitation can have a negative impact on yields. On the retail side, temperature can lead to unexpected changes in consumer behaviour such as substituting soft drinks with teas and coffees in an unusual cold spell. Extreme weather, such as heavy snow and rain can also disrupt supply chains and stop customers from shopping.

Weather derivatives offer several advantages. First, they enable farmers to transfer the financial risks of adverse weather conditions to capital market participants with more sophisticated risk management structures and which are better placed to adsorb short-term losses resulting from bad weather (Štulec, 2017). Second, the pay-out on weather derivatives is linked to variations from specified weather indices[4] rather than paying an indemnity based on incurred damages which reduces the farmer's administrative and transaction costs (Vedenov and Barnett, 2004). They can also be tailored to correspond to firms' risk appetites and hedging requirements and provide companies with flexibility in determining which risks to manage. This is especially true when considering that, unlike conventional insurance policies, the instruments can have multiple applications. Examples include compensating the holder for actual losses, off-setting opportunity costs and diversifying portfolios.[5] As a result, weather derivatives can have a material effect on overall risk exposure and firm value (Vedenov and Barnett, 2004; Pérez-González and Yun, 2013).

For example, Leggio (2007) finds that weather derivatives are useful tools to alleviate the negative consequences of low demand due to bad weather. Consequently, weather derivatives help to reduce revenue volatility by up

[3] Weather derivatives were introduced in the US in 1997. Weather derivatives differ from weather insurance, which is used to cover catastrophic events, e.g. hurricanes, floods and windstorms.
[4] For example, temperature-related weather derivatives compensate for deviations from the basis of 65Torres-GinerF per the Heating Degree Days (HDD) index and Cooling Degree Days (CDD) index. Precipitation-related products compensate for variations in precipitation.
[5] Weather derivatives can be used to speculate on the change of weather, a potentially risky undertaking.

to 80%. Using a natural experiment, Pérez-González and Yun (2013) find that the use of weather derivatives leads to an increase in firm value because the instruments enable the effective hedging of negative weather-related realisations. Similarly, based on an empirical study in the food retail sector, Štulec (2017) finds that weather derivatives can offset sales uncertainty. Consequently, holding weather derivatives tends to be viewed positively by stakeholders as a means of improving firms' risk profiles. Associated risk management disclosures (see IASB, 2014) can also provide stakeholders with an overview of how organisations hedge their financial risks leading to reduced information asymmetries and higher firm value. For example, Pierce (2015) shows that the use of hedge accounting significantly reduces earnings volatility. Steffen (2016) finds that derivative and hedging disclosures tend to improve investors' understanding of firm risk exposure. High-quality risk disclosures can also signal that firms are discharging accountability to their stakeholders leading to improved reputations and competitiveness and lower agency costs.

In addition to weather derivatives, Rial-Lovera et al. (2017) outline a range of techniques and strategies being used by the agricultural sector to address the increasing risks posed by climate change and weather variability. These are summarised in Table 1.

Although not exhaustive, Table 1 provides clear evidence of a consciousness of risk. The development of an array of risk management techniques shows that, on average,[6] food producers and retailers have accepted weather variability as an integral part of their experienced realities and are taking active steps to mitigate their risk exposure. This, we feel, is indicative of the operation of the risk society which frames business in terms of risk and the steps taken to mitigate risk. As the market for information on risk (such as weather forecasts, derivative indices and insurance data) expands and risk management practices become more established (Bates and Goodale, 2017) a type of risk discourse emerges in terms of which agricultural activity is framed, not only as an economic or biological activity, but one which is characterised by a type of 'risk ecology'.

2.4 Risk disclosure, climate change disclosures and greenhouse gas accounting

Risk management and internal control are a core feature of efficient corporate governance. In the context of a risk society, it is also becoming increasingly important for organisations to report on their key risks and mitigation strategies

[6] We use this caveat because there are still climate change sceptics among members of the farming community. In a study by Islam et al. (2013), 16% of the farmers surveyed were unconvinced about the reality of climate change. While small, this figure is still concerning given the direct impacts of climate change-related weather risks affecting farming in the UK.

Table 1 Agricultural mitigation strategies and techniques for dealing with climate change-induced weather risk

To deal with drought:	Greater use of water conservation techniques	Irrigation management	Greater drought resistant crop use	Changing planting/ harvest dates	Reduced tillage adoption
To deal with flooding:	Greater use of flood tolerant crops	Increase range of equipment	Adoption of soil conservation techniques (increased use of 'cover' crops)	Adoption of crop rotation	Changes in timing and quantities of fertiliser usage
To deal with high temperatures:	Crop varieties with longer duration	Sequential planting	Earlier sowing and later harvesting	Integrated pest management	Greater use of heat tolerant crops
To deal with low temperatures:	Shifts to crop varieties with different flowering and maturity dates				
To deal with changes in the length of growing seasons:	Adjusting planting and harvesting dates	Extend geographical range of suitable production areas	Integrated weed management		

for functional purposes and as a means of articulating their business in a discourse which resonates with a culture of risk awareness and contro

For example, there is a growing demand for risk disclosure by members of the investment community who feel that current reporting trends are inadequate for contextualising risk exposure and explaining how organisations are identifying and managing their risks (Solomon et al., 2000; ICAEW, 2011). These criticisms apply to both traditional financial issues as well as social and environmental metrics which are, increasingly, being understood as an integral part of an organisation's overall risk exposure (see, e.g. Edgley et al., 2010; ICAEW, 2011; Atkins and Maroun, 2015).

Per conventional finance theory, appropriate risk reporting can assist in reducing a company's cost of capital and creating a more stable business environment (Abraham and Cox, 2007; Rajab and Handley-Schachler, 2009). There is also a sense that risk management and reporting are becoming an integral part of the business ethos and defining feature of legitimate management control systems and reporting protocols. In support of this assertion is the codification of minimum levels of risk disclosures in terms of already institutionalised financial accounting standards. In US-GAAP, the use of and reporting on financial instruments as part of a firm's risk management strategy is outlined by SFAS 119, SAFS, 133 and FRR 48.[7] In the UK, where International Financial Reporting Standards (IFRS) are applied, IFRS 7 and IFRS 9 are applicable.

In addition to accounting standards, the *UK Corporate Governance Code* requires a board of directors to take responsibility for an organisation's risk governance and to ensure that key risks and steps taken to mitigate those risks are reported to stakeholders (FRC, 2018). These guidelines are complemented by recommended 'best practice' outlined by the IIRC (2013) and the GRI (2016). The academic community has also gone to considerable lengths to explain how management control systems (including those which deal with risks) form an integral part of the overall business management environment. They are essential for engaging with stakeholders, managing threats to legitimacy and ensuring business continuity (Melnyk et al., 2003; Alrazi et al., 2015; De Villiers et al., 2017). The net effect is considerable normative pressure on organisations to provide an account of their risk and to demonstrate how their business models are being aligned with a culture of risk identification and curtailment. One area which has received considerable attention is accounting for climate change.

The last decade has seen a rapid increase in climate change reporting (KPMG, 2015, 2017). An estimated 400 initiatives encourage or, in some cases,

[7] SFAS 119, Disclosures about Derivative Financial Instruments and Fair Value of Financial Instruments (1994), SFAS 133, Accounting for Derivative Instruments and Hedging Activities (1998), and the Financial Reporting Release (FRR) 48, Disclosure of Accounting Policies for Derivative Financial Instruments etc. (1997).

mandate companies to provide information on the costs, opportunities and risks associated with climate change. Most notable are the efforts of the Carbon Disclosure Project (CDP) which seeks to promote awareness about climate change and encourage the measurement of and reporting on interventions designed to reduce greenhouse gas emissions (CDP, 2017). The CDP compiles information from the world's largest companies to provide insights into the scope and quality of interventions taken by companies to lower their carbon footprint. In addition, the Financial Stability Board Task Force developed and published a series of voluntary recommendations in December 2016 which aim to:

> ... ensure that investors, lenders and other financial market participants receive consistent climate-related financial disclosures that are useful in understanding material climate-related risks and opportunities (Deloitte, 2017, p. 4).

Some firms disclose information on greenhouse gas (GHG) emissions voluntarily in order to gain competitive advantage (Rankin et al., 2011) or as part of a symbolic display designed to manage impressions (Borghei et al., 2016). In other cases, GHG disclosures can enhance environmental transparency. They are positively influenced by the presence of corporate sustainability officers and environmental committees (Peters and Romi, 2014) suggesting that some companies are starting to make a concerted effort to lower their carbon footprint. Carney (2015) takes a similar stance suggesting that climate change disclosure could have an emancipatory potential by making companies, investors and other stakeholders sensitive to environmental issues and aware of the total cost of their behaviour. More specifically,

> ..... a framework for firms to publish information about their climate change footprint, and how they manage their risks and prepare (or not) for a 2 degree world, could encourage a virtuous circle of analyst demand and greater use by investors in their decision making. It would also improve policymaker understanding of the sources of CO_2 and corporate preparedness (Carney, 2015).

This is especially true if climate change moves away from oversimplification of relationships in order to calculate the monetary cost of carbon and becomes more inter-disciplinary, context-specific and outcomes-focused (Isenhour et al., 2016).

In support of the emancipatory potential of the carbon accounting project is its ability to facilitate engagement with institutional investors and NGOs on climate change risks and opportunities. A dialogical type of accounting, in terms of which companies can seek advice from environmental groups or researchers and investors and other stakeholders can participate in discussions on the risks posed by climate change, promotes information sharing and the development of innovative strategies for combating climate change (Solomon et al., 2011; Atkins et al., 2015).

This is not to say that companies' risk reporting is free from limitations. Enslin et al. (2015) and Raemaekers et al. (2016), for example, find that current risk management and reporting practices:

> run the risk of appearing superficial due to the lack of meaningful forward-looking disclosure and the failure to link the possible impact of risks to the strategic objectives that could be affected (Enslin et al., 2015, p. 277).

Similarly, a recent report by Deloitte (2017) argues that, while there is a growing awareness of key environmental issues, like climate change, many organisations struggle to understand and act on the associated risk. Atkins and Atkins (2017, 2019) make a comparable point in connection with biodiversity and extinction of species. At the technical level, the difficulties of accounting for climate change, loss of species or ecological degradation should not be overlooked. Estimation methods necessary for reporting on GHG emissions, species under threat or changes in biomes are inherently complex and difficult to apply in practice (Milne and Grubnic, 2011; Jones and Solomon, 2013). The accounting infrastructure, necessary for collecting and analysing environmental data, is often under-developed and a single method or approach for accounting for environmental metrics is yet to be developed. Consequently,

> the existing surfeit of existing schemes and fragmented disclosures means a risk of getting 'lost in the right direction' (Carney, 2015).

3 Research method

The research subscribes to a social constructionist world view in terms of which language is seen as playing a key role in constructing organisational identify and revealing the operationalisation of a risk society mind-set (see Laine, 2010; Tregidga et al., 2014). As a result, data collection and analysis is carried out in an interpretive fashion.

The study employed a meaning-oriented content analysis of the weather-related disclosures produced by 15 FTSE100 companies involved in food retail and, in some cases, food production. The companies analysed re listed in Table 2.

We analysed all of the publicly available disclosures made by these companies in their annual/integrated and sustainability reports as well as on their websites. We focused on the periods which coincide with some of the most recently reported extreme weather events.

Table 2 Companies included in the analysis

ABF ASSOCIATED BRITISH FOODS PLC	Food producers	FTSE 100
CARR CARR'S GROUP PLC	Food producers	FTSE All-Share
CWK CRANSWICK PLC	Food producers	FTSE 250
DCG DAIRY CREST GROUP PLC	Food producers	FTSE 250
DVO DEVRO PLC	Food producers	FTSE All-Share
GNC GREENCORE GROUP PLC	Food producers	FTSE 250
PFD PREMIER FOODS PLC	Food producers	FTSE All-Share
TATE TATE & LYLE PLC	Food producers	FTSE 250
BOK BOOKER GROUP PLC	Food and drug retailers	FTSE 250
GRG GREGGS PLC	Food and drug retailers	FTSE 250
CLS MCCOLL'S RETAIL GROUP PLC	Food and drug retailers	FTSE All-Share
MRW MORRISON (WM) SUPERMARKETS PLC	Food and drug retailers	FTSE 100
OCDO OCADO GROUP PLC	Food and drug retailers	FTSE 250
SBRY SAINSBURY (J) PLC	Food and drug retailers	FTSE 100
TSCO TESCO PLC	Food and drug retailers	FTSE 100

3.1 Data collection

Data were collected from each corporate report and website and analysed systematically to ensure that all sections or parts of the reports and web pages were taken into account. In the first reading, the researchers considered the overall content and structure of the data sources, their scope and the context in which the reporting entities were operating. Next, each source was carefully reviewed to identify disclosures dealing directly or indirectly with the British weather.

The researchers focused on three specific types of content: information on the weather; disclosures outlining key risks and mitigation strategies and logistical arrangements or explanations of supply chains. For this purpose, no distinction was made between qualitative or quantitative disclosures. Narrative information, pictorial and graphical information were treated as having equal relevance. As well as food products, which are the main focus of this chapter because of the impact of climate change on farming, we also collected data on weather-related impacts on other items, such a clothing.

As each source was read, applicable content was highlighted and aggregated. This occurred until each section or subsection of the annual, integrated or sustainability reports had been reviewed. The same approach was followed for web content. The result was a log of the different weather-related disclosures categorised by company and source/location of the disclosure.

3.2 Data analysis

It was not our intention to determine the frequency of disclosures or to test for changes in the volume of quality of information being reported in a positivist sense. As a result, paragraphs, images, tables or subsections served as the unit of analysis rather than individual words or sentences. This avoided having to disaggregate data at a level which was not necessarily intended by the authors of the report content and mitigated the risk of a loss of meaning and context due to analysis taking place at a the granular, rather than conceptual, level (Krippendorff, 1989).

Data were analysed by considering how content, choice of language and tone construct meaning (adapted from Laine, 2010; Merkl-Davies et al., 2011; Atkins et al., 2018). This required each section of reports or web pages dealing with the weather to be considered in the context of how the respective organisation identified and understood risk and the actions being taken, if any, to manage those risks. The possibility of the disclosures forming part of a broader emancipatory agenda or compliance-orientated approach to reporting was also considered.

As is the case with all interpretive text analysis, data collection and analysis is inherently subjective. The researchers decided against using traditional inter-coder checks for validity and reliability as this runs the risk of reducing the analysis to a disclosure checklist. A more interpretive approach was used. The first round of data collection and analysis was being performed by two of the researchers to ensure consistency. A type of collective peer-review followed. This entailed a detailed discussion of each disclosure and how it was being interpreted as part of the broader risk society framework before including the information in the preliminary results. In the final stage, findings were presented to colleagues at two separate workshops to test inferences and ensure that results resonated with a broad audience.

The final results were grouped under broad theme headings (as per Section 4) which were derived interpretively by the researchers. Results are not organised by scientific method or objective data reduction. Their structure and sequence reflects the application of a subjective analytical model which 'progressed in hermeneutical manner' as the chapter was being developed (adapted from Laine, 2010; Merkl-Davies et al., 2011; Atkins et al., 2018).

4 Findings on weather risk disclosures

The data analysis revealed a series of themes covered by British food producers and retailers. Each is discussed in the following sections.

4.1 'The Great British weather': a traditional challenge for British food retailers

As discussed in Section 2, unpredictable weather is a historical feature of the British Isles. British farmers have battled with the weather for centuries developing products which are tailored to deal with the 'Great British weather'. Weather-related risk is, therefore, part of the local culture. For example:

> Designed with the great British weather in mind, a rain cover, foot muff and hood will keep your little one dry and warm. Whether you're in the city or out in the countryside, the multi-terrain pushchair is ready for **every eventuality** (Sainsbury's website, emphasis added).[8]

Similarly, the business model is adapted to traditionally unpredictable weather:

> That's why strawberries are grown under polytunnels ... which protect the delicate fruit from **whatever the British weather throws at them** (Sainsbury's website, emphasis added).[9]

The erratic weather which food producers and retailers have factored into their strategies and day-to-day operations is, however, becoming increasingly unpredictable and triggering alarm among industry leaders with Jonathan Church, a Tesco spokesman, commenting: 'Rapidly changing weather can be a real challenge' (Werdigier, 2009). How this abnormal variability is being incorporated in organisation's understanding of risk and, in turn, their strategies, management systems and reporting to stakeholders, is the key consideration of this research.

4.2 Acknowledgment that climate change is increasing weather risk

There is some mention in corporate disclosures of links between food/weather-related risks and climate change but this is sporadic. For example, in a section of their website dealing with 'How Sainsbury's is responding to global challenges' the retailer explains:

> The world is changing. Global warming, extreme weather and volatile prices are making it harder for farmers to produce the foods we love. So Sainsbury's is changing too (Fairly Traded: accessed 22/7/2017).

[8] Giving all parents a supporting Cuggl New nursery brand, Cuggl, launches exclusively at Argos (accessed on 23/7/2017) 19 July 2017 https://www.about.sainsburys.co.uk/news/latest-news/2017/19-07-2017.
[9] Behind the scenes of how we grow the perfect strawberry (accessed on 23/7/2017)11 July 2017. https://www.about.sainsburys.co.uk/news/latest-news/2017/10-07-17-strawberries.

The disclosure is very broad. Exactly how global warming is affecting the business is not outlined and the change being referred to is undefined. The company does provide some information about a new pilot programme aimed at supporting tea farmers in Africa but details are limited and the link between the initiative and climate change action is unclear.

Morrisons gives a more detailed account of the negative impact of climate change:

> Climate change is a direct result of human activity, from population growth, demand and consumption patterns increasing emissions into the atmosphere. It is already affecting the way we live today, we see this through severe weather patterns, rising temperatures and sea levels causing food and resource scarcity and devastation to communities (Morrisons website).

The business impact is explicated, albeit briefly:

> security of food supply; loss of food quality through changing climate patterns; food price inflation; loss of livelihoods for Morrisons suppliers; new markets; changing buying habits (Morrisons website).[10]

Under operational risks in the annual report and statements, ABF Ltd states that

> Our businesses rely on a stable supply of natural resources some of which are vulnerable to external factors such as natural disasters and climate change (ABF website).[11]

The company outlines the types of risk which it deals with, albeit in broad terms:

> We operate across a vast range of environments, which requires our businesses to adapt to local or national issues such as changes to regulation, increasing pressure on agricultural land, weather patterns and local social pressures ….. Our businesses are constantly responding to a range of physical, regulatory and financial risks associated with major environmental matters, including adapting to challenging weather patterns and temperatures, operating in water-stressed areas and responding to changing water and carbon-related regulations (ABF Ltd website).[12]

These high-level disclosures are supported by more detailed case-specific information which provides context and gives a better sense of the impact which weather conditions can have business operations. For example in 'The Impact of Changes in Weather' on the ABF website, the company explains:

[10] Corporate Responsibility Report 2016/17 (accessed on 8/7/2017). https://www.morrisons-corporate.com/Global/corporate/Morrisons_CR_Report2017_Interactive_FINAL.pdf.
[11] Annual reports and accounts 2016 (for the year ended 17 September 2016) (accessed on 19/5/2017). http://www.abf.co.uk/documents/pdfs/ar_cr_2016/2016_annual_report.pdf.
[12] Corporate Responsibility Report 2016 (accessed on 23/5/2017)http://www.abf.co.uk/documents/pdfs/ar_cr_2016/2016_corporate_responsibility_report.pdf.

> Our sugar operations account for over 82% of our total energy use. The impact of weather on the size and quality of the sugar beet and cane crops therefore directly affects the amount of energy needed to process those crops, and the efficiency with which we can do this. Our sugar factories are based in parts of the world with different climates and significant variability in local weather patterns. This year alone, our operations have managed the impacts of excess rainfall and flooding, droughts and hard frosts which have generally led to reduced crop yields. Through appropriate risk management and adaptability, our businesses are prepared to respond to these extremes in weather patterns (ABF Ltd website).[13]

Exactly how the company has responded is not dealt with. For example, details on how risk management plans are designed, implementation costs and measures of performance against targets are not provided. Similarly, the Chairman's statement includes the following generic comment on climate-related risk:

> Whether responding to challenging financial markets; integrating acquired businesses; disposing of businesses; dealing with the effects of unusual weather on our supply chains or retail demand .... Our people have responded with enthusiasm (ABF website).[14]

From a critical perspective, the company is dealing with environmental issues which are of increasing concern to stakeholders but without committing to the specific details necessary for holding the organisation accountable for its environmental performance. More optimistically, there is a sense of climate change being accepted as part of the corporate reality. There is no effort to deny the fact that climate change is taking place and is linked with increasingly erratic weather. Similarly, the risks arising from changing weather patterns are being acknowledged. These are a type of sunk cost which leave business with no realistic alternative but to engage 'enthusiastically'.

4.2.1 Weather risk, food production and crop accounting

There is some discussion on the companies' websites about the relationship between harvests, inflation and the weather. For example, Sainsbury states:

> ....we've seen a reduction in commodity prices, particularly in areas like fresh produce, as a result of a better winter, a less inclement winter, leading to better crops (Sainsbury's website).[15]

[13] Corporate Responsibility Report 2016 (accessed on 23/5/2017) http://www.abf.co.uk/documents/pdfs/ar_cr_2016/2016_corporate_responsibility_report.pdf.
[14] Annual Results Announcement Year ended 17 September 2016 (accessed on 28/5/2017). http://www.abf.co.uk/documents/pdfs/ar_cr_2016/annual_results_announcement_2016.pdf.
[15] First Quarter Trading Statement 2014 (transcript) (accessed on 6/8/2017) Wednesday 11th June 2014, 8:30. https://www.about.sainsburys.co.uk/~/media/Files/S/Sainsburys/documents/reports-and-presentations/2014/j-sainsbury-plc-q1-trading-statementtranscript.pdf.

Morrisons also discusses the impact of unpredictable weather on harvest and the related price effects:

> Poor weather, which affected harvests both in the UK and abroad, had a particularly negative impact on the price of some food commodities including potatoes and carrots, which increased by 23% and 18% respectively (Morrisons website).[16]

In most cases, the weather is operationalised as a business risk with material financial and operating consequences. Consider the following two examples dealing with apples, milk and lamb:

> If I could change anything about my job it would be the British weather, especially the hail. The damage it can do to top fruit is sad to see. I've seen fantastic crops destroyed in just five minutes of hail (Tesco website).[17]
>
> Without milk we would not have a business. Restricted milk supply could be caused by economic factors, weather, fuel availability or an epidemic which affects dairy cows. This could lead to lower sales and profits (Dairy Crest website).[18]
>
> ... lambs stay together as a group but their diet is based entirely on grass and forage. If grass is in short supply because of bad weather, they are fed other natural supplementary feeds to ensure the nutritional needs of the animal are met (Sainsbury's website).[19]

Other parts of the value chain also bear the brunt of a changing climate:

> Wet autumn conditions lowered wheat plantings to 70% of normal levels, reducing the demand for fertiliser and crop protection products although volumes are expected to pick up as spring planting resumes (ABF website).[20]

Weather changes also have demand-side implications. UK summers are expected to be warmer and leading to changing consumer patterns.

> As the warmer weather begins to emerge, Sainsbury's has seen a 58% increase in French rosé sales in the past four weeks, and is expecting bumper sales this weekend with temperatures expected to hit 23 degrees. When temperatures soar, we turn to rosé over white. A warm spell sees rosé sales up 30% in comparison to 20% for white wine (Sainsbury's website).[21]
>
> Supermarkets that stock meat and other barbecue items in anticipation of sunshine are often left with unsold food if the weather turns out rainy and cold instead. A

[16] Preliminary Results for the Year Ended 3 February 2013 (accessed on 8/7/2017). https://otp.tools.investis.com/clients/uk/morrison/rns/regulatory-story.aspx?cid=623&newsid=329302.

[17] British supplier support from Tesco's Master of Apples (accessed on 13/8/2017). https://www.tescoplc.com/news/blogs/topics/british-fruit-supplier-support-from-tesco-master-of-apples/ (accessed on 14 Sep 2016).

[18] Dairy Crest Group plc Strategic report 2017 (for the year ended 31 March 2017) (accessed on 19/6/2017). http://www.dairycrest.co.uk/~/media/Files/D/Dairy-Crest-Group/documents/annual-report-2017-strategic-report.pdf.

[19] Extract from: Animal Health & Welfare Report (accessed on 06/8/2017). https://www.about.sainsburys.co.uk/~/media/Files/S/Sainsburys/pdf-downloads/animal-health-and-welfare.pdf published in July 2017.

[20] Interim Report 2013 (Until 02 March 2013) (accessed on 26/5/2017). http://www.abf.co.uk/documents/pdfs/news/abf_interim_2013.pdf.

[21] 2017 is looking rosé as Brits' tastes change (accessed on 23/7/2017) 07 April 2017. https://www.about.sainsburys.co.uk/news/latest-news/2017/07-04-2017.

> temperature increase of 18 degrees generally triples sales of barbecue meat and increases demand for lettuce by 50 percent, Tesco said (Werdigier, 2009).

Variations in consumer purchases due to the weather can, however, result in volatile sales which companies need to manage.

> In 2007, the company bought an option to offset the risk that December would be warmer than expected. After seeing sales of jackets in December 2006 tumble by 30% because of unexpected warm weather in the Northeast, the company decided to buy a contract for the following winter to cover up to $10 million in potential losses if temperatures in December 2007 were warmer than normal (Morrison, 2009).

Logistical challenges are also documented:

> .... We rely on the ability of our employees to be able to get to their place of work .... This leaves us exposed to traffic congestion, road works, congestion charges and inclement weather, particularly snow, all of which could at times render deliveries difficult or even impossible (Ocado Group plc website).[22]

Dealing with its 'principal risks and uncertainties' ABF Ltd lists physical damage to operations and assets, availability and cost of agricultural raw materials, higher operating charges and the 'negative impact on the environment and the communities which depend on land used by our operations' as key risks arising from climate change.

Ocado has a similar message:

> We are vulnerable to fluctuations in the availability and price of food items. General economic conditions, unanticipated demand, problems in production or distribution, natural disasters, adverse weather conditions during the growing and harvesting seasons (Ocado Group plc website).[23]

4.2.2 Weather risk and non-food products

In addition to selling food, many of the retailers under review also sell non-food products. We found substantial disclosures relating to weather risks and their impacts on the latter. For example, Sainsbury and ABF warn:

> The Sainsbury's Group's and the Home Retail Group's businesses are subject to seasonal peaks, with higher sales and operating profits generated during certain peak trading periods. Weak sales during such peak trading periods, or as a result of extreme or unseasonal weather conditions, could adversely impact the Sainsbury's

[22] Offering memorandum (accessed on 9/7/2017). http://www.ocadogroup.com/~/media/files/o/ocado-group/financing-2017-disclaimer/offering-memorandum-382265.pdf.

[23] Offering memorandum (accessed on 9/7/2017). http://www.ocadogroup.com/~/media/files/o/ocado-group/financing-2017-disclaimer/offering-memorandum-382265.pdf.

> Group and the Home Retail Group, as applicable, and, following completion of the Acquisition, the Combined Group (Sainsbury's website).[24]
> Unseasonable weather and cautious consumer sentiment led to value declines in the clothing retail sector in some of our important markets, particularly the UK and Germany (ABF website).[25]

Similarly, Primark's trading update stated,

> Following a strong performance at the start of the financial year, trading was weaker in the weeks leading up to and over Christmas, as a result of unseasonably warm weather across northern Europe, resulting in like-for-like sales for the first half that were less than 1% below last year (ABF website).[26]

Climate change effects are not limited to the production and supply offood. The implications are broader. Erratic weather influences consumer confidence and behaviour in addition to the availability of food and non-food items. This means that diversified retailers, which appear would otherwise enjoy a type of natural hedger, remain exposed to the underlying financial and economic risks associated with climate change.

4.3 Acknowledgement of financial materiality of weather risk

Given the pervasive impact of climate change on retailers' business models, we found several instances where weather risks were disclosed, not just in terms of products, customers or operations, but also as direct and material financial concerns. For example,

> We are disappointed that our Grocery business reported materially lower sales in the quarter due to warmer weather; particularly in September (Premier Foods website).[27]
>
> The Sainsbury's Group's and Home Retail Group's sales are also sensitive to periods of extreme weather conditions. The Sainsbury's Group and the Home Retail Group may see a reduction of sales during periods of inclement weather due to reduced customer footfall. The number of customers visiting the Sainsbury's Group's or the home Retails Group's stores may also decline during periods of extreme weather conditions affecting the relevant local catchment area. Extreme weather conditions may also result in orders placed online being unable to be delivered. Prolonged unseasonal weather conditions, and/or temporary severe weather during the Sainsbury's Group's and the home Retail Group's peak trading periods could have a material adverse effect on the Sainsbury's Group's, the Home Retail Group's

[24] Extract from HRG acquisition- Prospectus dated 5 July 2016 (accessed on 23/7/2017). https://www.about.sainsburys.co.uk/~/media/Files/S/Sainsburys/documents/investors/prospectus-5-july-2016.pdf.
[25] Annual reports and accounts 2016 (for the year ended 17 September 2016) (accessed on 19/5/2017). http://www.abf.co.uk/documents/pdfs/ar_cr_2016/2016_annual_report.pdf.
[26] Interim Results Announcement 24 weeks ended 27 February 2016 (accessed on 28/5/2017). http://www.abf.co.uk/documents/pdfs/2016/abf-interim-results-announcement-2016.pdf.
[27] Trading update for the 13 weeks ended 1 October 2016 (accessed on 9/7/2017). http://www.premierfoods.co.uk/media/news-releases/Items/%E2%80%8BTrading-update-for-the-13-weeks-ended-1-October-2.

> and, following completion of the Acquisition, the Combined Group's businesses, results of operations, financial condition or prospects (Sainsbury's website).[28]

For some of the companies under review, climate change may be so material that it is being grouped with other high consequence risks:

> the occurrence of one or more natural disasters, such as floods, pandemic outbreaks, weather conditions, such as major or extended winter storms, terrorist acts or disruptive global political events, in the UK or in countries in which the Sainsbury's Group's or the Home Retail Group's suppliers are located, or other disruptions, could materially and adversely affect the Sainsbury's Group's the Home Retail Group's and, following completion of the Acquisition, the Combined Group's businesses, results of operations, financial condition or prospects (Sainsbury's website).[29]

4.3.1 Weather risk in food retailers' international food supply chain

The main focus of this chapter is on the 'not so great' British weather and the extent to which its increasing variability arising from climate change is being reported by UK food retailers. However, we also found substantial disclosures by the same companies dealing with climate change-induced weather risks for their supply chain. For example, Tesco deals with extreme weather and the impact on Spanish suppliers:

> This winter, the rains in Spain were quite a pain, as they caused floods that destroyed many of the crops. It was the worst weather to hit production in over 30 years. This unprecedented weather led to price rises and shortages with some products such as courgettes, peppers and leafy salads. In some cases over 40% of the crop was lost (Tesco plc website).[30]

ABF discussed the consequences of unpredictable weather on sugar supplied from Illovo, Africa's largest sugar producer with agricultural and production facilities spanning six countries:

> Illovo made good progress following last year's weather-related crop shortfalls and, with further recovery expected in the new season, sugar production in this financial year is expected to improve to 1.7 million tonnes compared with 1.4 million tonnes produced in the comparable months last year (ABF plc website).[31]
>
> In 2016, we abstracted 800 million m^3 of water, a decrease of 14% compared with 2015. In the main, this is due to Illovo's reduction in water abstracted for agricultural

28 Extract from HRG acquisition-Prospectus dated 5 July 2016 (accessed on 23/7/2017). https://www.about.sainsburys.co.uk/~/media/Files/S/Sainsburys/documents/investors/prospectus-5-july-2016.pdf
29 Extract from HRG acquisition-Prospectus dated 5 July 2016 (accessed on 23/7/2017). https://www.about.sainsburys.co.uk/~/media/Files/S/Sainsburys/documents/investors/prospectus-5-july-2016.pdf.
30 Lettuce tell you the news: Best of British Back in Baskets (accessed on 23/7/2017). https://www.tescoplc.com/news/blogs/topics/british-lettuce-tesco/.
31 Interim Results Announcement (24 weeks ended 4 March 2017) (accessed on 26/5/2016). http://www.abf.co.uk/documents/pdfs/2017/interim_results_2017_announcment.pdf.

> use because of the extreme dry weather in many parts of southern Africa. There was inadequate water in the rivers and other water sources to fulfil irrigation requirements (ABF plc website).[32]

Similar problems were experienced among Chinese and European suppliers,

> Our two beet factories in north China at Zhangbei and Qianqi, are operating well and will process a record beet crop. Adverse weather conditions experienced during beet harvesting and storage have adversely impacted sugar levels in the beet but, with the higher volumes and improved prices, this business is expected to deliver a much improved profit this year (ABF plc website).[33]
>
> In Spain, the northern campaign was delayed to maximise beet development from the reduced area under cultivation in the 2013 crop year. Although the campaign commenced well, adverse weather towards the end of the period resulted in challenging harvest conditions and an interrupted campaign at our La Beneza factory (ABF website).[34]

Climate change is a global problem and not specific to a single organisation. Unfortunately, environmental accounting provides an economic account of a firm's performance and is included as part of a sustainability or integrated report which is constrained by the boundaries of financial reporting conventions on the 'reporting entity' (see Cuckston, 2017, 2018). The above disclosures go some way to address this by showing how an organisation's business risk has geographic context. This is, however, a far cry from a consolidated ecological account of the risks posed by climate change in both anthropocentric and deep ecological terms.

4.3.2 Investor-company dialogue on weather risk

There were numerous mentions of weather-related risk in investor-company dialogue at quarter trading statement meetings. The content of these discussions resemble findings from earlier private climate change disclosures, which illustrate how investors are engaging directly with companies as part of a process of gaining a shared understanding of the dangers posed by changing weather conditions (Solomon et al., 2011). Consider, for example, the following two explanation where the weather is used to explain sales trends in different parts of the business to investors:

> It was an awkward summer in terms of some aspects of the seasonal business because it was a little bit wetter and colder in the first part and a bit warmer and drier

[32] Annual reports and accounts 2016 (for the year ended 17 September 2016) (accessed on 19/5/2017). http://www.abf.co.uk/documents/pdfs/ar_cr_2016/2016_annual_report.pdf.
[33] Pre Close Period Trading Update (27 February 2017) (accessed on 19/5/2017). http://otp.investis.com/clients/uk/associated_british_foods/rns/regulatory-story.aspx?cid=1464&newsid=847977.
[34] Associated British Foods plc announces its interim results for the 24 weeks ended 1 March 2014 (accessed on 26/5/2017). http://www.abf.co.uk/documents/pdfs/2014/interim_results_announcement_final.pdf.

> in the second part. That has probably been helpful to the Convenience business. So where there is a delta to the supermarket business, it is probably driven as much by the weather in August and the early part of September as anything else...The convenience business in particular is quite affected by weather, particularly in the big towns and, as I have already observed, it was probably not the most helpful summer for the grocery industry, but that would be particularly the case for the convenience stores. So I did see a slowdown in our convenience business, but still a great engine for growth (Sainsbury's website).[35]
>
> General merchandise has been slower than we have been used to and that is largely a reflection of a couple of things. Firstly, the weather was particularly unseasonal, or unseasonable over the summer. So the seasonal categories were not particularly buoyant in the year or in the first half (Sainsbury's website).[36]

There is a type of systematic dialogue taking place where the company is providing a detailed and neutral account of information to investors in order to explain sales trends and firm performance. The language is informal and the tone is neutral. Particularly interesting is the seamless incorporation of the weather as part of a standard business discourse used to inform investors. Consider also the following exchange between an institutional investor and Sainsbury's:

> Question: 'And we had adverse weather this year in the UK during Q1, so to what extent did that impact your non-food sales this quarter?' (Edouard Aubin, Morgan Stanley)
>
> Response: 'It sounds a bit like you're making excuses when you start talking about the weather, but we had already had the warmest week of the year this time last year. So there were certainly some seasonal weather effects in the quarter ..... so there's a few underlying factors in the numbers at which are probably on balance unhelpful but, as I say, blaming the weather **is the last refuge of scoundrel** so we won't do that this time!' (Mike Coupe, Sainsbury's website).[37]

The exchange points to the retailer's desire to please and reassure its investors, as found by earlier research on private reporting and engagement between companies, shareholders and analysts (Atkins et al., 2015b). At the same time, we see that the weather is not being used as a convenient excuse for poor performance. This view is reaffirmed when, in a different engagement, the same company representative downplays the impact of the weather on financial returns:

[35] First Quarter Trading Statement Tuesday 4 th July 2017 (transcript) (accessed on 6/8/2017). https://www.about.sainsburys.co.uk/~/media/Files/S/Sainsburys/documents/reports-and-presentations/2017/transcript-j-sainsbury-first-quarter-trading-statement-4-july-2017.pdf.

[36] Interim Results 2015 Analysts Wednesday 11 November 2015 (transcript) (accessed on 6/8/2017). https://www.about.sainsburys.co.uk/~/media/Files/S/Sainsburys/documents/reports-and-presentations/2015/transcript-j-sainsbury-interimresults-analysts-11-november-2015-v1.pdf.

[37] First Quarter Trading Statement Wednesday 10th June 2015, 08:30 BST (transcript) (accessed on 6/8/2017). https://www.about.sainsburys.co.uk/~/media/Files/S/Sainsburys/documents/reports-and-presentations/2015/q1-transcript-final.pdf.

on weather, you wouldn't be surprised to know that year-on-year there aren't that many different sunshine days and warm weather days. It just happened that they came a little bit later this year and caused a bit of a furore because of the timing. Actually it happened last year in May. So I would not call a lot of significance to the weather year-on-year, it is not that significant especially when you stretch it over a 16 week quarter that began at the back end of March. So don't get too carried away about that (Sainsbury's website).[38]

4.3.3 Weather risk mitigation strategies and innovations

Given the material risks posed by climate change and increased attention accorded to it by the investor community, it comes as no surprise that companies are disclosing information on their strategies for mitigating weather-related risks. For example,

> Sainsbury's will boost support for hundreds of thousands of farmers and workers - helping them to become more resilient in the face of escalating challenges from climate change to global competition, health and geo-political tensions. The plan includes the roll-out of new Sainsbury's Sustainability Standards across key crops and ingredients, the pilot of a new sustainable sourcing approach for tea farmers and the set-up of an expert advisory board (Sainsbury's website).[39]

Sainsbury could be criticised for departing from generally accepted Fair Trade conventions by introducing its own standards which may be biased or difficult for smaller farmers to implement. One could also argue that the plan is an under-developed and reactionary response to climate change. Nevertheless, the detailed disclosures suggest that this is more than just symbolic reporting aimed at reassuring stakeholders while avoiding substantive action (cf Borghei et al., 2016):

> The Sainsbury's Sustainability Standards, which build on and recognise existing certifications... have been piloted on prawns in Thailand and Belize and will now be piloted on other key crops such as tea, wheat, potatoes, sugar and bananas. One of their imminent applications is with tea farmers in Africa as part of a pilot on key lines of Sainsbury's tea, called Sainsbury's Fairly Traded - a new way of working that will enable tea farmers in Africa to strengthen their businesses and communities as they tackle ever growing challenges, such as the impact of climate change and associated droughts, soil erosion and crop diseases (Sainsbury's website)[40].

[38] First Quarter Trading Statement Tuesday 4 th July 2017 (transcript) (accessed on 6/8/2017). https://www.about.sainsburys.co.uk/~/media/Files/S/Sainsburys/documents/reports-and-presentations/2017/transcript-j-sainsbury-first-quarter-trading-statement-4-july-2017.pdf.

[39] Sainsbury's unveils ambitious new sourcing approach to give greater support to farmers (accessed on 23/7/2017). https://www.about.sainsburys.co.uk/news/latest-news/2017/23-05-2017.

[40] Fact Sheet: Sainsbury's Sustainability Standards. Fact Sheet: Fairly Traded Pilot, from.Sainsbury's unveils ambitious new sourcing approach to give greater support to farmers (accessed on 23/7/2017). https://www.about.sainsburys.co.uk/news/latest-news/2017/23-05-2017.

There are references to company partnering with different suppliers and running workshops to educate farmers on the plan's features and the suggested methods for becoming more resilient to adverse weather (see Atkins et al., 2018). In response to the possible criticism of departing from already established codes of best practice, stakeholders are reassured that Sainsbury's plan has been calibrated and references to pilot studies suggest that a carefully planned implementation strategy has been developed rather than an ad hoc environmental initiative. Key performance indicators, timelines and actions taken to date are not discussed in as much detail. Nevertheless, the disclosures point to a type of emancipatory accounting where weather-related risks are being internalised and the company is, at a minimum, taking the first steps to mitigate those risks (consider Maroun and Atkins, 2018; Atkins et al., 2019).

Other organisations are also working to develop and implement operating policies, codes of best practice and alternate product lines as a way of making their business models more resilient. The Farming Minister, David Heath, discusses how the introduction of a voluntary code and new milk price formula could help dairy farmers:

> The bad weather has been a real blow to dairy farmers, but the Code is a sign of what can genuinely be achieved through collaboration and overcoming differences (Dairy Crest website).[41]

Premier Foods have introduced a product range which reduces weather-related impacts on the business,

> ... the new ready-to-use range ... can be used across a wide variety of dishes, from soups to casseroles, to curries and risottos - making it less weather dependent than traditional stock (Premier Foods website).[42]

Similarly, Morrisons have introduced a new line of misshapen vegetables. What is especially interesting is that the environmentally responsible consumer is co-opted in this risk mitigation strategy, albeit indirectly, thereby creating a market for food items which may otherwise have been left to waste.

> To support producers we adjust our specifications when weather conditions lead to poor harvests and following customer feedback we have introduced a range of 'wonky' vegetables, such as parsnips, carrots and onions, which are cheaper in price and give customers the choice on which they want to buy (Morrisons website).[43]

[41] Farming Minister joins Dairy Crest to launch milk price formula in Parliament (accessed on 19/6/2017) http://www.dairycrest.co.uk/media/latest-news/news/2013/25-04-13.aspx.

[42] Tasty New additions to iconic Oxo range with first Ready-to-Use option and new Beef Stock Pot (accessed on 9/7/2017). http://www.premierfoods.co.uk/media/news-releases/Items/Tasty-new-additions-to-iconic-Oxo-range-with-first.

[43] CR programme-Food waste (site no date) (accessed on 8/7/2017). https://www.morrisons-corporate.com/cr/policy/food-waste/.

Another weather risk mitigation strategy described by ABF includes an EU-funded three-year Maragra Smallholder Sugar Cane Development Project (MSSDP) granted to Illovo in Mozambique. This project:

> ... enables smallholder farmers to increase their household food security by providing technologies that reduce vulnerability to climate shocks such as drought. The project introduced a new human-powered water pump that means smallholder farmers are less reliant on rainfall (ABF Ltd website).[44]

ABF's *Annual Reports Announcement* (2017) outlines other mitigation strategies implemented to address the climate-induced weather-related risks such as protecting infrastructure from floods.

Despite evidence on the increasing popularity of weather derivatives, we found very few disclosures in the UK food retail and production industries where these instruments were being used to hedge risks.[45] To confirm this finding, we reviewed the news coverage on Factiva, which is a global news database of just under 33,000 premium sources[46]. We also collected 1271 publications including 'weather derivatives' (or related terms) in their key words, titles or summaries/abstracts. Those articles mention '"weather derivatives' in the energy sector (35%), in the electricity/gas utility (30%), financial investment (14%) and insurance sector (13%). Only 2% and 0.6% of articles mention 'weather derivatives' in the agriculture and the retail/wholesale sectors respectively. There are three possible reasons for this finding: Firstly,

the lack of interest in whether derivatives may be due to the complexity of financial products Huault and Rainelli-Weiss, 2011). The use, valuation and accounting for weather derivatives require considerable expertise. The weather derivatives market is relatively new. Food retail and production sectors in the UK may be too conservative to roll out an extensive hedging strategy using weather derivatives. Secondly, weather derivatives may be too costly[47](Musshoff et al., 2011). Finally, weather derivatives can offset short-term losses but are not a substitute for structural changes required to ensure resilient business models capable of adapting to an unpredictable environment in the medium- to long-run.

[44] Corporate Responsibility Report 2016 (accessed on 23/5/2017). http://www.abf.co.uk/documents/pdfs/ar_cr_2016/2016_corporate_responsibility_report.pdf.
[45] We found only example where a company referred to a commodity price hedge in the financial statements.
[46] We review the press articles from Factiva until 7 March 2018.
[47] Basis risk is an important concern in the hedging effectiveness of weather derivatives, in which the value of derivatives might not move in line with that of the underlying exposure. For example, a farm may face the geographical basis risk if there is the difference between the site of production and the reference weather station.

4.3.4 Changing weather brings opportunities: Accounting for strawberries, asparagus and apricots

Bad weather is not always bad news:

> A wet autumn in 2012 lowered wheat plantings thereby reducing demand for fertiliser and crop protection products. However, the cool spring and warm summer of 2013 provided good growing conditions for autumn planted crops and spring cereals creating a better harvest potential than was previously expected (ABF website).[48]

Warmer winters and changes in rainfall are have a positive impact on some type of food production in the UK. Sainsbury's provides illustrations of weather-related potato, radish and apple accounting:

> It's great news that we're able to kick start the Jersey Royal season and due to the warmer weather over the winter it's a whole week earlier than last year (Sainsbury's website).[49]
>
> The milder winter has meant that the radish have grown much faster than usual. It's a real bonus for us to be able to increase the length of the British radish season and get them to Sainsbury's customers earlier than ever before (Sainsbury's website).[50]
>
> The recent hot weather is having a positive effect on the English apple crop, leading UK supermarket Sainsbury's has revealed, with the first apples of the season coming in 5% bigger than usual (Sainsbury's website).[51]

The agricultural sector has been developing plant hybrids which are more weather-resistant as a strategy for mitigating climate risks. Their research and development are starting to bear fruit as highlighted by Tesco which has invested heavily in cultivating new species of apricots:

> Apricots flourish in continental climates with cold winters, but without spring frosts. So traditionally they have grown in countries like Spain, France, Morocco, Turkey and Iran. Tesco were the first UK retailer to start selling English grown apricots and started working on a production partnership with one of the UK's largest stone fruit producers from Kent back in 2010.... In the last year UK demand for apricots has rocketed by an unprecedented 75 per cent, making it one of Britain's most in demand fruits. Apricot trees still need a lot of sunshine and as a result all the production at the moment is in the southern counties, with growers based either in Kent of the Isle of Wight. Until the late nineties English apricot production was not thought possible because of our cooler temperatures (Tesco website).[52]

[48] Annual Results Announcement Year ended 14 September 2013 (accessed on 30/5/2017). http://www.abf.co.uk/documents/pdfs/2013/annual-results-announcement.pdf.
[49] Sainsbury's is the first major retailer to stock Jersey Royals in 2017 (accessed on 23/7/2017). https://www.about.sainsburys.co.uk/news/latest-news/2017/08-03-2017.
[50] Sainsbury's customers get first bite of UK radish harvest (accessed on 23/7/2017),12 April 2017. https://www.about.sainsburys.co.uk/news/latest-news/2017/12-04-2017.
[51] Sunshine is the Core Ingredient in our First Apples of the Season (accessed on 25/7/2017), 6 September 2016. https://www.about.sainsburys.co.uk/news/latest-news/2016/06-09-2016-a.
[52] Great news for apricot fans as first British fruit of the year hits Tesco stores (Accessed on 23/7/2017). https://www.tescoplc.com/news/news-releases/2017/british-apricots/.

There are examples of 'asparagus' and 'strawberry accounting' which explain how UK farmers are benefitting from warmer and sunnier weather:

> The recent sunshine has brought the harvest forward this year, meaning that remarkably, customers will be able to enjoy British asparagus in March ... Tesco Produce Buying Manager David Daniels said: 'This is a brilliant opportunity for customers who love English asparagus, to enjoy it unusually early in the year. Thanks to the mild weather of late, our grower in Suffolk has been able to produce enough asparagus to supply stores across the UK, which is really quite a feat. (Tesco website).[53]
>
> The sunshine experienced across the UK has meant that our British strawberry crop has come on in bigger volumes than forecast. Growers from Herefordshire, Kent, Lancashire and Scotland have told us about a small crop flush brought on by the bonus British sunshine (Tesco website).[54]

Other crops, such as lettuce and cauliflower are also thriving due to the change in the UK's weather. Similarly, while climate change is adversely affecting some crop yields or livestock rearing, it simultaneously benefits farmers due to a price effect:

> The wet winter in the UK, particularly in the west country, adversely impacted livestock farming but excellent forage growing conditions during a dry spring and summer, as well as a softening commodity market, led to a period of more stable prices and increased confidence among farmers (ABF website).[55]

Similarly, a heat wave which can devastate crops can have an unexpected benefit for a diversified retailer:

> Our top five selling lines across the whole of Argos last week comprised of two fans, two paddling pools and play sand, with over 165,000 units of paddling pools and fans sold. Fans: This has been one of our busiest weeks ever for fans - yesterday we sold 92,000 units alone (Sainsbury's website).[56]

The changing weather can also be good news for NGOs or charities working with the homeless:

> There are times when unfortunately food surplus is unavoidable for a host of reasons such as unpredictable weather. In these instances, donating this surplus food to charity is our absolute priority, and we have been working hard to ensure we are

[53] Highly prized English asparagus to go on sale at Tesco - almost a month early (Accessed on 23/7/2017). https://www.tescoplc.com/news/news-releases/2017/english-asparagus-goes-on-sale/.
[54] Bumper British strawberries crop taken to help our farmers (accessed on 10/8/2017). https://www.tescoplc.com/news/blogs/topics/food-waste-bumper-british-strawberry-crop/ 15 Jun 2017.
[55] Annual Results Announcement for the Year ended 13 September 2014 (accessed on 28/5/2017). http://www.abf.co.uk/documents/pdfs/2014/annual_results_announcement.pdf.
[56] Argos customers are cooling down and making the most of the heatwave in the garden (accessed on 23/7/2017), 20 June 2017. https://www.about.sainsburys.co.uk/news/latest-news/2017/20-06-17-argos-weather.

> continually increasing the number of local food charities we work with (Sainsbury's website).[57]

The disclosure can be seen as a type of philanthropic accounting in the sense that Sainsbury's is explaining how the losses resulting from adverse weather conditions are being partially 'recouped' as part of their commitment to corporate social responsibility. This can be dismissed as a type of impression management rather than genuine altruism (see, e.g. Brown et al., 2006; Zadek et al., 2015). The monetary values are, however, relatively low and the disclosure suggests that the organisation is more concerned with explaining how it is managing efficiency and cost implications associated with adverse weather than trying to win stakeholder support with a superficial marketing campaign.

5 Conclusion

Our analysis provides interesting insights into how UK food retailers understand, internalise and react to the not-so-great British weather. Disclosures in their annual, integrated and sustainability reports and on company web pages represent a discourse of risk awareness which recognises the need to manage risk and, at the same time, expresses anxiety about the magnitude of the threat posed by climate change.

There is a sense that companies appreciate the operational and financial implications of erratic weather and that this environmental issue is not being seen as just a side effect of doing business. The weather is being incorporated as part of a broader business discourse on strategy development, operating resilience and risk mitigation. As part of this process, specific initiatives taken to reduce the impact which climate change is having on each organisation are discussed, albeit to different extent. Examples include building weather-resistant infrastructure, launching new products and, where applicable, identifying and capitalising on opportunities presented by the ever changing weather.

Contrary to international research findings, UK retailers are not using weather derivatives as part of a strategy for hedging environmental risks. The complexity and cost of this risk management model is a possible reason for this. An alternate explanation is that UK firms are taking a longer-term and more structural approach to managing weather-related risks. Hedging is effective for offsetting short-term losses. Re-engineering business models and changing operating practices are more appropriate for ensuring resilience in the long-term.

[57] Sainsbury's food surplus and food waste: How we are delivering a positive impact (accessed on 23/7/2017) Released: 19 September 2016. https://www.about.sainsburys.co.uk/~/media/Files/S/Sainsburys/documents/reports-and-presentations/2016/sainsburys-food-surplus-and-food-waste-figures-15-16-report.pdf.

In our opinion, the approach to risk management points to more than just symbolic reporting or impression management. Important details – such as cost estimates, implementation challenges and actions taken to date – are limited but efforts to formulate and act on specific plans points to a more emancipatory form of accounting based on action and proactive risk reduction. In support of this view are examples of companies forming partnerships with suppliers to tackle the effects of changing weather; active dialogue on the business impact of climate with investors and an awareness of the broad geographical context of weather-related risks. Perhaps most important is the realisation that the weather cannot be seen as just a part of the 'non-financial reporting' model. It is being internalised as a material business risk with operating, financial and reputational implications.

The risk discourse is strongly grounded in anthropocentric terms. This is to be expected given the capitalist context in which the UK retailers operate and the fact that annual, integrated and sustainability reports, while dealing with the information needs of stakeholders, are often targeted primarily at shareholders. This is not, in itself, a limitation. As explained by Gallhofer and Haslam (2019), accounting does not need to be radical to be emancipatory and it is possible to construct a comprehensive and shared understanding of environmental risk without having to dispense entirely with Capitalist constructs. A deep ecological perspective cannot, however, be overlooked.

In articulating the meaning and relevance of the weather, some retailers made subtle references to the 'not-so-great British weather' as a defining feature of national and cultural identity. This type of reporting can play an important role in enabling accounting's reflexive potential and ability to promote change. As illustrated by the prior research on aesthetic accounting, narratives, poems and pictures concerning nature can help to construct an 'archive' (or 'diary') which both preserves and reminds readers about what needs to be protected. The works by the seventeenth and eighteenth century writers such as Josselin (1649) and Evelyn (1685) are excellent examples. We are not proposing that companies become nature diarists but some narrative or pictorial information in the deep ecological tradition can provide context, allow stakeholders to understand risks more clearly and make vivid what we stand to lose if climate change is not halted.

Our findings make an important contribution to the environmental accounting research. They offer additional empirical evidence on a widening of accounting; how this can be used to construct new perspectives on environmental risk and how these can be managed. This is done in an original setting. There are no prior examples of weather accounting and how this meshes with a risk society logic. We execute the research in an interpretive (and normative) tradition where the aim is to explore and to inform rather than to test empirically. To this end, we offer the following recommendations.

First, weather accounting by UK companies is aligned with an emancipatory accounting framework but has a long way to go. As found by the earlier research on sustainability and integrated reporting, disclosures can be too generic and might not explain exactly how companies are changing their strategies and business processes (see McNally et al., 2017; Van Zijl et al., 2017). An emancipatory weather account gives a clear explanation of the risks faced (in anthropocentric and deep ecological terms), how the company plans to mitigate each risk and the resources required. Action plans should be laid out; implementation challenges must be discussed and reviews of progress against set targets are essential for proactive and transparent management and reporting (see Atkins et al., 2018; Maroun and Atkins, 2018).

Second, the boundaries of weather accounting need to be expanded. The prior research is usually content to deal with environmental management and reporting at the level of the firm. Climate change is a global problem which impacts every part of society. While reporting by individual firms is an important start, what is needed is an ecological-level account which contextualises the problem at the macro-level, outlines the steps taken by different stakeholders and explains which risks are not being addressed (see Cuckston, 2017, 2018).

Finally, academics and standard-setters need to focus on developing practical guidance for how companies can identify, assess and report on risks associated with social and environmental issues. Mainstream accounting research is preoccupied with examining capital market effects and much of the prior interpretive research focuses on theoretical development. There is little in the academic (and professional) literature which explains *how* to construct an emancipatory account. Possible areas for future research include experimenting with alternative formats for presenting information about the weather (see Sullivan and Hannis, 2017); outlining a conceptual model which can be used to organise an anthropocentric and deep ecological perspective on weather-related risks (see Atkins and Atkins, 2019) and developing a framework for how to identify these risks, partner with stakeholders and implement actions plans for effectively (see Atkins et al., 2018).

6 References

Abraham, S. and Cox, P. 2007. Analysing the determinants of narrative risk information in UK FTSE 100 annual reports. *The British Accounting Review* 39(3), 227–248.

Al-Htaybat, K. and von Alberti-Alhtaybat, L. 2018. Integrated thinking leading to integrated reporting: case study insights from a global player. *Accounting, Auditing and Accountability Journal* 31(5), 1435–1460.

Alrazi, B., De Villiers, C. and van Staden, C. J. 2015. A comprehensive literature review on, and the construction of a framework for, environmental legitimacy, accountability and proactivity. *Journal of Cleaner Production* 102, 44–57.

Ascui, F. and Lovell, H. 2011. As frames collide: making sense of carbon accounting. *Accounting, Auditing and Accountability Journal* 24(8), 978–999.
Atkins, J. and Atkins, B. 2016. *The Business of Bees: An Integrated Approach to Bee Decline and Corporate Responsibility*. London: Routledge.
Atkins, J. and Atkins, B. 2019. *Around the World in 80 Species. Exploring the Business of Extinction*. London: Routledge.
Atkins, J., Atkins, B. C., Thomson, I. and Maroun, W. 2015. 'Good' news from nowhere: imagining utopian sustainable accounting. *Accounting, Auditing and Accountability Journal* 28(5), 651–670.
Atkins, J., Barone, E., Maroun, W. and Atkins, B. 2017. Bee accounting and accountability in the UK. In: Atkins, K. and Atkins, B. (Eds). *The Business of Bees: An Integrated Approach to Bee Decline and Corporate Responsibility*. London: Routledge. 198–211.
Atkins, J. and Maroun, W. 2015. Integrated reporting in South Africa in 2012: perspectives from South African institutional investors. *Meditari Accountancy Research* 23(2), 197–221.
Atkins, J. and Maroun, W. 2018. Integrated extinction accounting and accountability: building an ark. *Accounting, Auditing and Accountability Journal* 31(3), 750–786.
Atkins, J., Maroun, W. and Atkins, B. 2019. How can accounting, integrated reporting and engagement prevent extinction? In: Atkins, K. and Atkins, B. (Eds). *Around the World in 80 Species. Exploring the Business of Extinction*. London: Routledge. 52–88.
Atkins, J., Maroun, W., Atkins, B. C. and Barone, E. 2018. From the big five to the big four? Exploring extinction accounting for the rhinoceros. *Accounting, Auditing and Accountability Journal* 31(2), 674–702.
Atkins, J. F., Solomon, A., Norton, S. D. and Joseph, N. L. 2015. The emergence of integrated private reporting. *Meditari Accountancy Research* 23(1), 28–61.
Bates, J. and Goodale, P. 2017. Making data flow for the climate risk market. *Television and New Media* 18(8), 753–768. View this article in WRRO.
Blackmore, C. 2014. Learning to change farming and water management practices in response to the challenges of climate change and sustainability. *Outlook on Agriculture* 43(3), 173–178.
Borghei, Z., Leung, P. and Guthrie, J. 2016. The nature of voluntary greenhouse gas disclosure: an explanation of the changing rationale: Australian evidence. *Meditari Accountancy Research* 24(1), 111–133.
Brockett, P. L., Wang, M. and Yang, C. 2005. Weather derivatives and weather risk management. *Risk Management and Insurance Review* 8(1), 127–140.
Brown, J. and Dillard, J. 2014. Integrated reporting: on the need for broadening out and opening up. *Accounting, Auditing and Accountability Journal* 27(7), 1120–1156.
Brown, W. O., Helland, E. and Smith, J. K. 2006. Corporate philanthropic practices. *Journal of Corporate Finance* 12(5), 855–877.
Buchling, M. and Maroun, W. 2019. Extinction accounting by the public sector: South African national parks. In: Atkins, K. and Atkins, B. (Eds). *Around the World in 80 Species. Exploring the Business of Extinction*. London: Routledge. 201–218.
Burt, T. P. and Ferranti, E. J. S. 2012. Changing patterns of heavy rainfall in upland areas: a case study from northern England. *International Journal of Climatology* 32(4), 518–532.
Carney, M. 2015. Breaking the tragedy of the horizon: climate change and financial stability, 29 September 2015 speech. Available at: www.bankofengland.co.uk/publications/Pages/speeches /2015/844.aspx.

CDP 2017. Climate change [Online]. *CDP*. Available at: https://www.cdp.net/en/climate [Accessed 10 December 2017].

Ceballos, G., Ehrlich, P. R. and Dirzo, R. 2017. *Biological Annihilation via the Ongoing Sixth Mass Extinction Signaled by Vertebrate Population Losses and Declines*. Proceedings of the National Academy of Sciences.

Cuckston, T. 2013. Bringing tropical forest biodiversity conservation into financial accounting calculation. *Accounting, Auditing and Accountability Journal* 26(5), 688–714.

Cuckston, T. 2017. Ecology-centred accounting for biodiversity in the production of a blanket bog. *Accounting, Auditing and Accountability Journal* 30(7), 1537–1567.

Cuckston, T. 2018. Making extinction calculable. *Accounting, Auditing and Accountability Journal* 31(3), 849–874.

de Brujin, K., Buurman, J., Mens, M., Dahm, R. and Klijn, F. 2017. Resilience in practice: five principles to enable societies to cope with extreme weather events. *Environmental Science and Policy* 70, 21–30.

Deloitte 2017. *Thinking Allowed. Climate-Related Disclosure: Integrating Climate-Related Information in the Annual Report*. London: Deloitte LLP.

De Villiers, C., Hsiao, P. K. and Maroun, W. 2017. Developing a conceptual model of influences around integrated reporting, new insights and directions for future research. *Meditari Accountancy Research* 25(4), 450–460.

Dillard, J. and Reynolds, M. 2008. Green owl and the corn maiden. *Accounting, Auditing and Accountability Journal* 21(4), 556–579.

Dischel, R. S. 2002. Introduction to the weather market: dawn to mid-morning. In: Dischel, R. (Ed). *Climate Risk and the Weather Market: Financial Risk Management with Weather Hedges*. London: Risk Books, 3–24.

Edgley, C., Jones, M. J. and Solomon, J. F. 2010. Stakeholder inclusivity in social and environmental report assurance. *Accounting, Auditing and Accountability Journal* 23(4), 532–557.

Enslin, Z., Bruwer, W. and Viljoen, C. 2015. Enhancing risk-related disclosure in South Africa: a study on guidelines and current practices. *Journal of Economic and Financial Sciences* 8(1), April, 261–280.

Fifka, M. S. 2013. Corporate responsibility reporting and its determinants in comparative perspective: a review of the empirical literature and a meta-analysis. *Business Strategy and the Environment* 22(1), 1–35.

Flower, J. 2015. The international integrated reporting council: a story of failure. *Critical Perspectives on Accounting* 27, 1–17.

Foucault, M. 1983. Afterword: the subject and power. In: Dreyfus, H. L. and Rabinow, P. (Eds). *Michel Foucault: Beyond Structuralism and Hermeneutics* (2nd ed.). Chicago: University Press, 208–226.

FRC 2018. The UK corporate governance code. Available at: https://www.frc.org.uk/getattachment/88bd8c45-50ea-4841-95b0-d2f4f48069a2/2018-UK-Corporate-Governance-Code-FINAL.pdf [Accessed 10 December 2018].

Freeman, M. C. and Groom, B. 2013. Biodiversity valuation and the discount rate problem. *Accounting, Auditing and Accountability Journal* 26(5), 715–745.

Gaia, S. and Jones, M. J. 2017. UK local councils reporting of biodiversity values: a stakeholder perspective. *Accounting, Auditing and Accountability Journal* 30(7), 1614–1638.

Gallhofer, S. and Haslam, J. 1996. Accounting/art and the emancipatory project: some reflections. *Accounting, Auditing and Accountability Journal* 9(5), 23–44.

Gallhofer, S. and Haslam, J. 2019. Some reflections on the construct of emancipatory accounting: shifting meaning and the possibilities of a new pragmatism. *Critical Perspectives on Accounting* 63, 1–18.

Giddens, A. 1990. *The Consequences of Modernity*. Cambridge: Polity Press.

Gömann, H. 2015. How much did extreme weather events impact wheat yields in Germany?: a regionally differentiated analysis on the farm level. *Procedia Environmental Sciences* 29, 119–120.

Gordon, C. (Ed.) 1980. *Power/Knowledge: Selected Interviews and Other Writings 1972-1977*. Harlow, Essex: Pearson Education.

Gray, R., Kouhy, R. and Lavers, S. 1995a. Corporate social and environmental reporting: a review of the literature and a longitudinal study of UK disclosure. *Accounting, Auditing and Accountability Journal* 8(2), 47–77.

Gray, R., Walters, D., Bebbington, J. and Thompson, I. 1995b. The greening of enterprise: an exploration of the (NON) role of environmental accounting and environmental accountants in organizational change. *Critical Perspectives on Accounting* 6(3), 211–239.

Gray, R. and Milne, M. J. 2018. Perhaps the Dodo should have accounted for human beings? Accounts of humanity and (its) extinction. *Accounting, Auditing and Accountability Journal* 31(3), 826–848.

Green, W., Taylor, S. and Wu, J. 2017. Determinants of greenhouse gas assurance provider choice. *Meditari Accountancy Research* 25(1), 114–135.

GRI 2016. Consolidated set of GRI sustainability reporting standards (2016). Available at: https://www.globalreporting.org/standards/gri-standards-download-center/?g=ae2e23b8-4958-455c-a9df-ac372d6ed9a8. Available at: https://www.globalreporting.org/reporting/g4/Pages/default.aspx [Accessed 10 February 2017].

Hopwood, A. G. 1987. The archaeology of accounting systems. *Accounting, Organizations and Society* 12(3), 207–234.

Huault, I. and Rainelli-Weiss, H. 2011. A market for weather risk? Conflicting metrics, attempts at compromise, and limits to commensuration. *Organization Studies* 32(10), 1395–1419.

IASB 2014. International financial reporting Standard 7: financial instruments: disclosure. In: *Pitermartisburg, Lexisnexus*. International Accounting Standards Committee Foundation Publications Department.

IIRC 2013. The international framework: integrated reporting. Available at: http://www.theiirc.org/wp-content/uploads/2013/12/13-12-08-THE-INTERNATIONAL-IR-FRAMEWORK-2-1.pdf [Accessed 1 October 2013].

Institute of Chartered Accountants for England and Wales (ICAEW) 2011. *Reporting Business Risks: Meeting Expectations*. London: ICAEW.

Intergovernmental Panel on Climate Change 2013. Climate change 2013: the physical science basis. Available at: http://www.ipcc.ch/report/ar5/wg1/ [Accessed 1 June 2015].

Isenhour, C., O'Reilly, J. and Yocum, H. 2016. Introduction to special theme section "accounting for climate change: measurement, management, morality, and myth". *Human Ecology* 44(6), 647–654.

Islam, M. M., Barnes, A. and Toma, L. 2013. An investigation into climate change scepticism among farmers. *Journal of Environmental Psychology* 34, 137–150.

Jonäll, K. and Sabelfeld, S. 2019. Accounting for survival of polar bears: an artic icon on thin ice. In: Atkins, K. and Atkins, B. (Eds). *Around the World in 80 Species. Exploring the Business of Extinction*. London: Routledge. 337–356.

Jones, M. J. 2003. Accounting for biodiversity: operationalising environmental accounting. *Accounting, Auditing and Accountability Journal* 16(5), 762–789.

Jones, M. J. and Solomon, J. F. 2013. Problematising accounting for biodiversity. *Accounting, Auditing and Accountability Journal* 26(5), 668–687.

Jorgensen, B. N. and Kirschenheiter, M. T. 2003. Discretionary risk disclosures. *The Accounting Review* 78(2), 449–469.

Khisty, C. J. 2006. Meditations on systems thinking, spiritual systems, and deep ecology. *Systemic Practice and Action Research* 19(4), 295–307.

Klein, A. M., Vaissière, B. E., Cane, J. H., Steffan-Dewenter, I., Cunningham, S. A., Kremen, C. and Tscharntke, T. 2007. Importance of pollinators in changing landscapes for world crops. *Proceedings. Biological Sciences* 274(1608), 303–313.

KPMG 2015. Currents of change. The KPMG survey of corporate responsibility reporting 2015. Available at: https://assets.kpmg.com/content/dam/kpmg/pdf/2016/02/kpmg-international-survey-of-corporate-responsibility-reporting-2015.pdf [Accessed 11 December 2016].

KPMG 2017. The road ahead. The KPMG survey of corporate responsibility reporting 2017. Available at: https://assets.kpmg.com/content/dam/kpmg/xx/pdf/2017/10/kpmg-survey-of-corporate-responsibility-reporting-2017.pdf [Accessed 20 November 2017].

Krippendorff, K. 1989. Content analysis. In: Barnouw, G., Gerbner, W., Schramm, T., Worth, L. and Gross, L. (Eds). *International Encyclopedia of Communication*. New York: Oxford University Press. 403–407.

Lajili, K. and Zéghal, D. 2005. A content analysis of risk management disclosures in Canadian annual reports. *Canadian Journal of Administrative Sciences / Revue Canadienne des Sciences de l'Administration* 22(2), 125–142.

Laine, M. 2010. Towards sustaining the status quo: business talk of sustainability in Finnish corporate disclosures 1987–2005. *European Accounting Review* 19(2), 247–274.

Lanka, S. V. 2019. An ecological auto-ethnography of a monarch butterfly. In: Atkins, K. and Atkins, B. (Eds). *Around the World in 80 Species. Exploring the Business of Extinction*. London: Routledge. 324–334.

Lash, S., Szerszynski, B. and Wynne, B. 1996. *Risk, Environment and Modernity: Towards a New Ecology*. London: Sage.

Leggio, K. B. 2007. Using weather derivatives to hedge precipitation exposure. *Managerial Finance* 33(4), 246–252.

Maraun, D., Osborn, T. J. and Gillett, N. P. 2008. United Kingdom daily precipitation intensity: improved early data, error estimates and an update from 2000 to 2006. *International Journal of Climatology* 28(6), 833–842.

Maroun, W. and Atkins, J. 2018. The emancipatory potential of extinction accounting: exploring current practice in integrated reports. *Accounting Forum* 42(1), 102–118.

McNally, M.-A., Cerbone, D. and Maroun, W. 2017. Exploring the challenges of preparing an integrated report. *Meditari Accountancy Research* 25(4), 481–504.

McNally, M.-A. and Maroun, W. 2018. It is not always bad news. *Accounting, Auditing and Accountability Journal* 31(5), 1319–1348.

Melnyk, S. A., Sroufe, R. P. and Calantone, R. 2003. Assessing the impact of environmental management systems on corporate and environmental performance. *Journal of Operations Management* 21(3), 329-351.

Mennicken, A. and Miller, P. 2012. Accounting, territorialization and power. *Foucault Studies* 13, 4-24.

Merkl-Davies, D., Brennan, N. and Vourvachis, P. 2011. Text analysis methodologies in corporate narrative reporting research. In: *23rd CSEAR International Colloquium*. St Andrews.

Met Office 2011. *Climate: Observations, Projections and Impacts*, Produced by the Met Office, © Crown Copyright (vol. 2011).

Milne, M. J. and Grubnic, S. 2011. Climate change accounting research: keeping it interesting and different. *Accounting, Auditing and Accountability Journal* 24(8), 948-977.

Milne, M. J., Tregidga, H. and Walton, S. 2009. Words not actions! The ideological role of sustainable development reporting. *Accounting, Auditing and Accountability Journal* 22(8), 1211-1257.

Morrison, J. 2009. Managing weather risk: will derivatives use rise? *Futures Industry* 14(1), 26-29.

Musshoff, O., Odening, M. and Xu, W. 2011. Management of climate risks in agriculture-will weather derivatives permeate? *Applied Economics* 43(9), 1067-1077.

Nicolov, S. 2019. Recovered species? The eastern North Pacific grey whale unusual mortality event. In: Atkins, K. and Atkins, B. (Eds). *Around the World in 80 Species. Exploring the Business of Extinction*. London: Routledge, 1999-2000.

Parker, L. D. 2005. Social and environmental accountability research: a view from the commentary box. *Accounting, Auditing and Accountability Journal* 18(6), 842-860.

Pérez-González, F. and Yun, H. 2013. Risk management and firm value: evidence from weather derivatives. *The Journal of Finance* 68(5), 2143-2176.

Peters, G. F. and Romi, A. M. 2014. Does the voluntary adoption of corporate governance mechanisms improve environmental risk disclosures? Evidence from greenhouse gas emission accounting. *Journal of Business Ethics* 125(4), 637-666.

Pierce, S. 2015. Does the accounting for derivatives affect risk and value? *SSRN Electronic Journal*. Available at SSRN. Available at: https://ssrn.com/abstract=2685896.

Raemaekers, K., Maroun, W. and Padia, N. 2016. Risk disclosures by South African listed companies post-King III. *South African Journal of Accounting Research* 30(1), 41-60.

Rajab, B. and Handley-Schachler, M. H. 2009. Corporate risk disclosure by UK firms: trends and determinants. *World Review of Entrepreneurship, Management and Sustainable Development* 5(3), 224-243.

Rankin, M., Windsor, C. and Wahyuni, D. 2011. An investigation of voluntary corporate greenhouse gas emissions reporting in a market governance system: Australian evidence. *Accounting, Auditing and Accountability Journal* 24(8), 1037-1070.

Rial-Lovera, K., Davies, W. P. and Cannon, N. D. 2017. Implications of climate change predictions for UK cropping and prospects for possible mitigation: a review of challenges and potential responses. *Journal of the Science of Food and Agriculture* 97(1), 17-32.

Russell, S., Milne, M. J. and Dey, C. 2017. Accounts of nature and the nature of accounts: critical reflections on environmental accounting and propositions for ecologically

informed accounting. *Accounting, Auditing and Accountability Journal* 30(7), 1426–1458.
Sibanda, M. and Mulama, M. 2019. Business contributions to extinction risk mitigation for black rhino in Laikipia, Kenya. In: Atkins, K. and Atkins, B. (Eds). *Around the World in 80 Species. Exploring the Business of Extinction*. London: Routledge. 192–200.
Siddiqui, J. 2013. Mainstreaming biodiversity accounting: potential implications for a developing economy. *Accounting, Auditing and Accountability Journal* 26(5), 779–805.
Solomon, A. and Clappison, M. 2019. Accounting for capitve belugas: a whale of a business. In: Atkins, K. and Atkins, B. (Eds). *Around the World in 80 Species. Exploring the Business of Extinction*. London: Routledge. 289–323.
Solomon, J. F., Solomon, A., Joseph, N. L. and Norton, S. D. 2011. Private climate change reporting: a discourse of risk and opportunity? *Accounting, Auditing and Accountability Journal* 24(8), 1119–1148.
Solomon, J. F., Solomon, A., Norton, S. D. and Joseph, N. L. 2000. A conceptual framework for corporate risk disclosure emerging from the agenda for corporate governance reform. *British Accounting Review* 32(4, December), 447–478.
Solomon, J. F., Solomon, A., Norton, S. D. and Joseph, N. L. 2011. Private climate change reporting: an emerging discourse of risk and opportunity? *Accounting, Auditing and Accountability Journal* 24(8), 1119–1148.
Solomon, J. F. and Thomson, I. 2009. Satanic Mills?: an illustration of Victorian external environmental accounting. *Accounting Forum*. Elsevier 2009, 74–87.
Spence, C. 2009. Social accounting's emancipatory potential: a Gramscian critique. *Critical Perspectives on Accounting* 20(2), 205–227.
Steffen, T. D. 2016. Improving financial statement footnotes: evidence from derivative and hedging disclosures. Available at SSRN. Routledge. Available at: https://ssrn.com/abstract=2739883 or http://dx.doi.org/10.2139/ssrn.2739883.
Stubbs, W. and Higgins, C. 2014. Integrated reporting and internal mechanisms of change. *Accounting, Auditing and Accountability Journal* 27(7), 1068–1089.
Štulec, I. 2017. Effectiveness of weather derivatives as a risk management tool in food retail: the case of Croatia. *International Journal of Financial Studies* 5(1), 2.
Sullivan, S. and Hannis, M. 2017. "Mathematics maybe, but not money": on balance sheets, numbers and nature in ecological accounting. *Accounting, Auditing and Accountability Journal* 30(7), 1459–1480.
TCFD 2016. Recommendations of the task force on climate-related financial disclosures, 14 December 2016. Available at: www.fsb-tcfd.org/publications.
Thomson, I. 2015. 'But does sustainability need capitalism or an integrated report' a commentary on 'The International Integrated Reporting Council: a story of failure' by Flower, J. *Critical Perspectives on Accounting* 27, 18–22.
Tregidga, H. 2013. Biodiversity offsetting: problematisation of an emerging governance regime. *Accounting, Auditing and Accountability Journal* 26(5), 806–832.
Tregidga, H., Milne, M. and Kearins, K. 2014. (Re)presenting 'sustainable organizations'. *Accounting, Organizations and Society* 39(6), 477–494.
Van Zijl, W., Wöstmann, C. and Maroun, W. 2017. Strategy disclosures by listed financial services companies: signalling theory, legitimacy theory and South African integrated reporting practices. *South African Journal of Business Management* 48(3), 73–85.

Vedenov, D. V. and Barnett, B. J. 2004. Efficiency of weather derivatives as primary crop insurance instruments. *Journal of Agricultural and Resource Economics* 29, 387–403.

Wang, M., Wen, M.-M. and Yang, C. C. 2010. Weather derivatives, price forwards, and corporate risk management. *The Journal of Risk Finance* 11(4), 358–376.

Werdigier, J. 2009. Tesco, British grocer, uses weather to predict sales. Available at: http://www.nytimes.com/2009/09/02/business/global/02weather.html [Accessed 30 January 2018].

Winterman, D. 2013. "Is the British weather unique in the world?". *BBC News Magazine*. Available at: http://www.bbc.co.uk/news/magazine-24305230 [Accessed 8 October].

Zadek, S., Evans, R. and Pruzan, P. 2015. *Building Corporate Accountability: Emerging Practice in Social and Ethical Accounting and Auditing*. London: Routledge.

Part 3

Protecting pollinators and insect biodiversity in agri-food supply chains

Chapter 15

Using data to assess the impact of agriculture on pollinators and pollinator services

Noa Simón Delso, BeeLife European Beekeeping Coordination, Belgium; and Walter Haefeker, European Professional Beekeepers Association, Germany

1 Introduction

The amount of human- and machine-generated data is growing exponentially and the farming and beekeeping sectors are no exception. Digital agriculture is becoming mainstream. Farmers are using data generated locally by their farm equipment as well as remotely by drones and satellites to more precisely manage their crops. Many beekeepers are adopting digital tools to manage their hives and are adding sensors such as hive scales to remotely monitor the status of their colonies. Professional beekeepers generate both hive management and business data. This growth in data creates the opportunity to use large amounts of already existing data to understand better the interaction between beekeeping and farming.

Scientists have often pioneered the use of digital tools in their research projects (including those related to pollinators) but, in most cases, data are gathered only for the duration of the project and often become orphaned after the funding runs out. Governments are starting to add data digitalization to

http://dx.doi.org/10.19103/AS.2023.0128.13
Published by Burleigh Dodds Science Publishing Limited, 2024.

the required reporting for receipt of subsidies and compliance with relevant regulations (e.g. on use of chemicals such as pesticides) whether from farmers or beekeepers. However, the data tend to be annual snapshots for administrative purposes and lack the granularity necessary to correlate farming and beekeeping practices with bee health. Reports of declining bee health have created a heightened sense of urgency to share bee-related data in order to better understand the multitude of factors impacting bees.

2 Current data on pollinators

Farming has a multitude of impacts on ecosystems. A wide range of species are potentially being impacted. Many studies have added to our understanding of these effects. However, the academic interest has not been evenly distributed.

Pollinators have long been of particular interest because the ecosystem services they provide are not only important to conservationists but also for agriculture. This priority is reflected in the regulation of pesticides in the European Union. In the stated protection goals of Regulation (EC) No. 1107/2009 concerning the placing of plant protection products on the market, only two species are mentioned by name: humans and honey bees (EC, 2009). In Section 3.8. Ecotoxicology, the requirements are clearly stated:

3.8.3.

> An active substance, safener or synergist shall be approved only if it is established following an appropriate risk assessment on the basis of Community or internationally agreed test guidelines, that the use under the proposed conditions of use of plant protection products containing this active substance, safener or synergist:

- will result in a negligible exposure of honeybees, or
- has no unacceptable acute or chronic effects on colony survival and development, taking into account effects on honeybee larvae and honeybee behaviour.

This particular focus on the honey bee has resulted in a significant body of research designed to meet risk assessment requirements (Maus et al., 2022). The availability of data from research projects is probably not proportional to the importance of honey bees relative to other species in the ecosystem such as solitary bees (Buchanan, 2023). In the current European Food Safety Authority (EFSA) Guidance, this has been somewhat rectified by extending the risk assessment to other bee species like bumble bees and solitary bees (EFSA, 2013). There is also an additional body of research focused on supporting the beekeeping sector. The records kept by beekeepers make the managed honey bee particularly interesting for data analysis (EFSA, 2021b).

In order to establish a baseline for risk assessment, the EFSA has published a scientific opinion in which the characteristics of a healthy managed honeybee colony were defined as follows (EFSA, 2016):

- an adequate size, demographic structure and behaviour;
- an adequate production of bee products (both in relation to the annual life cycle of the colony and the geographical location); and
- provision of pollination services.

The data requirements for risk assessment therefore are not limited to counting dead or alive colonies (EFSA, 2021a). Because honey bees are also the foundation of commercial enterprises and the subject of pollination contracts, data on their performance in the production of bee products or as pollinator for commercial crops have to be included. This means that data availability is significantly higher for the managed honey bee, even though other bee species maybe of equal or greater importance for our ecosystems.

The focus on the honey bee is a clear case of the 'streetlight effect', named after a well-known joke:

> A policeman sees a drunk man searching for something under a streetlight and asks what the drunk has lost. He says he lost his keys and they both look under the streetlight together. After a few minutes the policeman asks if he is sure he lost them here, and the drunk replies, no, and that he lost them in the park. The policeman asks why he is searching here, and the drunk replies, 'this is where the light is'.

The potentially high availability of data makes the mining of both farming and beekeeping data very attractive. However, it is important to be mindful of the extremely strong observational bias toward the managed honey bee.

3 Farm data standardization

Data records of farmers or beekeepers are typically stored in very different formats, depending on the systems being used for acquisition and processing. As a result, even if the different parties were willing to share their data, the technical difficulties associated with making sense of the raw data files from diverse systems would be insurmountable. A prerequisite for analyzing data from multiple sources is the standardization of the data formats and terminology. The standardization process also presents an opportunity to build trust among stakeholders (Wolfert, 2020).

The farm sector already has several, sometimes competing standards for interfacing with farm equipment. The goal here is the interoperability of products from different manufacturers. More recently, several standardization efforts have been launched to facilitate the exchange of data across farm management information system and cloud software (Wolfert, 2020). Here we have higher-level commercial data about field sizes and location as well as crops being seeded, fertilized, sprayed and harvested with great spatial and temporal granularity. The tools contain not only planned interventions for each field with dates. The equipment is also increasingly capable of recording the actual actions down to the square centimeter and the microsecond and delivering the information back to the farm management system.

The ISOBUS standard is designed to allow communication between farm equipment of different manufacturers. The data are closely linked to hardware (equipment) operations but can be mined for higher-level information. The ISOBUS messages instructing spraying equipment to open or close a valve contain very precise information about spraying activity. Enriched with other data about time and position, even such specific data can contribute to a clear picture of farming activity.

Much higher levels are farm management data, which can be combined in platforms like the AgGateway, which uses already standardized data formats such as AgXML. Parallel efforts like AgroEDI and AEF also have been successful in standardizing some of the agricultural data (Aspexit, 2021). Readers can find further information on these data platforms via the following links:

- AgGateway: https://www.aggateway.org/Home.aspx;
- AgXML: https://www.newswire.com/news/agxml-management-committee-announces-publication-of-v4-166075;
- AgroEDI: https://proagrica.com/news/proagrica-and-agro-edi-europe-announce-strategic-partnership-to-facilitate-traceability-and-data-exchange/; and
- AEF: https://www.aef-online.org/about-us/about-the-aef.html.

The potential for assessing the impacts on the ecosystem using the information contained in these systems cannot be overstated.

There is increasing concern, that major players like Monsanto (now part of Bayer) and John Deere are building big data platforms to further cement their market dominance. This may further shift power away from farmers (Carbonell, 2016). The EU Commission's concerns about Digital Farming in the Bayer Monsanto Deal are described via this link: https://www.euractiv.com/section/agriculture-food/news/vestager-bayer-monsanto-deal-must-preserve-digital-farming/.To counter this trend, open source initiatives have sprung up to provide an alternative to closed corporate systems. FarmOS, a web-based application for farm management, planning and record keeping, is an example of this (https://farmos.org).

4 Data standardization in beekeeping

In the beekeeping sector, standardization is still in its infancy. An effort is underway to take stock of the existing data islands, finding commonalities and proposing standard representations with more and more coverage. Because of the current focus on bee health, standardization efforts have started with colony data (see e.g. EFSA, 2021a,b). However, there are many others in the beekeeping sector where data exist and could benefit from standardization.

Beekeepers need to track their hives: location, health status, age of queen, honey production, etc. Beekeepers need to track their hive products: inventory, sales, lot numbers, customers, etc. Beekeepers need to coordinate certain activities with other beekeepers and veterinary authorities such as the treatment of bee diseases like American foulbrood (AFB) and control of parasites like the varroa mite or the small hive beetle. There are specific recording requirements for organic certification. There are, e.g. products beekeepers use as winter feed and pollen supplements, where data on quantity and composition exist and sharing would be useful for scientific research as well as for organic certification.

A lot of information about bee genetics is being recorded in queen breeding programs. Beekeepers sometimes track the performance of specific queens in their hive management software. This information is potentially useful to find correlations between impacts and genetic fitness of the colonies. There is a worldwide effort to breed bees with varroa-sensitive hygiene (a behavioral trait in which bees remove pupae infected by the varroa mite) (Danka et al., 2016). Reducing the varroa mite as a factor may increase the resilience of the colonies in other areas or negatively impact genetic fitness in other areas not covered by the breeding programs. As bee diseases, pathogens and parasites spread geographically, standardized records on different strains as well as treatments and their effectiveness can help veterinary services develop and implement effective measures.

There are many projects monitoring environmental data as well as contaminants and residues in bee products. The results are difficult to compare because of a lack of standardization of both data and collection methods. The global COLOSS project has developed the 'Bee Book' to harmonize the scientific methods for studying the honey bee (see e.g. https://www.northernbeebooks.co.uk/products/dietemann-ellis-neumann-the-coloss-beebook-vol-1/). This effort would benefit greatly if each method described would also have an associated data standard. The COLOSS Project is, e.g. discussed in Delso et al. (2022). Further information on the COLOSS project can be found via the following website: https://coloss.org/.

Increasingly, bee research is also aided by modeling. These models need real-world data for validation as well as calibration.

The beekeeping sector is not just important for pollination but also a source of bee products as food (Haefeker, 2021) or medicine (e.g. apitherapy: https://apitherapy.org/). Pollen collected by beekeepers is also essential as feed for bumblebees commercially produced for pollination in agriculture. The data on the authenticity of bee products (using analytical techniques such as nuclear magnetic resonance imaging) as well as market data (prices, trade flows) are essential tools to combat fraud in the supply chain and should be maintained in a standard format globally. The trade data implicitly contain information about different impacts on the beekeeping sector resulting from changes in agricultural practices or other factors such as climate change or bee diseases.

A lot of information about bee genetics is also being recorded in queen breeding programs. Maintaining the data in a standardized format would not only facilitate the sharing of information among breeders but also preserve the records, which often are maintained by a single volunteer. As with scientific projects, there is a big problem with orphan data in proprietary formats. Having records in a consistent format over long periods of time is of increasing importance for understanding biodiversity trends and the effects of climate change on pollinators.

There are many projects monitoring environmental data as well as contaminants and residues in bee products. The results are difficult to compare because of a lack of standardization of both data and collection methods. Increasingly, bee research is also aided by modeling. These models need real world data for validation as well as calibration (Yoko et al., 2021).

In 2017, the EFSA initiated the EU Bee Partnership. Stakeholders from the beekeeping and agricultural sector to explore the potential of sharing bee-related data in the EU (EFSA, 2018). It was decided to focus on standardization first and to create the Bee Hub as a proof of concept to demonstrate the feasibility of this approach (Simón Delso et al., 2021). The EU Bee Partnership Prototype Platform was tasked with collecting and exchanging standardized data on pollinator health and other relevant data within the framework of the Internet of Bees Project.

The work developed during the prototyping of the platform funded by EFSA has included the design and implementation of data models for the data integrated into the Prototype Platform (https://bee-ppp.eu). A procedure for data and metadata integration has been created and implemented, including validation of the data quality. The platform was enriched with datasets coming from existing projects and public databases. Additionally, several algorithms for exploring the data and extracting ready-to-use information have been proposed in this phase providing examples and avenues for future development. Finally, a reporting feature was added, and some reports were produced, to illustrate the type of information that the analyses of the platform could generate in the future for the benefit of stakeholders, institutions and more sustainable bee health and beekeeping in the EU.

The EU efforts are a major contribution to the global standardization effort of the Apimondia Bee XML working group, which is developing and improving standardization for bee health data by sharing expertise and efforts to better communicate data (https://beexml.org/about/). Bee XML is a self-describing data format that allows the exchange of data on bees and beekeepers (Haefeker, 2022).

5 Data sharing in farming

The potential to use the 'big data' toolkit to transform existing data from different sources into actionable information remains unrealized not only because of the obvious lack of standardization but also because there often is an unwillingness to share data. Even if the data format is well defined, the information actually may not be reliable because of inconsistent data entry or because of hesitancy to fully share the data.

Since tracking activities are essential to any well-run farm and beekeeping operation there is a strong incentive for operators to keep accurate and current data. It does not make sense for the operator to lie to himself. But if third parties like tax authorities and regulators also demand access to the data, this is no longer the case.

Both the farming and the beekeeping sectors currently have strong incentives against the sharing of data, which need to be overcome by policy changes as well as anonymization techniques already well understood in computer science.

Digital agriculture and smart farming are being widely adopted in the sector. But at the same time, there is an ongoing debate about ownership of the data. In many cases, the farmer may have access to the data, but they actually belong to the manufacturer of the equipment or the developer of the app or cloud service the farmer is using. Major tasks like seeding and spraying are frequently performed by contractors using their own equipment. Does the farmer or the contractor own the data generated by the machine used to perform the talk on the farmer's field? Before data can be shared, the question of data ownership needs to be resolved.

Even before the advent of digital agriculture, efforts to gather more precise data on pesticide use have faced fierce resistance from farmers and the chemical industry. A case in point is the German Ministry of the Environment, which unsuccessfully tried to institute monitoring of pesticide use in 2004 (UBA, 2005; Lahl, 2006) (for background, see https://www.pan-europe.info/old/Resources/Reports/Towards_pesticide_use_reduction_in_Germany.pdf). There was immediate resistance by conservative members of the German Parliament and others (DBD, 2005). The project was compared to the methods of the Stasi - the secret police in former East Germany (LBW, 2004). A massive lobbying campaign succeeded and the project had to be abandoned. The federal agency published a report in 2006 explaining the decision (APL, 2006):

> The Federal Environment Agency has not continued the 'unannounced field observations' of the research project 'Recording misuse in the application of plant protection products and deriving suggestions for improvement for future enforcement activities in the plant protection sector' since October 2005. Although the first interim results of the project are not yet representative, they indicate a misapplication rate of about 50%.

The fact that big data has such a great potential to show the connection between farming practices and environmental impacts may be a strong incentive to withhold data or prevent the sharing of data by farmers through restrictive licensing terms (Bronson and Knezevic, 2016).

In 2012, the EFSA conducted a risk assessment of already authorized plant protection products using new data including those from actual use in the field. This ultimately resulted in the European Commission placing severe restrictions in

the use of plant protection products and treated seeds containing neonicotinoids (clothianidin, imidacloprid and thiamethoxam) in 2013 to protect honeybees. More information on neonicotinoids can be found via the following link: https://www.hsph.harvard.edu/news/press-releases/study-strengthens-link-between-neonicotinoids-and-collapse-of-honey-bee-colonies/.

The chemical industry challenged these actions in the European Court of Justice (ECJ). The industry argued that it was such a long and expensive process to get a plant protection product authorized, that they had to have legal certainty to be able to market the products after authorization to recover their investment. The ability of the Commission to restrict or revoke the authorization based on new data was seen as a major threat to the business model of the industry. In 2021, the ECJ reaffirmed a lower court ruling from 2018 which established that 'the Commission may review the approval of an active substance at any time' (ECJ, 2020, 2021; Stokstad, 2018).

Since 'new data' are the basis for initiating and conducting such a review, sharing data may be seen as extremely problematic not just by the manufacturers but also by the users, whose farming practices rely on the availability of plant protection products. As long as there is a strong possibility that the data may reveal less-than-perfect farming practices and unforeseen problems with plant protection products, data may be very difficult to obtain.

6 Data sharing in beekeeping

Incentives and disincentives may result in a hesitancy to share data or to record 'alternative facts' in the dataset, rendering it useless for everybody. Systems need to be designed to have built-in protection for personal and business data which do not depend on trusting current or future authorities.

Different government agencies and institutions as well as research projects have an interest in maintaining data on the location and movement of bee hives as well as data on bee losses and other monitoring data. Such data collection efforts tend to run into serious difficulties in the beekeeping sector. This is especially true, when they are based on the assumption, that bees are just another farm animal. But bees are not flying cows.

While farm animals can be individually tagged and tracked, individual honey bees can be considered more as part of a superorganism where individual members are less significant than the ways they interact as a hive. The same bee hive, e.g. may have replaced 100% of its individuals during a season. Colonies can be combined and split, which requires a completely different type of record keeping.

Beekeepers currently have very little incentive to participate in such a system. Participation as a condition for receiving government support is not happening, because the money actually received by beekeepers is too insignificant. Participation as a condition for placing products on the market does not work for the many hobby beekeepers who distribute their products to friends and family. Beekeepers are also very skeptical of the notion that participation may result in useful expert assistance from government agencies. There is simply not enough practical beekeeping expertise at government agencies to be perceived as sufficiently valuable assistance (Cazier et al., 2018).

While the incentives to participate are weak at best, the disincentives to participate in a traceability system are very strong. Beekeepers are aware that producers in other farming sectors like dairy, pigs and poultry, where there is strong regulation and a high degree of monitoring and traceability to meet safety and quality standards, do not seem to benefit from this system (e.g. in terms of a price premium for their products) and, indeed, seem to struggle financially in meeting these requirements and making a return. There is no desire for the beekeeping sector to follow this example. Participation is frequently viewed as a loss of privacy and independence. In many countries, there are tax consequences with arbitrary thresholds. In addition, beekeepers may be forced into a farmers' insurance system to compensate for the declining number of active farmers paying into the pool. Draconian measures without sufficient compensation in the case of small hive beetle (*Aethina tumida*) (SHB) eradication efforts have reduced trust that authorities will act in beekeepers' best interests.

Given this situation, participation under very specific conditions still may make sense, if it is part of a system used by beekeepers for their own needs for record keeping and hive management.

There are a number of commercial and private hive management systems available for beekeeping operations, which are capable of tracking everything necessary to record key information about bees and, in some cases, also bee products. But this information is the private property of the beekeepers. Some of the information in these systems can be used by the beekeeper to generate reports for veterinary purposes or the organic certification of a beekeeping operation. While the data belong to the beekeeper in theory, the implementation of some cloud-based hive tracking systems may not prevent third parties from accessing the information.

7 Solutions to data sharing

So what could a data-sharing environment that farmers and beekeepers would be likely to participate in look like?

Assuming data are not owned by the provider of the equipment or the service, there is a way to keep information collected by farmers or beekeepers secure through strong encryption. If only the farmer or the beekeeper has the key to the encrypted data, even a cloud-based system can be designed to maintain a high level of privacy. This is the principle that allows Apple to prevent the FBI from getting at iPhone users' data even if presented with a court warrant.

At the same time, it is necessary to clearly define limited needs for data to be shared. For many purposes, it may not be necessary to work with personally identifiable data. Good aggregate and anonymized data may be sufficient. Alerts and notifications can be transmitted to beekeepers and farmers without the need to know any personal data. It is possible to design the system in such a way that it provides anonymized aggregate data, which could be useful for scientific, regulatory and veterinary purposes.

A good example for this can be found in current navigation systems for cars. Modern systems can show more than a map. They can inform the user about traffic jams in real time. But how does the navigation system know where the traffic is not flowing? The system gets the aggregate data from the mobile phone providers. The movement of mobile phones from cell tower to cell tower captures the traffic flow. In order to get this very useful information, it is completely unnecessary to know the identity of phone users. It is also not necessary to require drivers to carry a mobile phone. And the anonymous aggregate tracking of the movement of mobile phones does not result in speeding tickets for individual drivers. The navigation example shows that it is possible to have very accurate real-time tracking of the movement of cars without invasion of privacy of the drivers.

The same approach could be taken for tracking the movements of hives in beekeeping as well as other farming activities as long as the spatial resolution does not allow the de-anonymization of the data. To analyze the effect of certain farming practices on bees, it is may not necessary to work with the actual geography, but simply share proximity values. This approach has successfully been implemented by Chile's Agriculture and Livestock Service to manage contamination risks for GMOs (SAG, 2011).

Data sharing could also be designed to provide strong incentives for farmers and beekeepers to participate. The data-sharing platform could be designed to allow farmers to contact the beekeepers prior to the application of plant protection products or veterinary services to reach beekeepers in case of an outbreak of AFB or SHB in an area. It would be easy to build a system which notifies all beekeepers with hives within a certain radius. Essentially a message goes out on the whole network. When the user is logged in, the hive location data are decrypted using the private key of the account allowing the system to perform a data match against the outstanding bulletins. This approach means it is not necessary to know any personal data to be able to reach out to everybody using the system.

By offering a system which is extremely useful to farmers, beekeepers and government agencies, if it has guaranteed privacy built into the architecture, valuable aggregate information can be generated. Such a system would actually provide more accurate data than a mandatory system because the encryption, anonymization and aggregation make sure that there are no negative consequences for the individual beekeeper or farmer. This removes any incentive to provide false data. Beekeepers and farmers are not the only actors. Sharing of data owned by equipment manufacturers and service providers may have to be mandated. This is especially true if the service is provided by the chemical industry which may have a conflict of interest if data highlights problems with plant protection products they sell.

8 Conclusion

Data collection efforts present unique challenges in both the farming and the beekeeping sector. The incentives to participate are too insignificant while strong disincentives exist both for operators and suppliers. But another approach using encryption, anonymization and aggregation may actually address some of the concerns resulting in more participation and more accurate data than even a mandatory system. Systems need to be designed to have built-in privacy protections which do not depend on trusting government initiatives. Commercial suppliers of farm management and hive tracking systems need to be included in the process. The possible role of authorities could be to promote standardized ways for the exchange of data in all systems used by farmers and beekeepers. All systems should be open source to allow the community to understand and get comfortable with what is being processed.

9 Competing interests

Noa Simón Delso, Scientific director and project manager of BeeLife, which was a party to the proceedings on the side of the European Commission in the European Court of Justice case C-499/18 P.

Walter Haefeker is Honorary President of the European Professional Beekeepers Association and the initiator of the beeXML Standardization effort. He is also a honorary member of the German Professional Beekeepers Association, which was a party to the proceedings on the side of the European Commission in the European Court of Justice case C-499/18 P.

Both authors have been or continue to be affiliated with the European Bee Partnership, the beeXML standardization effort and related projects.

10 Where to look for further information

- https://beexml.org.
- https://www.efsa.europa.eu/en/supporting/pub/en-1423.

11 References

APL 2006. *Anwendung von Pflanzenschutzmitteln in der Landwirtschaft: Ergebnisse von Untersuchungen des Umweltbundesamtes und Vergleich mit Erkenntnissen der Länder* (vol. 2006). https://www.umweltbundesamt.de/sites/default/files/medien/publikation/long/3557.pdf.

Aspexit 2021. Standards and data exchange in agriculture. Available at: https://www.aspexit.com/standards-and-data-exchange-in-agriculture/.

Bronson, K. and Knezevic, I. 2016. Big data in food and agriculture. *Big Data and Society* 3(1). https://doi.org/10.1177/2053951716648174.

Buchanan, S. 2023. The role of habitat conservation and restoration in protecting pollinators in agricultural landscapes. In: Kevan, P. and Willis Chan, S. D. (Eds). *Promoting Pollination and Pollinators in Farming*. Burleigh Dodds Science Publishing, Cambridge, UK.

Carbonell, I. M. 2016. The ethics of big data in big agriculture. *Internet Policy Review* 5(1). https://doi.org/10.14763/2016.1.405.

Cazier, J., Haefeker, W. and Hassler, E. 2018. Data sharing risks and rewards. *Bee Culture Magazine*.

Danka, R. G., Harris, J. W. and Dodds, G. E. 2016. Selection of VSH-derived "Pol-line" honey bees and evaluation of their Varroa-resistance characteristics. *Apidologie* 47(3), 483. https://doi.org/10.1007/s13592-015-0413-7.

DBD 2005. *Deutscher Bundestag Drucksache* (vols. 15/4935). https://dserver.bundestag.de/btd/15/049/1504935.pdf.

Delso, N. S., Mędrzycki, P., Sgolastra, F., Fourrier, J., Azpiazu, C., Tosi, S. and Hatjina, F. 2022. Apitox: the COLOSS task force to investigate the impact of pesticides on bees. *Bee World* 99(1), 32. https://doi.org/10.1080/0005772X.2021.2016291.

EC 2009. Regulation (EC) 1107/2009 of the European Parliament and of the council of 21 October 2009 concerning the placing of plant protection products on the market and repealing council directives 79/117/EEC and 91/414/EEC, n.d. https://eur-lex.europa.eu/legal-content/EN/TXT/?uri=celex%3A32009R1107.

ECJ 2020. Court of Justice of the European Union. 2020. Opinion of advocate general Kokott delivered on 17 September 2020. Case C-499/18 P Bayer CropScience AG and Bayer AG v European Commission., 2020.

ECJ 2021. *Judgment of the court (First Chamber) 6 May 2021 (*) (Appeal - Regulation (EC) No 1107/2009 - Articles 4 and 21 - Criteria for Approval - Review of Approval - Plant Protection Products - Implementing Regulation (EU) No 485/2013 - Active Substances Clothianidin and Imidacloprid - Seeds Treated with Plant Protection Products Containing Those Active Substances - Prohibition of Non-professional Use - Precautionary Principle).* Court of Justice of the European Union.

EFSA 2013. EFSA guidance document on the risk assessment of plant protection products on bees (*Apis mellifera*, *Bombus* spp. and solitary bees). *EFSA Journal* 11(7), 3295.

EFSA 2016. EFSA Panel on Animal Health and Welfare (AHAW): assessing the health status of managed honeybee colonies (HEALTHY-B): a toolbox to facilitate harmonized data collection. *EFSA Journal* 14(10), 4578. https://doi.org/10.2903/j.efsa.2016.4578.

EFSA 2018. Terms of reference for an EU Bee Partnership. *EFSA Supporting Publications* 15(5) (EFSuppl.3). https://doi.org/10.2903/sp.efsa.2018.EN-1423.

EFSA 2021a. Analysis of background variability of honey bee colony size. Technical Report. *EFSA Supporting Publications* 18(3), EN-6518.

EFSA, EFSA Scientific Committee, More, S., Bampidis, V., Benford, D., Bragard, C., Halldorsson, T., Hernández-Jerez, A., Bennekou, S. H., Koutsoumanis, K., Machera, K., Naegeli, H., Nielsen, S. S., Schlatter, J., Schrenk, D., Silano, V., Turck, D., Younes, M., Arnold, G., Dorne, J. L., Maggiore, A., Pagani, S., Szentes, C., Terry, S., Tosi, S., Vrbos, D., Zamariola, G. and Rortais, A. 2021b. A systems-based approach to the environmental risk assessment of multiple stressors in honey bees. *EFSA Journal. European Food Safety Authority* 19(5) (EFSuppl.2), e06607. https://doi.org/10.2903/j.efsa.2021.6607.

Haefeker, W. 2021. Pollen supplements and substitutes in the EU feed market: a product/market survey for bees and other animal species. *EFSA Supporting Publications* 18(2), EN-6461. https://doi.org/10.2903/sp.efsa.2021.EN-6461.

Haefeker, W. 2022. Collaboration platform for the standardization of the exchange of data about bees and beekeepers. *BeeXML.org*. Available at: https://beexml.org/about/(accessed 7.4.22).

Lahl, U. 2006. Anwendungen von Pflanzenschutzmitteln in der Landwirtschaft. *Umweltwissenschaften und Schadstoff-Forschung* 18(1), 41. https://doi.org/10.1065/uwsf2005.08.101.

LBW 2004. Landtag von Baden-Württemberg, Drucksache 13 / 3492. https://www.landtag-bw.de/files/live/sites/LTBW/files/dokumente/WP13/Drucksachen/3000/13_3492_D.pdf.

Maus, C., Alix, A. and Schmehl, D. R. 2022. Assessing the impact of pesticides on pollinators. In: Kevan, P. and Willis Chan, S. D. (Eds). *Promoting Pollination and Pollinators in Farming*. Burleigh Dodds Science Publishing, Cambridge, UK, 183.

SAG 2011. *Chile's Agriculture and Livestock Service (SAG): New Tools for Chilean Beekeepers to Meet Demanding European GM Standards (FreshFruitPortal.com.pdf).* https://www.freshfruitportal.com/news/2011/12/09/new-tools-for-chilean-beekepers-to-meet-demanding-european-gm-standards/.

Simón Delso, N., Sušanj, G. and Salazar Abello, A. 2021. The EU Bee Partnership (EUBP) Prototype Platform: data model description. *EFSA Supporting Publications* 18(7) (EFSuppl.3). https://doi.org/10.2903/sp.efsa.2021.EN-6694.

Stokstad, E. 2018. European Union expands ban of three neonicotinoid pesticides. *Science*. https://doi.org/10.1126/science.aau0152.

UBA 2005. *Feldbeobachtungen Sowie Pflanzen- und Bodenuntersuchungen Anwendungen von Pflanzenschutzmitteln in der Landwirtschaftlichen Praxis Begründung für die Notwendigkeit Einer Sachverhaltsaufklärung*. https://www.umweltbundesamt.de/sites/default/files/medien/publikation/long/3566.pdf.

Wolfert, S. 2020. Improving data standards and integration for more effective decision-making in agriculture. In: Armstrong, L. (Ed.). *Improving Data Management and Decision Support Systems in Agriculture*. Burleigh Dodds Science Publishing, Cambridge, UK, 17.

Yoko L, D., Nuno, C., Per, K., Joana, A., Jørgen A, A., Mette G, B., Marianne, B., Silvia, C., Julie, F., Geoff B, G., Annika S, J., Birgit, L., Sara, L., Alice, P. M., da Silva Antonio, A., Beate, S., Peter Borgen, S. and José Paulo, S. 2021. Research project on field data collection for honey bee colony model evaluation. *EFSA Supporting Publications* 18(7) (EFSuppl.3). https://doi.org/10.2903/sp.efsa.2021.EN-6695.

Chapter 16

Developing and implementing plans to conserve insect biodiversity in agricultural landscapes

Scott Longing, Texas Tech University, USA; and Brendan Kelly, Texas Tech University/Texas A&M AgriLife Research, USA

1 Introduction

As abundant and mobile biotic organisms, insects perform essential functions that benefit humans. Indirect benefits to humans include their roles in sustaining wild plant communities and feeding herbivorous and insectivorous wildlife, while direct benefits include other *ecosystem services* provided to humans such as pollination, pest control, decomposing dung, and filtering carbon and nutrients from freshwater systems. Losey and Vaughn (2006) estimated insects provided a value of US$57 billion annually in the US, attributed to four ecosystem services:

- Pollination;
- Pest control;
- Dung burial; and
- Food for wildlife.

This chapter focuses on insect biodiversity conservation in agricultural landscapes, reflected in the numerous beneficial insects occurring in these systems. Among the beneficial insects, pollinators represent a logical point

http://dx.doi.org/10.19103/AS.2023.0128.14

of focus because of their importance in crop pollination and the capacity for agricultural producers to facilitate direct positive impacts through land management that includes strategies to promote resources for pollinators. Moreover, improving resources for pollinators in agricultural landscapes supports other ecosystem services such as pest control, improved water quality, and mitigation of soil erosion to enhance rural aesthetics (Wratten et al., 2012).

Insects are integral to global agricultural production, with pollination services provided by insects valued at US$212 billion annually (Gallai et al., 2009). Although the production of global crop commodities such as corn, wheat, and rice do not depend on insect pollinators to produce yields, most crops worldwide, especially numerous specialty crops - including most fruits, nuts, and vegetables - require pollinators for improving and stabilizing yields (Klein et al., 2007). Without insect pollinators in agriculture, human diets would simply not contain the important macro- and micro-nutrients or diversity of foods made possible by our six-legged friends.

While insects are a vital component to human nutrition, the agricultural systems that require their inputs create stressors to populations through farming practices that affect pollinator resources (i.e. where they live/nest and what they eat) or their health. Land use and other factors affecting habitat and nutritional resources, pesticide exposure, climate variability and extremes, drought, and disease have affected insect biodiversity worldwide, especially for pollinators (Potts et al., 2010). Landscapes dominated by agricultural production have intensified many of these stressors. Pollinator 'health' in agriculture has mostly been investigated for honey bees and other commercially available, managed bees because they are essential in crop pollination. For honey bees, the effects of many of these stressors have been thoroughly investigated; for example, pesticide exposure and mortality from acute or chronic toxicity or sublethal effects has been investigated in relation to colony collapse disorder in honey bees (vanEngelsdorp et al., 2009). In addition to managed bees, conserving native wild bees and their habitat resources on farms can benefit production (e.g. Cusser et al., 2016) and simultaneously support achieving program conservation targets for biodiversity and natural resource conservation. However, the effects of environmental stressors on most native wild bee fauna are less known compared to what is known for honey bees. From a natural history perspective this is unfortunate considering insects (mostly bees) pollinate over 75% of all flowering plant species and are vital to sustainable agriculture (Ollerton et al., 2011). At a systems level, needs exist to both protect crop pollination services (of species known to provide pollination) and to protect the pollinator biodiversity of the agroecosystem, which regulates the vegetation communities and other wildlife that provides positive system feedbacks such as pest control and other important ecosystem services (Wratten et al., 2012).

The need to sustainably intensify the production of food and fiber for an expanding human population requires innovation at every level, including natural resources and biodiversity conservation. It also requires coordinated efforts from all sectors of the industry. This chapter introduces some contemporary issues related to beneficial insects in agriculture - mostly the insect pollinators and bees - and introduces new concepts linking biodiversity to consumer markets to advance insect conservation within sustainable agriculture. We frame approaches for conservation in the intensified agricultural landscapes of the Southern High Plains grassland region of the mid-continental United States. This region has undergone intensified agriculture since the mid-twentieth century, when use of underground water for irrigation created a sharp increase in row crop production resulting in extensive grassland fragmentation and agricultural intensification.

2 Pollinator conservation: the contemporary context

The term 'pollinator conservation' brings connotations of honey bee hives and monarch butterflies, or perhaps bumble bees. In the insect world, these are the 'charismatic megafauna', of which without their involvement insect conservation certainly would not have advanced as we have seen in recent decades. While it is the well-known insects that garner attention and link the human dimension to conservation of the natural world, a vast number of other insects perform critical functions that affect the production of agricultural commodities - functions that contribute to food webs and support higher trophic levels from herbivores to top predators. Without insects serving in this intermediate level of food chains, either through consumption of plant materials or through serving as prey for insectivorous species, a vast biodiversity would be compromised. From an anthropogenic perspective, losses of insects and silencing of songbirds were the basis of Rachel Carson's Silent Spring, which documented the broad effects of the organochlorine insecticide DDT among food webs as the pesticide bioaccumulated in fatty tissues of organisms and up food chains, ultimately affecting the health of predatory raptor offspring.

When considering innovations in insect conservation in agriculture, something that should be included in considerations of integrating pollinator conservation includes the diversity of the global agricultural landscape. The global agricultural landscape is vast - it is estimated to cover over a third of land surface worldwide - and diverse, with about a third of the total area dedicated to growing crops (FAO, 2021). The crops grown range from large industrial commodity farms, producing crops such as corn and soy, to smaller family subsistence farms. The landscape includes intensively managed specialty crops and local produce, all of which could directly benefit pollinators and insect biodiversity. Since in many cases these farms access urban markets, through production they can provide resources for pollinators in areas where resources have been compromised because of urbanization; they serve as additional green spaces in urban areas for pollinator resources, such as protected soils for nesting and flowering vegetation to provide nectar and pollen. Furthermore, beekeeping has become a popular activity in some urban areas and can be supported by locally abundant flowering vegetation provided by small urban and rural farms. In contrast, a commodity production system might farm considerably more acreage, but also could manage more diverse landscapes. For example, a cotton farmer in the Southern High Plains might farm an entire section, but also could manage the entire watershed of a playa lake for conserving freshwater quality and migrating waterfowl habitat. Therefore, prescriptions for pollinator habitat management depend on the different characteristics of farming operations and include acreage, habitat diversity and crop type, inputs and IPM practices, and the quality of wild or marginal lands containing resources. However, it should be noted that farm (crop) diversification is not exactly a fix for sustaining wild pollinators (St. Clair et al., 2020), whereas incorporation of natural habitats and the re-connectivity and improvements in pollinator forage quality should be primary goals.

While the need to conserve biodiversity needs no extended explanation - it is our view that all species are equally deserving of life - it should be noted that biodiversity conservation might not be a reasonable argument for crop pollination in that, in most communities, only a small portion of pollinators might contribute to the movement of crop pollen to benefit crop yields (Kleijn et al., 2015). We support this view (albeit more studies are needed to address regional fauna in crops and push-pull mechanisms among wild vegetation) and its integration with insect conservation in agriculture. Emerging research is showing how species traits such as body size and foraging distance (and potentially short dispersal of specialists) influence pesticide exposure (Longing et al., 2020). Furthermore, between 25% and 35% of native bee species in the U.S. specialize on a particular plant or group of plants and therefore represent a diverse fauna that would not likely visit crop flowers/resources (https://jarrodfowler.com/bees_pollen.html). While contributions of (ambient) native bee communities to crop pollination have not been investigated widely, Cusser et al. (2016) found that contributions of native bees in Texas cotton - supported by heterogeneous wild habitat on farms - reduced pollen limitation and increased seed cotton weight by 18%, resulting in an US$108 per acre increase in crop value.

3 Threats to pollinators

Agricultural production interacts with insect communities and broader wildlife biodiversity in multiple ways. Pollination services are provided to crops where crops receive inputs from managed pollinators (e.g. commercially available honey bees) or insect pollinators supported in managed or wild habitats. Where risks to pollinators can be mitigated, crops can even provide mass flowering resources for pollinators to assist populations and rear young (Todd et al., 2016), resources that help sustain populations especially when other wild flowering vegetation is scarce such as during drought periods. Furthermore, depending on resource availability in the agricultural system, push-pull mechanisms can influence pollinator foraging. However, the effects of mass flowering crops on behaviors such as competition among wild bees and risk to pesticide exposure are not fully understood. Accordingly, threats to pollinators should include understanding relative hazards among farm habitats utilized to produce crops and habitats set aside for wildlife that do not receive inputs such as pesticides and fertilizers.

How environmental stressors influence pollinators in agricultural landscapes depends on the life history traits of the species and its habitat resource use, including degree of generalization or specialization in pollinator diets. Generalist bees will forage across diverse flowering vegetation while more specialized bees will forage within a particular plant genus or on one plant species. Polylecty, oligolecty, and monolecty describe foraging specialization in terms of the number of different types of plants visited, respectively in descending order. Specialized bees (oligolectic and monolectic) are a significant component of native pollinator biodiversity. In U.S. arid regions specialized bees relying on specific host plants can comprise over 30% of the total number of species. Unfortunately, conservation actions to enhance and restore resources for pollinators on farms usually involve the planting of mixed flowering vegetation with not much consideration for these host plants supporting specialized bees. It also appears that conventional sampling methods to assess bee biodiversity might not detect this specialized fauna (authors unpubl. data).

The close association with plant resources (i.e. pollen and nectar and other tissues), foraging efficiency, and need to provision food for raising young creates an inherently strong associations between bee populations and habitat resources in lands managed for crop production. In intensified agricultural settings where landscape change is extensive, crop land cover can dominate leaving a heavily fragmented natural landscape where fauna must adapt or perish locally (i.e. localized extirpation of a sub-population) or throughout their distributional range, leading to species level extinctions. Sometimes natural land cover can become non-dominant land cover. For example, in the Southern High Plains region across a seven-county region, approximately 30% of the existing land cover (2017) was in grassland land cover, and this portion was in various sizes and parcels providing the habitat template from which pollinators forage for resources. This is also the wild land that could be managed to support resources for pollinators and pollinator biodiversity, but strategies for promoting all bee biodiversity (not just the common or easily collected species) remain undeveloped.

Pollinators are not only affected by physical landscape change that alters the dispersion of resources across landscapes via extensive habitat fragmentation, but populations also face acute and chronic threats that act to further stress populations, such as pesticide exposure to non-target organisms and resources (Krupke et al., 2012; Longing et al., 2020; Peterson et al., 2021), contaminated runoff, and loss of soil structure to sustain nesting resources for bees, among many others. In farming systems, habitat degradation and pesticide exposure represent dual major threats. Organisms occurring in these systems, whether bee pollinator or predatory lacewing, must adapt to disjunct and variable resources across landscapes, while being in proximity to farming practices that could be acutely or chronically disruptive and/or toxic (i.e. agrochemical exposure). How both habitat fragmentation and pesticide exposure affect wild bees at the population level remains largely unknown, yet a vast biodiversity of specialized pollinators could be affected if their resources disappear.

These threats to insect biodiversity in agriculture–habitat degradation and pesticide exposure–can compromise food and nesting resources among the widespread changes to soil and natural land cover (Fig. 1). To combat exposure of pollinators to pesticides, Sponsler et al. (2019) provide a multi-faceted socioecological approach to risk mitigation. Accordingly, we support the advancement of practices related to integrated pest management that place importance on mitigating pesticide exposure through the advancement of technologies aimed at reducing environmental contamination resulting from widespread pesticide application (e.g. neonicotinoid seed treatments, Krupke et al., 2012; Stewart et al., 2014). Douglas et al. (2020) determined that pesticide exposure hazard has increased because of the ubiquity of agrochemical-treated seed use in U.S., a practice that has reduced reliance on early season foliar applications but one that can lead to episodic contamination of natural systems (Krupke et al., 2012; Fig. 1).

Agriculture is not limited to the production of agricultural crops, but also includes livestock operations including confined animal operations and feedlots for cattle. These areas represent a relatively small portion of the total landscape but are intensively managed to control pests and disease. As a result, some agrochemicals used on feedlots and confined animal operations can affect non-target substrates and fauna locally. Peterson et al. (2017) found numerous chemicals used in feedlots occurring on flowering vegetation surrounding feedlots, which could affect foraging pollinators. In the same region, artificial cavity nests to attract wild bees were placed in habitats proximal and distal to confined livestock operations (the distal 'control' sites were still adjacent to cotton and other row crops) and nest substrates were collected to determine

Figure 1 Pesticide-treated seed on soil surface following planting in an intensified agricultural region in the U.S. (top). A red harvester ant carrying a treated seed. Non-target insects such as these seed collectors and other flying/foraging insects can be exposed to friable pesticide dusts from the seeds and planting process (Krupke et al., 2012). Fragmented native grasslands as a result of intensified agriculture (bottom). Pollinators and other wildlife depend on the mosaic of resources/patches and land management that supports restoration and conservation of native soils and vegetation.

agrochemical concentrations (Peterson et al., 2021). In this study, all substrates tested (mud, leaves, brood, pollen) contained at least one of 23 agrochemicals, mud contained 21 of 23 agrochemicals analyzed, and >70% of leaf particles collected by female bees and used to construct brood chambers contained at least one agrochemical. The chemical moxidectin was found in all nesting substrates from artificial nests at feed yard sites (Peterson et al., 2021), showing that veterinary pharmaceuticals should be considered as additional potential stressors in agricultural regions. Livestock operations represent an additional component of the agricultural landscape where non-target effects from agrochemicals could compromise pollination services and pollinator health. Furthermore, while pesticides can be detrimental to non-target organisms, the effects of other agrochemicals such as fungicides and herbicides are not well understood. An incomplete understanding of the synergistic toxicities of these agrochemical mixtures on pollinator health hinders our ability to determine the full consequences to crop production and the environment, as foraging insects can translocate agrochemicals among different resource substrates (Peterson et al., 2021).

4 Measures to support pollinators

A diverse plant population creates a healthy insect habitat and supports insect biodiversity. Only a couple of hundred species of the several hundred thousand known plant species are domesticated as a source of food for humans, and far fewer than that are produced in managed agricultural landscapes (FAO, 2019). This lack of flora diversity leads to concerns that agricultural landscapes lack a suitable habitat for supporting diverse insect populations. This is of heightened concern for specialized pollinators that rely on floral structures that are not present in the prevailing crops adopted in each region. For example, declines in milkweed plants as a result of herbicide tolerant crops has occurred extensively in the mid-continental agricultural region in the U.S. (Hartzler, 2010), becoming one of the stressors that influences declining monarch populations.

New plant species are in the process of domestication for use in production agriculture, but the acreage is limited (Al-Babili and Beyer, 2005; Raymer, 2002; Hartnell et al., 2022). While these may improve floral diversity on some specialty farms, newly domesticated crops do not appear to be a major contributor to floral diversity needed to support insect biodiversity in a global context. Furthermore, the lack of flora diversity may be the most acute in commodity farming, where only four crops contribute around 50% of global crop production (FAO, 2019). In order to meet the global caloric demand, commodity farming often requires large acreages dedicated to a single crop across a region. However, the target crop is not the only insect habitat in an agricultural landscape. Agricultural landscapes are complex ecosystems that depend on insect populations for productivity (Altieri, 1999). Nesting bees may use bare soil and 'weed' species may provide resources to support pollinators. More research is needed to understand the interactions between farm management and diverse insect populations, and ways to monitor improvements in conditions. Moreover, where habitat resources are known for a particular species or group of species, lands can be managed to promote native biodiversity locally and with practices aimed to promote populations of both bees and plants (e.g. native plant seed harvest to sow on proximal conservation lands and farms).

In the U.S., a major aim in the 2015 Pollinator Health Action Plan (White House, 2015) is to enhance and restore resources for pollinators. Essentially, this could include either active ecological restoration where landowners modify current resources through soil or vegetation management, or passive ecological restoration where simple conservation actions could include identifying and protecting parcels of land on properties that contain resources for generalist or specialist pollinators. For the latter, harvesting and sharing specific seeds to support wild bee specialists could be an effective approach, yet this is not typically practiced despite a push for using ecotype seed in habitat restoration. Ecotypes seeds are representative of native plant populations in the targeted restoration area and are adapted to the regional climate and environment. Such a strategy could involve identifying specialized ecotype plants (of specialist taxa or species of concern), implementing habitat conservation to protect the resource, collecting seed, monitoring biodiversity, and sharing seed among local landowners interested in the conservation of native pollinator biodiversity. Such a conservation action could immediately impact native bee biodiversity and given that between 25% and 30% of native bee communities in the U.S. consist of specialists, could be an effective conservation action for farmers (in that the totality of native biodiversity is being considered) and one that might bolster consumer markets in support of agricultural and environmental sustainability. Cultural implications from such conservation actions could be easily integrated in this approach, as the co-evolution of bees and flowering plants, especially considering domination of landscapes by humans, represents a biological element and interaction standing the test of time while exposed to past and present environments. This could involve critical evaluation of the environments that influences seed germination, or the number of different bees visiting a particular plant. Furthermore, how the seed disperses and germinates (in consideration of habitat fragmentation, contemporary soil disturbances, and climate) and how the specialized bees respond to improved resource conditions on farms represents a potentially unique conservation and monitoring system.

4.1 Integrated pest management

Chemical control is a widely adopted agricultural practice and currently plays a critical role in meeting global caloric and nutritional demands. Many insecticides that do not meet organic labeling requirements can be better managed through Integrated Pest Management (IPM). IPM is an environmentally sensitive continuum of pest management widely used across many agricultural landscapes that uses chemical control only after the proper use of action thresholds, field monitoring, and insect pest prevention using alternative cultural, biological, and physical methods of control (Kogan, 1998).

Integrated Pest Management frames plant and animal protection in terms of multiple synergistic cultural, mechanical, biological and chemical control methods. The practice of pest control aims to tolerate pests until numbers reach a critical threshold (i.e. an economic injury level threshold, EIL) (Kogan, 1998). In many cases, chemical control is used to suppress populations and to mitigate crop losses when pest densities are unmanageable using other suppressive tactics. By design, agrochemicals cause mortality in insect pests and in many cases these chemicals are toxic to non-target organisms including beneficial species, where sublethal and lethal effects can occur in non-target organisms such as honey bees (Long and Krupke, 2016).

As part of integrated pest management, seed treatments involve the coating of crop seeds with a mixture of agrochemicals to control early season pests on plant seedlings. While this practice may reduce early-season foliar applications of insecticides and fungicides, the widespread use of seed treatment technology in crop growing regions does not depend on pest densities or thresholds. Seed treatments are a prophylactic insecticidal treatment that ensure early season pest control (also reducing the number of pesticide applications) but do not depend on pest numbers or thresholds. Therefore, while it is a tool in the IPM toolbox, seed treatment is applying pesticides without immediate knowledge of pest numbers in the production field. Another potential problem (and research need) with this practice in relation to insect conservation is that pesticides are introduced into the environment broadly and systematically (in rows and fields), thereby contacting soils broadly and with potential soil contamination (see Krupke et al., 2012) and with unknowns about how confounding factors such as subsurface irrigation might further affect conditions. This is an important consideration because over 75% of all bee species are soil nesters and depend on soil conditions to raise young.

Insect biodiversity conservation depends on a well-functioning IPM system because insect biodiversity conservation is a fundamental component of the IPM continuum. A healthy and diverse agricultural ecosystem minimizes the need for pest control, increasing farm profitability through reducing agrochemical inputs or other suppresive pest control tactics. When chemical controls are implemented, applications are selected to be minimally disruptive to non-target insect populations. An accurate assessment of the potential impact a chemical control has on insect biodiversity helps IPM agents integrate that option into an environmentally sensitive management plan. An application that exists but is rarely used involves 'Environmental EILs' (E-EIL) where an environmental cost is added to the numerator, thus increasing the EIL (i.e. number of pest individuals to trigger pesticide application). Pollinator conservation actions would fit with E-EILs, since the number of bee species, number of pest predators, density of a specialist plant, or the proportion of farmland in conservation resulting from conservation improvements could be added to the total costs in EIL calculations, thereby lowering the EIL threshold (i.e. establishing a higher tolerance for higher pest numbers/densities because of increasing costs to the environment). However, relationships of E-EIL metrics and production or conservation outcomes should be quantified to effectively implement this approach, while some have argued against this approach because environmental costs are already considered in pesticide registration processes (Hutchins and Gehring, 1993).

4.2 The role of pest-resistant transgenic traits

Transgenic technologies provide management options for producers by introducing agronomically important traits from other species, such as bacteria. The most common transgenic technologies used in the management of agronomic crops are herbicide tolerances (HT) and insecticidal properties introduced through a *Bacillus thuringiensis* trait, known as Bt (Benbrook, 2012). The insecticidal properties provided by the Bt trait control lepidopteran larvae and other chewing insects by physically disrupting the midgut, causing death.

Insecticidal properties imparted by Bt traits provide value to the farmer by increasing yield and yield stability (Edgerton et al., 2012). However, the full impact of transgenic traits on insect biodiversity in agronomic fields is not well understood. Furthermore, while transgenic traits may lead to a reduction in the need for chemical control of insect pests, they do not eliminate the use of chemicals (Benbrook, 2012). Many insect pest populations are not susceptible to the cry protein produced by the Bt trait, and susceptible insect populations may build resistance to the technology. Controlling these resistant insect populations requires the use of Bt chemicals to keep populations below EILs. A full assessment of the potential impact of any agronomic practice on insect biodiversity requires an accounting of the complex insect ecosystem network, including impacts to biodiversity associated with transgenic traits for pest control.

4.3 Organic crop production and insect biodiversity

The use of insecticides or other chemical control of insect pests can pose a threat to ecosystem health and insect biodiversity. Organic regulations place limits on the types of chemicals allowed for use in pest management. Organic production began as a philosophical movement and was formalized into a government program in the Organic Foods Production Act of 1990 (Treadwell et al., 2003). These programs provide that farmers meeting a given set of government regulated standards can label their production as organic. Farmers that can certify their crops as organic receive a market premium for their harvest.

Organic farming practices are designed to reward farming practices that promote environmental health, ultimately leading to more sustainable agricultural ecosystems. Promoting environmentally conscious farming practices through organic agriculture has had a positive impact on environmental health and has helped conserve insect biodiversity through reducing inputs of conventional agrochemicals, but insect biodiversity is not a direct target of organic production.

While consumers pay the organic premiums, organic production places additional demands on the producer by limiting their management options. Only management tools that meet the organic standards are allowed. The use of many widely adopted agricultural management tools, such as transgenic technologies, is prohibited in organic systems.

4.4 Commodity farms as a driver of change

Commodity farms are larger scale operations that produce industrial raw materials such as food, fiber, and oil. When raw materials produced on commodity farms are used in industries that might otherwise use petroleum-based raw materials, they play an important role in generating a more sustainable economy with fewer greenhouse gas (GHG) emissions driving climate change.

In a circular bioeconomy, products made from biological commodities help to trap atmospheric carbon in a cycle, replacing petroleum feedstocks (Pascoli et al., 2022). In the United States alone, it is believed a billion tons of biomass can potentially be produced as a feedstock for industrial needs (Langholtz et al., 2016). Because commodity farms typically occupy large tracts of contiguous land, they also might contain concomitant portions of relatively large tracts of marginal or wild lands to support native biodiversity.

Carbon neutral or carbon negative industrial raw materials can even be produced with the right crop and management practices (Therasme et al., 2021). Seen through this context, trapping greenhouse gases from the atmosphere through crop production helps to promote insect biodiversity by mitigating climate change. However, each crop and production area is unique, and the conservation practices needed for effective engagement with the circular bioeconomy are varied. Furthermore, resources of pollinators might vary across production systems with establishment of flowering vegetation not the only beneficial conservation action. Improving soil disturbance regimes, providing clean freshwater, and leaving dead snags for bee habitat are all passive conservation actions that benefit pollinators. All these actions should improve pollinator biodiversity locally because of improvements in habitat resources, and this could be a target for establishing restoration trajectories for pollinators in agricultural landscapes. Furthermore, while there are plenty of chemical options for suppressing pest populations in urban and small farms (just go in any 'big box' home and garden store), large commodity farms use a disproportionate larger quantity of agrochemicals to produce crops and therefore are a key stakeholder in agricultural sustainability involving pollinator conservation and threats to biodiversity.

5 Incentivizing pollinator conservation practices

Biodiversity conservation depends on identifying and promoting farming practices that support productivity across agricultural landscapes while building habitats needed for diverse insect communities. Some existing programs that are designed to promote sustainable, or environmentally friendly, production practices by paying market premiums to farmers may indirectly promote insect biodiversity (e.g. organic crop production). However, the same markets have not identified direct biodiversity endpoints that could be used to offset environmental costs. We believe that identifying such endpoints could serve as both a direct conservation action to promote biodiversity on farms while being a specific target for enhancing the human connection via natural history education and ecological restoration of native populations in agricultural grasslands.

Agricultural markets are often characterized by a lack of market differentiation. Harvest occurs around the same general time for a region, leaving farmers to bring their harvest to market alongside many other farmers with the same or similar product. This leaves farmers little opportunity to differentiate themselves in the market. Most farm income is established through maximizing yield, with some premiums available for crop quality traits in different crops. These premiums tend to be a much smaller part of the farm revenue and are rarely the primary driver in farm management.

Many conservation programs build on this system by attempting to provide farmers with a way to earn a premium for their production through market differentiation. This market differentiation is typically a label applied to consumer goods that is earned based on metrics or programmatic requirements that are meaningful to consumers. These requirements are not necessarily tied to crop quality or yield. In the case of insect pollinators, direct contributions related to the regulation of plant reproduction via pollination provide a vehicle to link tangible outcomes and processes in agriculture with consumers. For example, the conservation of a biological target that positively relates to a high functional biodiversity on farms (because of ecological restoration or farm conservation to support biodiversity) might suggest optimal production because of ecosystem multifunctionality – the functional biodiversity sufficient to sustain natural systems (Soliveres et al., 2016) and potentially provide positive feedbacks to crop production (Cusser et al., 2016). This could be a potentially relatable concept for consumer markets that include environmental and biological conservation as a strategy, while simultaneously promoting natural histories of regional biodiversity in sustainable agricultural production.

5.1 Expanding environmental stewardship

The premiums paid to farmers for organic production help to promote environmental stewardship in agricultural landscapes. However, organic regulations are restrictive, limiting widespread adoption. Environmental stewardship and insect biodiversity conservation should not be limited to only those acreages that are able to meet organic production regulations.

Many other types of environmental stewardship systems attempt to broadly create market differentiation for the producer. Each producer and farm may focus on different aspects of the environment ranging from carbon sequestration to water quality (De Pinto et al., 2010; Mol and Oosterveer, 2015). Each one of these systems is forced to balance managing the regulation of agricultural landscape management practices with the desired environmental impact of the program.

It is likely that many of these programs, if not all, have a positive impact on insect biodiversity. Programs even exist that promote pollinator health. In the USDA Natural Resource Conservation Service, conservation practices are available to enhance resources for pollinators in working farms (Quigley et al., 2019; Ritten et al., 2017). However, specific biological targets for monitoring related to these improvements are generally not considered; It is assumed that increasing floral abundance and diversity influences all pollinator biodiversity, when it is likely that while generalist bee resources might be met, a large diversity of specialized bees might not benefit from such actions. Moreover, while calls to enhance resources for pollinators in the U.S. (White House, 2015) has been met with some successes (Sweikert and Gigliotti, 2019), most goals remain unmet 'given that most conservation policy is treated at farmer actions and not outcomes (Ehmke et al., 2015). This appears to be a missed opportunity, given the important role insects play in a productive agricultural landscape. Accordingly, developing biodiversity conservation targets in farmlands, given what we know specifically about bee communities and specialized fauna, represent promising targets.

5.2 Sustainability markets

Modern sustainability markets attempt to address the need for a more flexible environmental credit label through different traceability systems. For example, mass balance chain of custody systems promotes agricultural practices by creating a market for credits that are earned through meeting a set of programmatic requirements (Mol and Oosterveer, 2015). When a mass-balance credit is earned, the harvest and the environmental credit can be traded in separate marketing systems, each following a different chain of custody.

The flexibility of this approach allows buyers to source raw materials that meet their production needs while also paying premiums to those farmers using production practices that are meaningful to them. Mass-balance systems promote many environmentally and culturally significant production practices ranging from reduced chemical usage and the payment of a fair living wage to farm workers. Insect biodiversity or indicator species (i.e. bioindicators) could be integrated into an existing mass-balance system or implemented as its own credit market. Bioindicators can represent environmental, ecological or biological conditions that relate to the specific indicator species or guild. Improving the quality and connectivity of host plants for specific pollinators represents a potential bioindicator system to relate insect biodiversity conservation with consumer markets.

5.3 Insect biodiversity conservation markets

Insect biodiversity conservation requires a similar approach. In many cases it is not practical to target individual insects or insect populations; some insects are difficult to observe or detect and it is not practical to observe the full range of insect species in an environment. It is also not enough to expect agricultural conservation programs to promote insect biodiversity without some assessment of their impact or ability to monitor change. A proactive insect biodiversity conservation approach needs to be based on the diverse network of insect communities and their habitats.

An insect biodiversity credit market needs to balance promoting farm management practices with measured programmatic impacts on insect biodiversity. Factors such as insect size, camouflage, and their mobility make detecting insects a challenge with existing technologies, and these difficulties can increase with fauna that require specialized resources. Specialized insect nodes in the insect ecosystem services network may provide a method for measuring the direct benefit of an insect biodiversity credit (Fig. 2). And where specific insect resources are known, such specialized nodes can serve as bioindicators in the case that farming practices aimed at preserving specialist bee populations exist as a conservation umbrella - i.e. preserving the landscape to promote a particular plant species to support specialized bees may concurrently (passively through conservation or actively through sowing host plant seed) conserve broader wildlife and natural resources. Linking bioindicators to these important outcomes (i.e. broader biodiversity and ecosystem services) remains a research need to advance conservation in agriculture. Moreover, the usefulness of a bioindicator would depend on its relationship with other local state variables of the community. For example, do densities of the bioindicator positively relate to generalist bee diversity that could contribute to crop pollination? If the answer is yes, the specialized native bee would represent a valuable indicator species for monitoring as well as a tangible biological target to integrate the human dimension in promoting native wildlife.

Figure 2 An example of a regionally endemic plant supporting specialized native bee biodiversity in urban and rural areas of historic grasslands of the Southern High Plains (USA). Two native bees (Hymenoptera: Halictidae: Rophitinae, *Sphecodosoma* cf. *pratti* females shown here) specialize on this plant, with species co-evolving in grasslands and influenced in unknown ways by intensified agriculture and urbanization. Such a model bee-plant system could highlight native bee biodiversity and potentially serve as a biological target in farm ecological restoration and conservation initiatives, yet conventional sampling typically results in low detections of these native specialist bees.

The sage-grouse conservation system provides some insight as to how monitoring specialized insect nodes might work. The sage-grouse is a bird species with a unique nesting habit that requires sagebrush habitats in the western United States. A loss and fragmentation of sage-grouse habitat led to population decline (Baruch-Mordo et al., 2013). Despite sage-grouse populations being recognized as threatened, the species was unable to be listed under the Endangered Species Act due to political pressure.

A voluntary conservation system was developed that incentivizes landowners that manage their land in a way that promotes sage-grouse populations (Baruch-Mordo et al., 2013). However, the sage-grouse is difficult to observe directly, thus *detection* can be low. The birds are only prominent when they group together during their showy mating rituals called leks (Wann et al., 2019). The obligate relationship of the sage-grouse and the sagebrush allows the program to focus on promoting sage-grouse populations through sagebrush management and monitoring of leks. In the case of specialized bees, occurrences of a host plant in bloom usually corresponds to observations of bees foraging on the host plant and therefore a potentially higher expected probability of detection when compared to larger organisms with wider ranges. Determining occupancy and detection probabilities of specialized bees could support investigations of environmental drivers and threats to this fauna while bolstering monitoring strategies aimed at improving this biodiversity.

Moreover, while both generalist and specialist insects contribute to ecosystem services, specialist insects are the more fragile part of localized insect biodiversity networks. They are often the first insect ecosystem services lost when insect biodiversity is threatened – that attributed to the maintenance of plant-pollinator networks (Burkle et al., 2013). Monitoring the insect network should focus on these specialist insect hubs, where direct measurements resulting from conserving particular species could be related to broader biodiversity and supportive conservation practices. If the pollinator can be shown to always be present with the plant (i.e. high detection probabilities), the plant population provides an indicator measurement for that insect biodiversity network hub. The more bees a particular host plant supports at a specific location or farm the higher the importance of that biodiversity network hub (i.e. the plant or plant patch).

Specialized bee-plant biological indicators (i.e. specialized hubs) provide a measurable and immediate conservation target; Farmers may identify plants of a specialized bee on their land, set aside this land for conservation where applicable, and build pollinator habitat around that while learning about how to sustain this unique regional biodiversity in tandem with active ecological restoration (e.g. seed harvest). This transfers the action-based policy to tangible outcomes aimed at not only the functional roles of insects, but also the cultural, aesthetic value, and environmental safety value of conserving a unique component of regional biodiversity at the nexus of crop production and wild systems (Katumo et al., 2022). These specialized bee-plant systems could provide a robust model for monitoring, having adapted to regional climates over millenia a robust system for monitoring, as the plant-bee system has adapted to regional climates over millennia. Pardee et al. (2022) found that wild bees with a narrow diet breadth (not necessarily rare) increased in relative abundances during periods of low rainfall, suggesting the robustness of native specialized bee-plant systems for monitoring conservation actions in agriculture. Furthermore, Kremen and M'Gonigle (2015) found that small scale restoration on farms involving enhancing wildflower resources in small patches of marginal lands supported bees that were more specialized in floral and nesting resources and less mobile. This bolsters the concepts of monitoring specialized taxa where host plant resources are known and plants can be conserved or restored, thereby providing tangible outcomes by following the 'if you build it, they will come' philosophy for the conservation of specialized fauna.

5.4 Consumer education

Many ecological conservation systems are based on protecting threatened or endangered species. Waiting for insects to become threatened risks losing critical ecosystem services and creating a fragile insect biodiversity network. A consumer driven insect diversity credit system is a proactive system that does not require a species of insect to become threatened or endangered before action is taken.

The success of a proactive consumer driven insect biodiversity conservation system depends on consumers that are willing to pay a premium to farmers willing to adopt production practices that promote insect biodiversity. These premiums depend on consumers that are aware of the important ecosystem services provided by insect.

University entomology departments and extension services play a central role in consumer education and driving translational research opportunities in insect biodiversity conservation. Elementary through high school coursework provides students with limited exposure to a curriculum dedicated to insect ecosystem networks. Despite the important role university systems play in educating consumers about the important ecosystem services insects provide in agriculture, entomology programs at major universities are often underfunded and typically do not address biodiversity targets in agriculture, aside from those species directly known to influence a crop.

One important tool for consumer engagement involving biodiversity conservation is citizen science. Smartphone apps place the power of scientific observation in the hands of anyone on a neighborhood walk. These apps currently focus on observation and identification of different insects, but they can also be used to engage the public in conversations about important ecosystem services different insect communities play in urban or rural communities. Such applications also have a means of connecting people. Where biodiversity is measured, such activities can have important measurable impacts regarding the distributions of extant fauna, supporting research to advance species level biodiversity conservation (Chandler et al., 2017). Pollinators represent a potential flagship species - monarch butterflies and bumble bees are well-known flagship species and their plight has been documented. Such approaches linking actions linking humans to pollinator biodiversity natural histories and conservation actions prepares a new generation to advance sustainable agriculture.

6 Conclusion

While concerns about pollinator decline and its effect on agriculture and human lives have increased, an understanding of community and population level effects and the ecosystem services provided by the system remains incomplete, with advances hindered by policy focused on farmer actions but not outcomes. While benefits to pollinator communities from conservation practices on farms have been extensively documented, how conservation practices tie to biodiversity *conservation*, and how that in turn can be used to support crop production or consumer markets, has not been explored.

A potential biological target for monitoring wild land conservation on farms includes determining umbrella or focal species that could serve as targets relevant to regional natural histories and related to broader biodiversity or ecosystem services. Where risks can be mitigated, specialized bees occurring in agricultural grasslands are potential conservation targets where host plant associations are known and where seed can be expanded in ecological restoration. Furthermore, in supporting this fauna through mitigating risks of pesticide exposure (and other potentially detrimental actions) and in enhancing host plants for specific biodiversity, a restoration trajectory aims towards a system (i.e. vegetation, soil, fauna, disturbance) that supports the reproductive strategies and life histories of co-evolved grasslands species while highlighting the importance of insect biodiversity supported by agriculture.

Limitations of consumer markets in promoting conservation in agriculture include nontangible outcomes and potential long timeframes to assess the effects of conservation actions. Here, we provided background and ideas to support advances in insect conservation in agriculture, based on insect biodiversity. Considering the monetary value of ecosystem services such as pollination, as well as the extended ecosystem service benefits from wild land preservation on farms, the link between humans and nature in agroecosystems is strong and supporting of the result in agriculture - crop production to feed the human population. To bolster this link we present new ideas related to pollinator biodiversity and consumer market tools aimed at a synthesis of environmental stewardship - the conservation of specialized biodiversity in agriculture as a bioindication and monitoring tool. Such actions benefit the protection of the ecotypes of the region, the animals and plants that were here long before humans, while advancing environmental stewardship and an agroecosystem management approach that places value on genetic, species and ecosystem level biodiversity. This approach nurtures the linkage between humans and agriculture, advancing processes and outcomes in feeding and clothing a growing human population in a sustainable, environmentally responsible manner among stakeholders that value the diversity of life and the ecosystem services it provides.

7 Where to look for further information

Resources are available online to assist in initiatives for invertebrate conservation in agroecosystems, which supports ecosystem services including pollination and pest control. These important components of integrated pest and pollinator management seek to protect resources and ecosystem services while benefiting crop production.

- Pollinator Partnership (https://www.pollinator.org/).
- The Integrated Crop Pollination Project (https://icpbees.org/).
- Xerces Society for Invertebrate Conservation (https://www.xerces.org/).

Native bee pollen specialists requiring environmental conditions that sustain populations over time (via conservation actions related to land disturbance, farming practices, seed sources, crop-type, etc) could be a promising focal group in agricultural grasslands. Several other faunal groups have been evaluated for their potential role in serving as bioindicators in agricultural (earthworms (Paoletti, 1999), syrphid flies (Burgio and Sommaggio, 2007), invertebrate biodiversity (Paoletti, 1999), and birds (Padoa-Schioppa et al., 2006). For native bees occurring in the United States in North America, J. Fowler synthesized information related to the diverse pollen specialist bees. Where population conditions are known (including host-plant conditions) and where conservation or ecological restoration is implemented, these plant-pollinator specialized systems might serve as potential bioindicators and monitoring tools (https://jarrodfowler.com/bees_pollen.html).

A pioneering book on the importance and control of insects related agricultural production and natural processes influencing crop production and conservation in farm environments is by T.R. New–Invertebrate Conservation and Agricultural Ecosystems (2005, Cambridge University Press, 354 pp).

8 References

Al-Babili, S. and Beyer, P. 2005. Golden rice-five years on the road-five years to go? *Trends in Plant Science* 10(12), 565–573.

Altieri, M. A. 1999. The ecological role of biodiversity in agroecosystems. *Agriculture, Ecosystems and Environment* 74(1-3), 19-31.

Baruch-Mordo, S., Evans, J. S., Severson, J. P., Naugle, D. E., Maestas, J. D., Kiesecker, J. M., Falkowski, M. J., Hagen, C. A. and Reese, K. P. 2013. Saving sage-grouse from the trees: a proactive solution to reducing a key threat to a candidate species. *Biological Conservation* 167, 233-241.

Benbrook, C. M. 2012. Impacts of genetically engineered crops on pesticide use in the US--the first sixteen years. *Environmental Sciences Europe* 24(1), 1-13.

Burgio, G. and Sommaggio, D. 2007. Syrphids as landscape bioindicators in Italian agroecosystems. *Agriculture, Ecosystems and Environment* 120(2-4), 416-422.

Burkle, L. A., Marlin, J. C. and Night, T. M. 2013. Plant-pollinator interactions over 120 years: loss of species, co-occurrence, and function. *Science* 339(6127), 1611-1615.

Chandler, M., See, L., Copas, K., Bonde, A. M. Z., Lopez, B. C., Danielson, F., Legind, J. K., Masinde, S., Miller-Rushing, A. J., Newman, G., Rosemartin, A. and Turak, E. 2017. Contributions of citizen science towards international biodiversity monitoring. *Biological Conservation* 213, 280-294.

Cusser, S., Neff, J. L. and Jha, S. 2016. Natural land cover drives pollinator abundance, leading to reductions in pollen limitation in cotton agroecosystems. *Agriculture, Ecosystems and Environment* 226, 33-42.

De Pinto, A., Magalhaes, M. and Ringler, C. 2010. *Potential of carbon markets for small farmers*. International Food Policy Research Institute (IFPRI).

Douglas, M. R., Sponsler, D. B., Lonsdorf, E. V. and Grozinger, C. M. 2020. County-level analysis reveals a rapidly shifting landscape of insecticide hazard to honey bees (Apis mellifera) on US farmland. *Scientific Reports* 10(1), 797. https://doi.org/10.1038/s41598-019-57225-w.

Edgerton, M. D., Fridgen, J., Anderson, J. R., Ahlgrim, J., Criswell, M., Dhungana, P., Gocken, T., Li, Z., Mariappan, S., Pilcher, C. D., Rosielle, A. and Stark, S. B. 2012. Transgenic insect resistance traits increase corn yield and yield stability. *Nature Biotechnology* 30(6), 493-496.

Ehmke, M., Jones-Ritten, C., Shogren, J. and Panchalingam, T. 2015. Integrating ecological and economic considerations for pollinator habitat policy. *Choices* 30, 1-7.

FAO 2019. *The State of the World's Biodiversity for Food and Agriculture*, Bélanger, J. and Pilling, D. (Eds). Food and Agriculture Organization Commission on Genetic Resources for Food and Agriculture Assessments, Rome.

FAO 2021. *FAO Statistical Yearbook: World Food and Agriculture*. Food and Agriculture Organization, Rome.

Gallai, N., Salles, J.-M., Settele, J. and Vaissiere, B. E. 2009. Economic valuation of the vulnerability of world agriculture confronted with pollinator decline. *Ecological Economics* 68(3), 810-821.

Hartnell, G., Lemke, S. and Aulbach, C. 2022. Composition of a low erucic acid, low fiber field pennycress (Thlaspi arvense L) grain referred to as CoverCress™ developed through breeding and gene editing. *bioRxiv, 2022-06*.

Hartzler, R. G. 2010. Reduction in common milkweed (Asclepias syriaca) occurrence in Iowa cropland from 1999 to 2009. *Crop Protection* 29(12), 1542-1544.

Katumo, D. M., Liang, H., Ochola, A. C., Lv, M., Wang, Q. F. and Yang, C. F. 2022. Pollinator diversity benefits natural and agricultural ecosystems, environmental health, and human welfare. *Plant Diversity* 44(5), 429-435. https://doi.org/10.1016/j.pld.2022.01.005.

Klein, A. M., Vaissière, B. E., Cane, J. H., Steffan-Dewenter, I., Cunningham, S. A., Kremen, C. and Tscharntke, T. 2007. Importance of pollinators in changing landscapes for world crops. *Proceedings. Biological Sciences* 274(1608), 303-313.

Kleijn, D., Winfree, R., Bartomeus, I., Carvalheiro, L. G., Henry, M., Isaacs, R., Klein, A. M., Kremen, C., M'Gonigle, L. K., Rader, R., Ricketts, T. H., Williams, N. M., Adamson, N. L., Ascher, J. S., Báldi, A., Batáry, P., Benjamin, F., Biesmeijer, J. C., Blitzer, E. J., Bommarco, R., Brand, M. R., Bretagnolle, V., Button, L., Cariveau, D. P., Chifflet, R., Colville, J. F., Danforth, B. N., Elle, E., Garratt, M. P. D., Herzog, F., Holzschuh, A., Howlett, B. G., Jauker, F., Jha, S., Knop, E., Krewenka, K. M., Le Féon, V., Mandelik, Y., May, E. A., Park, M. G., Pisanty, G., Reemer, M., Riedinger, V., Rollin, O., Rundlöf, M., Sardinas, H. S.,Scheper, J., Sciligo, A. R., Smith, H. G., Steffan-Dewenter, I., Thorp, R., Tscharntke, T., Verhulst, J., Viana, B. F., Vaissiere, B. E., Veldtman, R., Westphal, C., Potts, S. G. 2015. Delivery of crop pollination services is an insufficient argument for wild pollinator conservation. *Nature Communications* 6, 7414. https://doi.org/10.1038/ncomms8414: 1038/ncomms8414.

Kogan, M. 1998. Integrated pest management: historical perspectives and contemporary developments. *Annual Review of Entomology* 43, 243-270.

Kremen, C. and M'Gonigle, L. K. 2015. EDITOR'S CHOICE: Small-scale restoration in intensive agricultural landscapes supports more specialized and less mobile pollinator species. *Journal of Applied Ecology* 52(3), 602-610.

Krupke, C. H., Hunt, G. J., Eitzer, B. D., Andino, G. and Given, K. 2012. Multiple routes of pesticide exposure for honey bees living near agricultural fields. *PLoS ONE* 7(1), e29268. https://doi.org/10.1371/journal.pone.0029268.

Langholtz, M. H., Stokes, B. J. and Eaton 2016. 2016. Billion-ton report: advancing domestic resources for a thriving bioeconomy. *Law and Medicine*. Oak Ridge National Laboratory, Oak Ridge, managed by UT-Battelle. LLC for the United States Department of Energy 1: Economic availability of feedstock, 1-411.

Long, E. Y. and Krupke, C. H. 2016. Non-cultivated plants present a season-long route of pesticide exposure for honey bees. *Nature Communications* 7(1), 11629.

Longing, S. D., Peterson, E. M., Jewett, C. T., Rendon, B. M., Discua, S. A., Wooten, K. J., Subbiah, S., Smith, P. N. and McIntyre, N. E. 2020. Exposure of Foraging Bees (Hymenoptera) to Neonicotinoids in the U.S. Southern High Plains. *Environmental Entomology* 49(2), 528-535.

Losey, J. E. and Vaughn, M. 2006. The economic value of ecological services provided by insects. *BioScience* 56(4), 311-323.

Mol, A. P. and Oosterveer, P. 2015. Certification of markets, markets of certificates: tracing sustainability in global agro-food value chains. *Sustainability* 7(9), 12258-12278.

Ollerton, J., Winfree, R. and Tarrant, S. 2011. How many flowering plants are pollination by animals? *Oikos* 120(3), 321-326.

Padoa-Schioppa, E., Baietto, M., Massa, R. and Bottoni, L. 2006. Bird communities as bioindicators: the focal species concept in agricultural landscapes. *Ecological Indicators* 6(1), 83-93.

Paoletti, M. G. 1999. The role of earthworms for assessment of sustainability and as bioindicators. *Agriculture, Ecosystems and Environment* 74(1-3), 137-155.

Paoletti, M. G. 1999. Using bioindicators based on biodiversity to assess landscape sustainability. In: *Invertebrate biodiversity as bioindicators of sustainable landscapes* (pp. 1-18). Elsevier.

Pardee, G. L., Griffin, S. R., Stemkovski, M., Harrison, T., Portman, Z. M., Kazenel, M. R., Lynn, J. S., Inouye, D. W. and Irwin, R. E. 2022. Life-history traits predict responses of wild bees to climate variation. *Proceedings. Biological Sciences* 289(1973), 20212697. https://doi.org/10.1098/rspb.2021.2697.

Pascoli, D. U., Aui, A., Frank, J., Therasme, O., Dixon, K., Gustafson, R., Kelly, B., Volk, T. A. and Wright, M. M. 2022. The US bioeconomy at the intersection of technology, policy, and education. *Biofuels, Bioproducts and Biorefining* 16(1), 9-26.

Peterson, E. M., Wooten, K. J., Subbiah, S., Anderson, T. A., Longing, S. D. and Smith, P. N. 2017. Agrochemical mixtures detected on wildflowers near cattle feed yards. *Environmental Science and Technology Letters* 4(6), 216-220.

Peterson, E. M., Thompson, K. N., Shaw, K. R., Tomlinson, C., Longing, S. D. and Smith, P. N. 2021. Use of nest bundles to monitor agrochemical exposure and effects among cavity nesting pollinators. *Environmental Pollution* 286, 117142. https://doi.org/10.1016/j.envpol.2021.117142.

Potts, S. G., Biesmeijer, J. C., Kremen, C., Neumann, P., Schweiger, O. and Kunin, W. E. 2010. Global pollinator declines: trends, impacts and drivers. *Trends in Ecology and Evolution* 25(6), 345-353.

Quigley, T. P., Amdam, G. V. and Harwood, G. H. 2019. Honey bees as bioindicators of changing global agricultural landscapes. *Current Opinion in Insect Science* 35, 132-137.

Raymer, P. L. 2002. Canola: an emerging oilseed crop. *Trends in New Crops and New Uses* 1, 122-126.

Ritten, C. J., Bastian, C., Shogren, J. E., Panchalangam, T., Ehmke, M. D. and Parkhurst, G. 2017. Understanding pollinator habitat conservation under current policy using economic experiments. *Land* 6(3), 57. https://doi.org/10.3390/land6030057.

Sponsler, D. B., Grozinger, C. M., Hitaj, C., Rundlöf, M., Botías, C., Code, A., Lonsdorf, E. V., Melathopoulos, A. P., Smith, D. J., Suryanarayanan, S., Thogmartin, W. E., Williams, N. M., Zhang, M. and Douglas, M. R. 2019. Pesticides and pollinators: a socioecological synthesis. *Science of the Total Environment* 662, 1012-1027.

St. Clair, A. L., Zhang, G., Dolezal, A. G., O'Neal, M. E. and Toth, A. L. 2020. Diversified farming in a monoculture landscape: effects on honey bee health and wild bee communities. *Environmental Entomology* 49(3), 753-764.

Stewart, S. D., Lorenz, G. M., Catchot, A. L., Gore, J., Cook, D., Skinner, J., Mueller, T. C., Johnson, D. R., Zawislak, J. and Barber, J. 2014. Potential exposure of pollinators to neonicotinoid insecticides from the use of insecticide seed treatments in the mid-southern United States. *Environmental Science and Technology* 48(16), 9762-9769.

Sweikert, L. A. and Gigliotti, L. M. 2019. Evaluating the role of Farm Bill conservation program participation in conserving America's grasslands. *Land Use Policy* 81, 392-399.

Therasme, O., Volk, T. A., Eisenbies, M. H., Amidon, T. E. and Fortier, M. O. 2021. Life cycle greenhouse gas emissions of ethanol produced via fermentation of sugars derived from shrub willow (Salix ssp.) hot water extraction in the Northeast United States. *Biotechnology for Biofuels* 14(1), 52.

Todd, K. J., Gardiner, M. M. and Lindquist, E. D. 2016. Mass flowering crops as a conservation resource for wild pollinators (Hymenoptera: Apoidea). *Journal of the Kansas Entomological Society* 89(2), 158-167.

Treadwell, D. D., McKinney, D. E. and Creamer, N. G. 2003. From philosophy to science: A brief history of organic horticulture in the United States. *Hortscience* 38(5), 1009–1014.

vanEngelsdorp, D., Evans, J. D., Saegerman, C., Mullin, C., Haubruge, E., Nguyen, B. K., Frazier, M., Frazier, J., Cox-Foster, D., Chen, Y., Underwood, R., Tarpy, D. R. and Pettis, J. S. 2009. Colony collapse disorder: A descriptive study. *PLoS ONE* 4(8), e6481. https://doi.org/10.1371/journal.pone.0006481.

Wann, G. T., Coates, P. S., Prochazka, B. G., Severson, J. P., Monroe, A. P. and Aldridge, C. L. 2019. Assessing lek attendance of male greater sage-grouse using fine-resolution GPS data: implications for population monitoring of lek mating grouse. *Population Ecology* 61(2), 183–197.

White House. 2015. *National Strategy to Promote the Health of Honey Bees and Other Pollinators*. Pollinator Health Task Force, Washington, DC.

Wratten, S. D., Gillespie, M., DeCourtye, A., Mader, E. and Desneux, N. 2012. Pollinator habitat enhancement: benefits to other ecosystem services. *Agriculture, Ecosystems and Environment* 159, 112–122.

Index

www.ingramcontent.com/pod-product-compliance
Ingram Content Group UK Ltd.
Pitfield, Milton Keynes, MK11 3LW, UK
UKHW021825100926
8556IPUK00003B/169

9 781801 463515